Interacting Bose-Fermi Systems in Nuclei

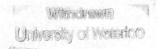

ETTORE MAJORANA INTERNATIONAL SCIENCE SERIES

Series Editor:

Antonino Zichichi

European Physical Society

Geneva, Switzerland

(PHYSICAL SCIENCES)

Recent volumes in the series:

Volume 5 **PROBING HADRONS WITH LEPTONS**
Edited by Giuliano Preparata and Jean-Jacques Aubert

Volume 6 **ENERGY FOR THE YEAR 2000**
Edited by Richard Wilson

Volume 7 **UNIFICATION OF THE FUNDAMENTAL PARTICLE INTERACTIONS**
Edited by Sergio Ferrara, John Ellis, and Peter van Nieuwenhuizen

Volume 8 **CURRENT ISSUES IN QUANTUM LOGIC**
Edited by Enrico G. Beltrametti and Bes C. van Fraassen

Volume 9 **ENERGY DEMAND AND EFFICIENT USE**
Edited by Fernando Amman and Richard Wilson

Volume 10 **INTERACTING BOSE–FERMI SYSTEMS IN NUCLEI**
Edited by F. Iachello

Volume 11 **THE SOLUTION OF THE INVERSE PROBLEM IN GEOPHYSICAL INTERPRETATION**
Edited by R. Cassinis

Interacting Bose-Fermi Systems in Nuclei

Edited by

F. Iachello

University of Groningen
Groningen, The Netherlands
and
Yale University
New Haven, Connecticut

Plenum Press · New York and London

Library of Congress Cataloging in Publication Data

Main entry under title:

Interacting Bose-Fermi systems in nuclei.

(Ettore Majorana international science series. Physical sciences; v. 10)
"Proceedings of the second specialized seminar on interacting Bose-Fermi systems in nuclei, held June 12-19, 1980, in Erice, Sicily"—Verso of t. p.
Bibliography: p.
Includes index.
1. Bosons—Mathematical models—Congresses. 2. Fermions—Mathematical models—Congresses. 3. Nucleon-nucleon interactions—Congresses. I. Iachello, F. II. Series. III. Title: Boson-Fermi systems.
QC793.5.B622I56 539.7'21 81-4319
ISBN 0-306-40733-7 AACR2

Proceedings of the second specialized seminar on Interacting Bose–Fermi Systems in Nuclei, held June 12–19, 1980, in Erice, Sicily

A Division of Plenum Publishing Corporation
233 Spring Street, New York, N.Y. 10013

Printed in the United States of America

PREFACE

During the week of June 12-19, 1980, a group of 42 physicists from 12 countries met in Erice, Sicily, for the second specialized seminar on "Interacting Bose-Fermi Systems in Nuclei". The countries represented were Belgium, Denmark, France, the Federal Republic of Germany, Israel, Italy, Japan, Mexico, the Netherlands, Poland, Sweden and the United States of America. The seminar was sponsored by the Italian Ministry of Public Education (MPI), the Italian Ministry of Scientific and Technological Research (MRST), the North Atlantic Treaty Organization (NATO) and the Regional Sicilian Government (ERS).

The purpose of this second seminar was twofold. First, to conclude the discussion on the Interacting Boson Model (IBM) already initiated at the first seminar of this series, held in Erice, June 6-9, 1978, and second to begin a discussion of the Interacting Boson-Fermion Model (IBFM). Thus, this book is divided into two parts. The first part, devoted to the Interacting Boson Model, begins with the results of calculations of properties of even-even nuclei (Casten, Federmann, Barrett, Clement, Lieb). It then continues with a study of the microscopic structure of the model. This study is at present a subject of considerable controversy and both the "pro" point of view (Arima, Otsuka) and the "contra" point of view (Broglia) are contained in this book. Possible extensions of the model to include other degrees of freedom, such as J=4 coupled pairs, additional J=0,2 modes, etc., discussed by Van Isacker, Barrett, Gelberg, Arima and Bäcklin, are included here.
One of the most interesting developments in the interacting boson model has been the study of its classical limit and the close relationship of this limit with the Bohr-Mottelson picture. This study, which was stimulated by some previous work of Gilmore and Feng, is presented in the contributions of Dieperink, Gilmore and Ginocchio.

The second part of the book is dedicated to the Interacting Boson Fermion Model (IBFM). This is the major new development occurred in this field since the first seminar, and it opens the possibility to study collective spectra of odd-even nuclei within the framework of an algebraic approach. Several aspects of this interacting boson-fermion model were presented at the seminar and are contained in this book. Among these, worth noting are the possible existence of Bose-Fermi symmetries (Iachello, Wood, Cizewski), the results of extensive calculations in medium mass and heavy nuclei (Scholten, Von Brentano, LoBianco, Wood) and the connection with the microscopic, shell-model, structure (Talmi, Scholten, Vitturi). The book is concluded by the remarks of Feshbach.

In addition to the contributions explicitly mentioned above, and directly connected with the interacting boson model, there were several other contributions, on related subjects. These contributions are also contained in the present book.

I wish to take this opportunity to thank all the participants and contributors who made the seminar a very enjoyable one. I also wish to thank Professor A. Zichichi, Director of the Centre for Scientific Culture "Ettore Majorana", who made it possible that this Seminar could take place in Erice, Professor G. Preparata, Director of the Seminars, and Dr. A. Gabrieli, for their help and collaboration. A final thank to Marianne Christen and Pauline DiGioia who helped me in many occasions. Without them, the preparation of this book would not have been possible.

F. Iachello
November 1980
Groningen, The Netherlands

CONTENTS

Part I. THE INTERACTING BOSON MODEL

STATUS OF EXPERIMENTAL TESTS OF THE IBA

Richard F. Casten

Physics Department
Brookhaven National Laboratory
Upton, New York 11973

INTRODUCTION

This review will survey the status of experimental tests of the IBA including a discussion of both IBA-1 and IBA-2. Overall, to date, the IBA has enjoyed enormous success, both in individual nuclei and as a general scheme for the unified treatment of entire regions. As the number of tests of the model has grown recently, the first significant discrepancies have also appeared and will be discussed in some detail. The review is limited to even-even nuclei. Even so, it is impossible to discuss all relevant work within the short space available. The growing pace of experimental work in this area is attested by the rather different contents of this review and an earlier one[1] this year: the reader is, therefore, also referred to that review where it was possible to treat some of the relatively smaller number of tests in more detail and also for its survey of tests of the IBA for odd mass nuclei.

STRUCTURE AND CHARACTERISTICS OF THE IBA: THE SYMMETRIC TRIANGLE

The IBA (see refs. 2-10) is an attempt to treat nuclei away from closed shells by grossly truncating the shell model space. This is illustrated in fig. 1: the IBA proceeds by replacing a general shell model Fermion space by a much smaller one restricted to include only correlated Fermion pairs coupled to spin 0 or 2. This truncated Fermion Hamiltonian is then mapped, by a specific procedure, onto a boson space. In distinction to boson expansion models, the essential approximation is thus in the Fermion space and not in the transition to a Boson Hamiltonian.

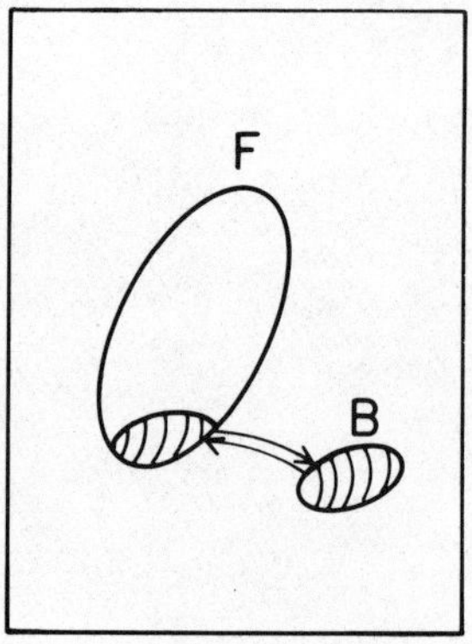

Fig. 1. Schematic illustration of the basic approximation scheme of the IBA in terms of Fermion and Boson spaces.

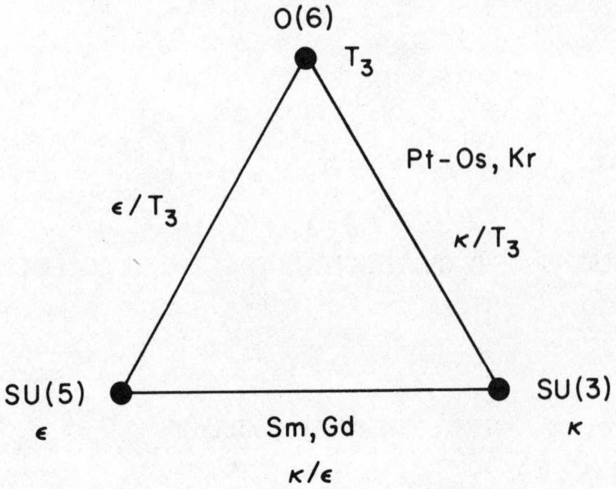

Fig. 2. A symmetry triangle the IBA-1. The essential coefficients describing each symmetry and transition leg are indicated.

The IBA can be considered on two levels, IBA-1 (refs. 2-4) and IBA-2 (refs. 6-8). The former further truncates the nuclear structure problem by neglecting the distinction between proton and neutron bosons. This is clearly a gross approximation, with consequences that will be different for different observables. The advantages of the IBA-1 are its simplicity, the natural emergence of limiting symmetries, the ease of calculation of transitional regions and the close correspondence to geometrical models. A general form[11] for the IBA-1 Hamiltonian is:

$$H=\varepsilon n_d+\kappa Q\cdot Q+\kappa' L\cdot L+\kappa'' P\cdot P+T_3[(d^+\tilde{d})^3(d^+\tilde{d})^3]^0_0+T_4(d^+\tilde{d})^4(d^+\tilde{d})^4]^0_0 \quad (1)$$

where Q, L and P are combinations of boson s and d creation and destruction operators, representing quadrupole, spin dependent and pairing interboson interactions. The IBA has an inherent group structure associated with the s,d boson approximation. The decomposition into subgroups yields three symmetries or coupling schemes characterized by analytic solutions, simple selection rules, geometrical analogues, and many parameter-free predictions. Two of these limits, SU(5) and SU(3), were recognized immediately[2] as corresponding, in the limit of large boson number, to the familiar vibrator and symmetric rotor models, while the third, the O(6) limit, was shown[3,5,12] to correspond to a γ-unstable picture. Each of the limits arises from the dominance of a specific term in eq. 1: the ε term for SU(5), the $Q\cdot Q$ term for SU(3) and the T_3 term (although $P\cdot P$ is also important) for O(6). The symmetries, and the relevant dominant coefficients in eq. 1, are displayed in the form of a symmetry triangle in fig. 2. The characteristics of each symmetry have been discussed frequently[1,2,9,10] and will not be repeated here; many, though, are evident in fig. 3, discussed below. In addition to the simplicity of the limiting cases, the IBA is particularly powerful as an elegant means of treating complex transition regions in surprisingly simple fashion. Thus, given that each limit is characterized by the dominance of a specific term in eq. 1, direct transitions between two limits correspond to the transition legs of the symmetry triangle and for the most part can be treated in terms of a single parameter, specifying the position along that leg, namely the ratio of the coefficients in eq. 1 which correspond to the two limits involved. Thus, the O(6) → SU(3) transition can be calculated primarily as a function of the single parameter κ/T_3, and the SU(5) → SU(3) transition corresponds to a changing ratio of κ/ε. Of course, more complex situations can also be treated simply by various combinations of the parameters of eq. 1 and this comment serves to highlight one of the most attractive features of the IBA, its generality.

Since the IBA-2 distinguishes protons and neutrons, it should be inherently superior. It is also more microscopic in that the parameter variations may be predicted from shell model considerations. On the other hand, the limiting symmetries and geometrical analogues are less apparent in IBA-2. It has been shown[9,10] that a simple approximation to the most general[6-8] neutron-proton boson Hamiltonian is

$$H = \varepsilon(n_{d_\pi} + n_{d_\nu}) - \kappa\, Q_\pi \cdot Q_\nu + V$$

where (2)

$$Q_{\pi(\nu)} = (d^+ s + s^+ \tilde{d})^{(2)}_{\pi(\nu)} + \chi_{\pi(\nu)}\, (d^+ \tilde{d})^{(2)}_{\pi(\nu)}$$

The terms have been discussed in the literature. (V is a Majorana term that has little effect on existing calculations and will not be discussed further.) Almost all tests of the IBA-2 have been based on eq. 2 or a minor extension of it to include a diagonal neutron-neutron interaction of the form $\frac{1}{2} \sum_L C_L (2L+1)^{\frac{1}{2}} [(d^+d^+)^L (dd)^L]^0$.

The appeal of the IBA-2, as represented by eq. 2, is that it has only four parameters and, at least in the single j shell approximation, their behaviour can be predicted[7-9] microscopically. Furthermore, both ε and κ are expected to be nearly constant across a major shell. χ_π and χ_ν should be large and negative at the start of a shell and large and positive near its end. For a single j shell, the expression for χ_ν(or χ_π) is $\chi = \chi_0\, (\Omega_\nu - 2N'_\nu)/\sqrt{(\Omega_\nu - N_\nu)}$ where Ω_ν is the neutron shell degeneracy, 1/2(2j+1), and N'_ν (N_ν) is the number of bosons relative to the beginning of the shell (to the nearest closed shell). (This behaviour of χ_ν is illustrated below in fig. 7.) Finally, for series of isotopes (isotones), $\chi_\pi(\chi_\nu)$ should be constant.

THE Pt-Os-W REGION: O(6) NUCLEI AND THE O(6) → SU(3) TRANSITION LEG

The most striking early prediction of the IBA was that of a new, third, symmetry, the O(6) limit, which existed, theoretically, on the same footing as the two familiar SU(5) and SU(3) limits. This symmetry differs considerably from the others, having unique characteristics such as the absence of a 0^+ state close in energy to the 2_2^+ and 4_1^+ levels, 0^+-2^+-2^+ sequences characterized by strong intracascade B(E2) values, and entire sequences of higher lying levels that repeat the sequence near the ground state, but which, though fully collective, are electromagnetically isolated from the latter by virtue of different values of a new quantum number σ and the E2 selection rule, $\Delta\sigma=0$. This implies, incidentally, that those higher levels are inaccessible via Coulomb excitation or inelastic scattering. They have been excited via the "feeding" reaction, (n,γ).

Given the prediction of this new symmetry, its almost simultaneous empirical discovery[3] in ^{196}Pt gave substantial confirmation to the IBA and, indeed, encouraged renewed testing of it. (In fact, the largest group of tests[5,12-14,16-20] have indeed centered on the Pt-W region, regarded as undergoing an O(6) → rotor transition.)

A comparison of the low lying levels and E2 transitions of ^{196}Pt with the three symmetries is shown in fig. 3. Since there is a precise one-to-one correspondence between the empirical levels and those of the O(6) limit, and because all allowed E2 transitions were found to be dominant and all forbidden ones were weak or unobserved, the diagrams for ^{196}Pt and the O(6) limit in fig. 3 are identical and therefore combined. The patterns of transitions in the different limits are clearly entirely different. Several decay branches are unique to the O(6) limit and each of the other limits has a number of transitions that are forbidden in O(6). The thickened transition arrows highlight these characteristic selection rules and, of course, serve to demonstrate that ^{196}Pt is a good O(6) nucleus.

In addition to branching ratios, the inherent collectivity of each level and the collective relationships are important and therefore the study of absolute B(E2) values is an essential test of the model as well. Recently, Bolotin et al.[14] measured a number of lifetimes in ^{196}Pt. By normalizing to the $2_1^+ \rightarrow 0_1^+$ transition, the B(E2) values thus deduced may be compared to the parameter-free predictions of the O(6) limit. The comparison is shown in Table 1 and is indeed remarkable. Note that it includes the high lying 0_2^+ and 4_2^+ levels.

The recognition of the O(6) structure for ^{196}Pt provides a

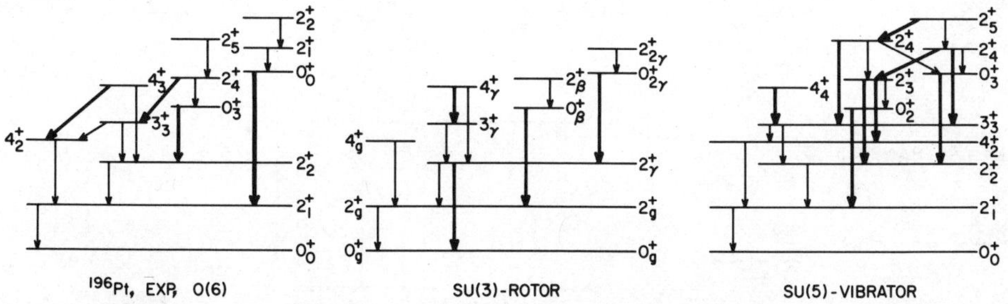

Fig. 3. The levels and strong E2 transitions in ^{196}Pt and the allowed E2 transitions, in each symmetry, between the same levels. (There are no reasonable analogues for the highest two 2^+ levels in SU(3).) Infinitesimal symmetry breaking is permitted to show the favored decay routes for the 0^+_β and $0^+_{2\gamma}$ levels in SU(3) and the uppermost 0^+ level in O(6). For ^{196}Pt-O(6) the thicker arrows are transitions that are forbidden in both other limits: for SU(3) and SU(5) they are transitions that are forbidden in O(6). The J^π subscripts are: the τ quantum numbers for O(6) where the E2 selection rule is $\Delta\tau = \pm 1$, the familiar β, γ or 2γ labelling of the rotor in SU(3) and the phonon or d boson number in SU(5).

Table 1. Absolute B(E2) values (e^2b^2) in ^{196}Pt normalized to the $2^+_1 \rightarrow 0^+_1$ transition

I_i	2_1	2_2	2_2	4_1	0_2	0_2	4_2	4_2	4_2	6_1
I_f	0_1	2_1	0_1	2_1	2_2	2_1	4_1	2_2	2_1	4_1
Exp[14)] (ΔExp)	.264 (.011)	.318 (.023)	Not obs.	.409 (.022)	.142 (.077)	.022 (.010)	.193 (.097)	.177 (.025)	.003 (.001)	.421 (.116)
O(6)	.264	.346	0	.346	.352	0	.167	.184	0	.352

benchmark for a new treatment[5] of the heretofore thought-to-be extremely complex Pt-Os region. Since the lighter Os isotopes resemble good rotors this sequence of nuclei corresponds to the O(6)→SU(3) transition leg in the symmetry triangle which can be treated very simply in terms of a single smoothly varying parameter κ/T_3. This has been done using the code PHINT[11] and the results[1,5] show that the IBA provides a remarkably accurate delineation of this region. In fact, well over a 100 E2 branching ratios, many rapidly varying, were predicted with only five significant discrepancies. Interestingly, four of these occurred in the K=4 bands in Os. Recently it was discovered that there had been an error in the code PHINT that only affected transitions between odd and even spin levels and only in nuclei deviating from the O(6) limit. All calculations of ref. 5 were then redone[13] with the identical parameters as before.

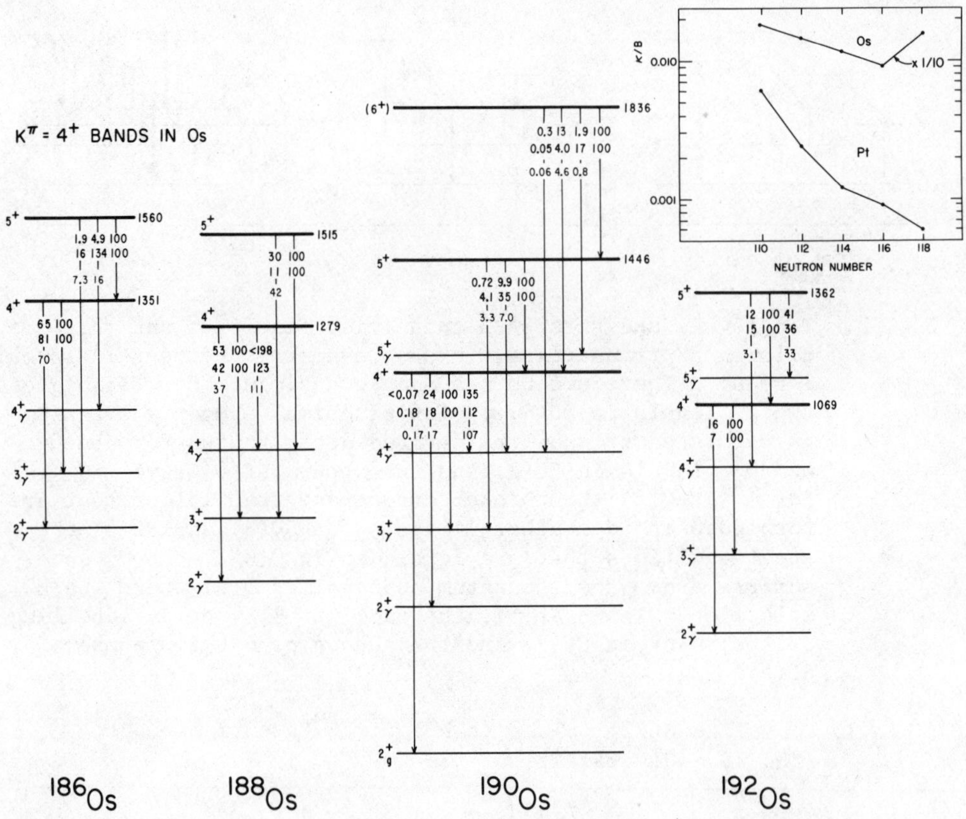

Fig. 4. The K=4 bands in Os. On the transition arrows relative B(E2) values are given: top row (empirical), second row (original calculations[5]), third row (all changes arising in the new calculations[13]). The inset shows the parameter values κ/B (B∝T_3).

It was found that the only significant changes occurred in the K=4 bands and that they occurred for just those cases of prior disagreement and that they were in the direction toward alleviating the discrepancies. The empirical results, the old, and the new, calculations for these bands are shown in fig. 4. The improvement is clearly evident, no major discrepancies remaining. The same calculations predict absolute B(E2) values as well. Figure 5 shows some of these. Again, the agreement is impressive and shows the developing structural changes across the O(6) → SU(3) transition.

Spurred by these results and by the goal of improving the agreement for level energies and quadrupole moments which are not

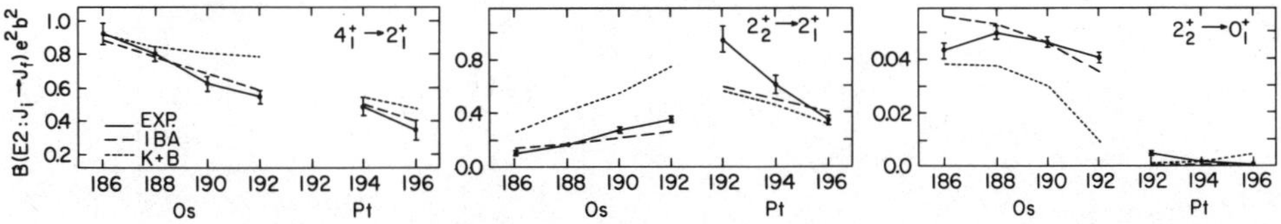

Fig. 5. Absolute B(E2) values, normalized to $2_1^+ \rightarrow 0_1^+$ transition and compared with IBA (ref. 5) and Kumar and Baranger (K+B) (ref. 15) calculations.

expected to be well predicted for O(6) nuclei in IBA-1, Bijker et al.[16] have recently carried out IBA-2 calculations for these same nuclei. These results are reported in detail in another contribution to this conference. In general, as expected, the IBA-2 predictions for E2 transitions differed little from the already excellent IBA-1 values, but level energies changed appreciably, greatly diminishing differences with experiment. Figure 6 shows a small sample of the results. The one involving 3_1^+ state is especially interesting since the $3_1^+ \rightarrow 2_1^+$ transition is forbidden in all three limits and reaches finite, albeit small, proportions only in transition regions. Along with two or three other similar branching ratios, (such as $(4_2^+ \rightarrow 2_1^+)/(4_2^+ \rightarrow 2_2^+)$ and $(5_1^+ \rightarrow 4_1^+)/(5_1^+ \rightarrow 4_2^+)$) whose numerators are $\Delta\tau=2$ in O(6), interband in SU(3) and two phonon changing in SU(5), these branching ratios provide a nice signature of transition regions. As seen in fig. 6, the empirical (and predicted) values start off small near $^{192-196}$Pt, then grow for the lighter isotopes and are substantially larger in Os. The plot on the left in fig. 6 shows the excellent agreement in most cases for the branching ratio from the 2_2^+ level and the striking and important disagreement for ^{194}Os which will be discussed below, following presentation of the IBA-2 calculations for W.

The W calculations, by Duval and Barrett,[17] continue the analysis of the O(6) → rotor transition and are also discussed in detail at this conference. We present a summary of some of the comparisons for energies, E2 transitions and quadrupole moments in figs. 7,8. Again, given the simplicity of the calculations, the smooth behaviour of the parameters and the vast amount of data interpreted by these calculations, the agreement is excellent. Also of interest, and typical of the capabilities of IBA-2 calculations, are new predictions for a number of yet-to-be-measured B(E2) values and branching ratios.

It is instructive to consider the parameter values (see fig. 7) in some detail. As expected, ε and κ are nearly constant, while χ_ν roughly follows the predicted (single j shell) trend but does

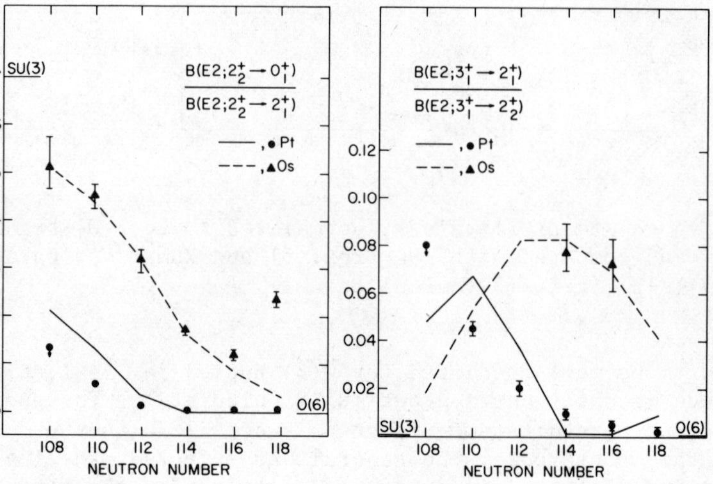

Fig. 6. Comparison of the data for Pt and Os with IBA-2 calculations (ref. 16). The 0(6) and SU(3) limiting values are also shown.

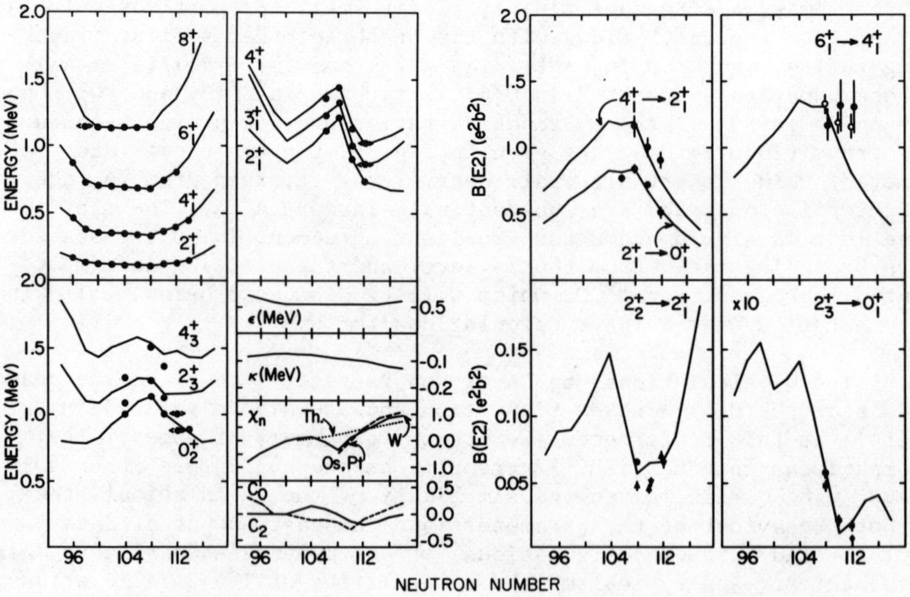

Fig. 7. Comparison of IBA-2 calculations of Duval and Barrett[17] with the data for W. The second box at bottom summarizes the parameter values used. Also shown for comparison are the χ_ν values of Bijker et al.[16] for Pt, Os and the predicted[9,10] trend for a single j shell.

drop a bit faster and cross zero before mid shell. Since actual shells have detailed microstructure, with fairly distinct subshells, one might expect that the overall trend in χ_ν would be a superposition of several negative-to-positive $\chi_\nu(j)$'s, each shifted sequentially to the right and each with a more vertical slope (than the single overall j χ_ν plotted) because of the lower degeneracy of each subshell. This is exactly what appears to be happening. Duval and Barrett attribute it to the closure of the $i_{13/2}$ shell. This kind of analysis implies that one may use the extracted parameter values from IBA-2 calculations to gain information on the local shell structure. Finally, as shown in fig. 7, it is striking, and indeed comforting, that the calculations[16] for Os-Pt, carried out independently, arrived at almost identical values for χ_ν (and also for the other parameters, ε, κ, and χ_π (see ref. 16)).

We can now return to a consideration of the discrepancy for the 2_2^+ branching ratio in ^{194}Os mentioned earlier. The striking point is the sudden empirical rise toward the SU(3) value of 0.7 and this is also reflected in a sudden upward jump in E_{2_2}, also not predicted. In ref. 18, this was discussed in the context of a phase transition, in ^{194}Os, to an oblate deformation, such that ^{194}Os has a larger, not smaller, β_2 value than ^{192}Os. The fact that this branching ratio was well reproduced in the IBA-1 calculations of ref. 5 merely reflects the enforced parameter change in κ/T_3 (see inset, fig. 4) and is ad hoc. In a proper IBA-2 calculation, within a single shell, such changes are not expected, and, therefore, with the smooth parameter values demanded by Bijker et al.,[16] it is clearly impossible to reproduce the ^{194}Os result. Just as clearly, this result could have been reproduced by adopting a χ_ν value similar to that for ^{188}Os or ^{190}Os. To do so would, in the context of the above discussion, probably reflect further microscopic subshell effects.

The IBA-2 also predicts quadrupole moments. Baktash et al.,[19] have summarized the comparisons with theory for both $Q(2_1^+)$ and $Q(2_2^+)$ for W, Os and Pt. We show their results in fig. 8. Overall, the agreement for the IBA is reasonable although clearly not as good as other approaches in Pt where the predicted magnitudes for $Q(2_1^+)$ and $Q(2_2^+)$ are too low. (The disagreement for the 2_2^+ state in ^{184}W is difficult to assess since this is the only nucleus where the empirical and expected approximation $Q(2_2^+) \approx -Q(2_1^+)$ does not hold and where no model predicts the empirical value.) The Pt discrepancy, however, must be considered a major one for the IBA in this mass region and, as yet, remains unexplained.

Before leaving this region it is important to site another significant test of the IBA. Cizewski et al.[20] have carried out (t,p) reaction studies leading to the even Pt and Os nuclei. It is safest to consider ground state transitions only since only for these can one normally neglect the difficulties associated with

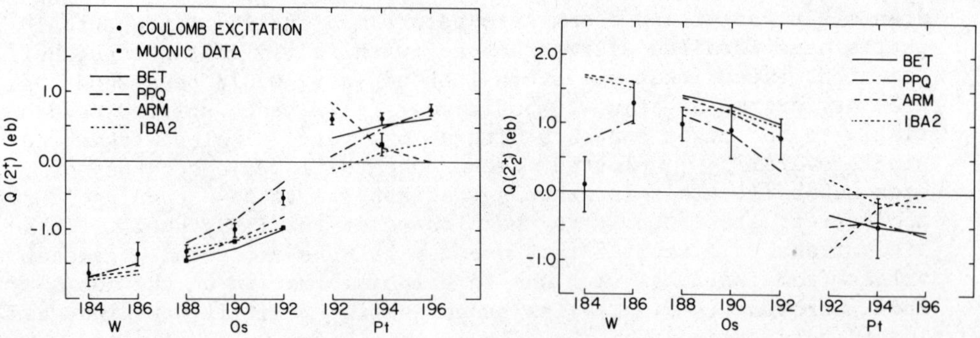

Fig. 8. Empirical quadrupole moments for W, Os, and Pt compared with the IBA (refs. 16,17), the boson expansion theory (BET), the asymmetric rotor model (ARM), and the calculations of Kumar and Baranger (PPQ). Taken from ref. 19.

multi-step reaction mechanisms. Figure 9 shows the IBA predictions for the O(6) and SU(3) limits compared with the data. While the theoretical differences are small, so are the experimental error bars and the agreement with the O(6) limit is clearly both excellent and preferred. This is one of the few tests of the IBA for reaction processes and is important because it again makes use of the elegant closed form analytic expressions available in the limiting symmetries. It is interesting that the differences in slopes of the theoretical cross sections arise from the different effects of finite

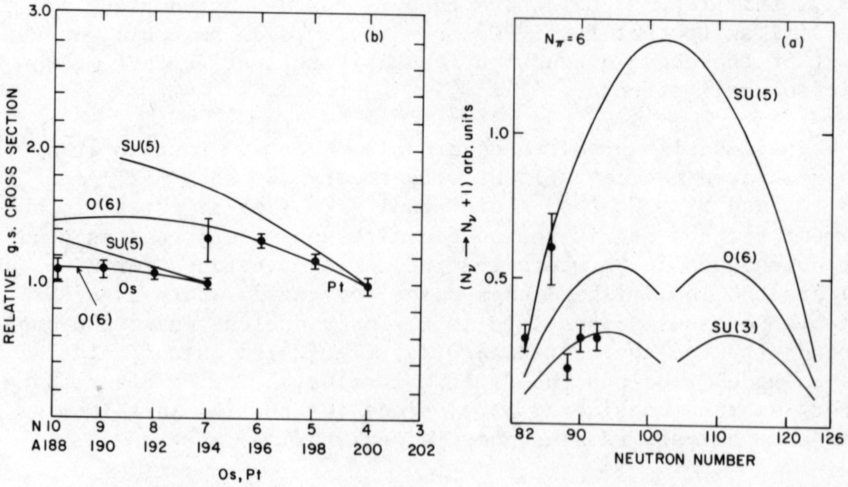

Fig. 9. Ground state (t,p) cross sections compared with IBA calculations. Left: Os, Pt (data from ref. 20). Right: Sm (ref. 2).

boson number in the two limits which in turn arise from the different structure of the SU(3) and O(6) wave functions, in particular the different values for $<n_d>$, the expectation value of the number of d bosons.

THE Sm REGION: THE SU(5) → SU(3) TRANSITION LEG

Some of the earliest extensive IBA-1 and IBA-2 calculations were carried out for the Sm transition region. Again, this transition can be calculated in IBA-1 as a function of the single parameter, κ/ε, which should grow monotonically along the transition leg of fig. 2. Figure 10 displays examples of the IBA-1 results and comparisons with other models. The IBA predictions are in rather good agreement with the empirical trends. Some discrepancies in details appear, as for the decay of the 2_3^+ level for ^{150}Sm, and that of the 3_1^+ state in ^{154}Sm. On the other hand, it is clear from both the $B(E2(2_2^+ \rightarrow 0_1^+)/B(E2:2_2^+ \rightarrow 2_1^+)$ ratio and the $B(E2:2_2^+ \rightarrow 0_1^+)$ value that the TWK calculations do not account for the transition to a deformed character: this probably arises, as indeed pointed out in ref. 21 itself, from poor convergence of their boson expansion as one departs further from the spherical basis. The closer agreement with experiment for the IBA reflects its greater facility in treating broad sequences of nuclei on the same footing. The

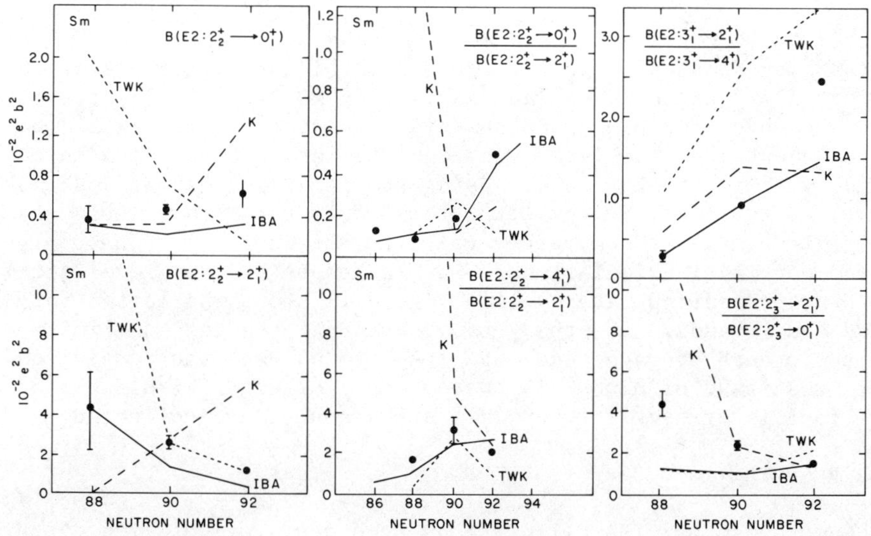

Fig. 10. Empirical (refs. 21 and 4) absolute B(E2) values and branching ratios in Sm compared with IBA-1 calculations[4] (solid lines), the BE calculations[21] (TWK) and those of Kumar (K) (ref. 15).

calculations of Kumar and Baranger depart from the data particularly in the more vibrational region. Other comparisons of these same three models with the data are given in ref. 1.

The same detailed IBA calculations provide reasonable agreement with measured (t,p) and (p,t) cross sections as recently summarized by Saha et al.[22] and in ref. 1. It is more instructive, however, to consider a simpler analysis. Equations 3 and 4 give simplified IBA expressions,[2] approximately correct for large N, for ground state (t,p) or (p,t) cross sections in the SU(5) and SU(3) limits.

$$S_o(N_\nu \rightarrow N_\nu+1) \rightarrow \alpha_\nu^2(N_\nu+1)(\Omega_\nu-N_\nu) \qquad SU(5) \qquad (3)$$

$$S_o(N_\nu \rightarrow N_\nu+1) \rightarrow \alpha_\nu^2 \frac{(N_\nu+1)}{3}(\Omega_\nu-5/3N_\nu) \qquad SU(3) \qquad (4)$$

In a rapid SU(5) → SU(3) transition, as occurs in Sm, one therefore expects a sudden drop of a factor of at least three in cross section. Figure 9 compares the (t,p) ground state cross sections with the analytic expressions for the three limits: the sudden phase change in Sm is nicely reflected by the IBA prediction. This same result can be argued from geometrical models as resulting from the poor overlap of deformed and spherical states at the transition point: the IBA provides in effect an alternate phrasing of the same physical effect, and one that arises automatically and naturally.

THE A∿80 REGION

The Kr isotopes have been studied in a systematic set of calculations[23] by Kaup and Gelberg. This was, historically, one of the earliest IBA-2 calculations and envisioned an O(6) → SU(3) transition such that ^{78}Kr had an intermediate character. New data[24] on ^{78}Kr have extended the level scheme up to rather high spins in the ground, γ and octupole bands. Using the IBA-2 parameters deduced in ref. 23 for ^{78}Kr, Hellmeister et al.[24] have extended those calculations to provide a comparison with the new data. They have also carried out IBA-1 calculations. Both attempts fit the level energies and γ-ray transition rates well. Figure 11 shows the latter for the ground and γ bands. For the γ band, the B(E2) values and limits span two orders of magnitude. For the ground band the comparison raises the issue of high spin states and thus we defer discussion of it to a later section where all such results are collected.

OTHER RESULTS

Numerous other IBA calculations, too many for detailed discussion, have been done. Many are mentioned in ref. 1 and others are contained in contributions to this conference. Three recent ones should, however, at least be cited. These are IBA-2 calculations of the Pd and Ru nuclei[25] and of the Ba, Ce, Xe region,[26] and IBA-1

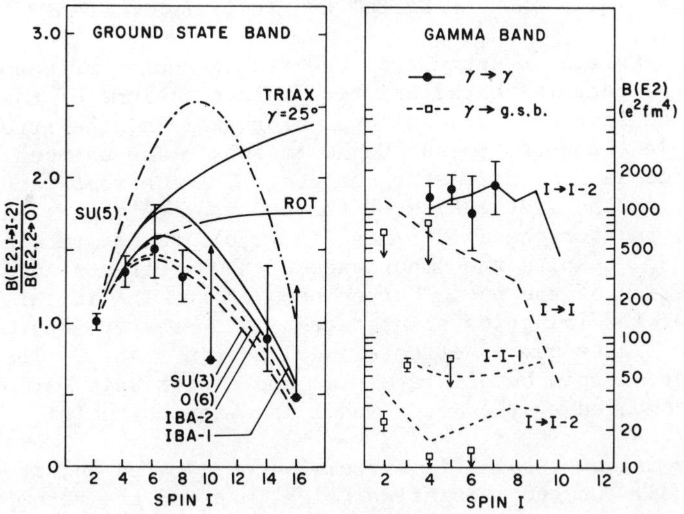

Fig. 11. Absolute B(E2) values for the ground and γ bands in ^{78}Kr. The g.s.b. calculations are compared to all three IBA limits and to two detailed calculations as well as to the rotational and triaxial (γ=25°) model. (From ref. 24)

calculations for Er (ref. 27). The first is important because the region treated borders on, on the one side, the most rapid spherical-deformed transition region known (A∿100), which in turn is near a major subshell closure and, on the other, the Cd nuclei with their enigmatic quintuplets of states at an energy expected for the two phonon triplett. The Ba, Ce, Xe calculations are interesting because they highlight the subtle differences in these neighboring nuclei: with decreasing neutron number in the 50-82 neutron shell, Xe goes from O(6)-like to SU(5)-like, while, further from the Z=50 closed shell in Ba, and more so in Ce, there is an intermediate region approaching the SU(3) limit. This shows up most obviously in the $B(E2{:}2_2^+ \rightarrow 0_1^+)/B(E2{:}2_2^+ \rightarrow 2_1^+)$ ratio which is zero in O(6) and SU(5) but 0.7 in the rotor. The maximum calculated value occurs in each element for N=66 (midshell) and has the values 0.06 for Xe, 0.16 for Ba, and 0.3 for Ce. Unfortunately the data are sparce and the calculations pointedly illustrate the need for extensive new data, in relatively uncharted mass regions. Finally, the most extensive comparison of the IBA in a single nucleus, and the first involving as many as 16 bosons, is currently underway by Warner et al.[27] in ^{168}Er. Preliminary results in a perturbed SU(3) framework yield good results for energies and for intra-to-interband branching ratios for the ground, γ, and two 0^+ bands, and qualitative agreement for the only other positive parity bands (both K=2) below 2 MeV.

HIGH SPIN STATES AND THE EFFECT OF FINITE BOSON NUMBER

The explicit treatment of the finite number of bosons in the IBA introduces certain characteristic features. Some of these are implicit in the foregoing discussions. For example, the structural changes in the W-Pt, Sm and Ce-Xe regions arise as much from an N dependence as from parameter changes. These effects are particularly striking in the systematics of the decay of 0^+ states in Os and Pt (ref. 5) and for the differences in (t,p) ground state cross sections in fig. 9. Despite the importance of these effects and the abundant confirmation of the basic correctness of the IBA in this respect, at least for low lying low spin states, even greater interest has centered on finite number effects in high spin states. These are of two types, a spin cutoff (at I=2N) and a high spin falloff in B(E2) values compared to the rotor model in deformed nuclei.

The issue, though, is extremely complicated and frequently misunderstood. On the theoretical side, many of the attempted comparisons to date have utilized only the SU(3) limit of the IBA, which is usually an unrealistic assumption. Others have dealt with nuclei involving backbending wherein the possibility of intruder states outside the IBA must be carefully considered. On the experimental side, the experiments (generally Coulomb excitation) designed to excite the high spin states are difficult and the extraction of B(E2) values requires a very complete and careful analysis of side band yields that deplete strength from the ground band. At the time of the survey of ref. 1, it was not felt that the experimental results, except for ^{78}Kr, were sufficiently reliable. Now, however, more sophisticated data analyses have been completed and reliable empirical results for some actinide nuclei,[28] for ^{194,196}Pt (ref. 29), for ^{156}Dy (ref. 30), as well as ^{78}Kr are now available.

On the left in fig. 11 the B(E2) values for the ground band in ^{78}Kr, normalized to the $B(E2:2^+ \to 0^+)$ value, are plotted against spin. There is apparently a falloff, most clearly defined for the $14^+ \to 12^+$ transition. Various IBA calculations are shown. All but the SU(5) case agree with the empirical falloff. The two geometrical model predictions, for the symmetric and triaxial rotors, of course display no falloff. In order to assess the significance of this result it is necessary to discuss the up-bend occurring in ^{78}Kr near spin 10^+. Lieb and coworkers[31] have studied the $(g_{9/2})$ system in ^{79}Rb. They obtain three results: the odd particle does not polarize the core, that is, the B(E2) values for the $13/2 \to 9/2$ and $17/2 \to 13/2$ transitions are just what would be expected for a rotor of constant shape based on the $B(E2:2^+ \to 0^+)$ value in ^{78}Kr , there is no evidence for backbending in ^{79}Rb and yet there is a falloff in B(E2) values. One can conclude[31] then that backbending in ^{78}Kr is due to the $\pi(g_{9/2})^2$ pair and that the falloff in B(E2) values there (and in ^{79}Rb) is not due to a change in structure associated with a crossing band but is a reflection of the finite particle number effects predicted in the IBA.

Turning to more strongly deformed nuclei, one finds that in ^{232}Th(N=12), ^{234}U(N=13), and ^{236}U(N=14) the ground bands extend[28] to spins 28+, 28+ and 30+, in each case beyond the predicted IBA cut-off, and that there is no apparent sudden change in level patterns or evidence for a drop in B(E2) values below the rotor ones. In ^{156}Dy, as well, no falloff in B(E2) values at high spins is observed.[30] These results have important implications, but care must be taken in assessing them. It is somewhat naive to take the extreme SU(3) limit as representing the IBA predictions. As shown in fig. 12, other limits, such as SU(5), have very different B(E2) values which, in fact, rise well above the rotor model predictions, especially for large boson numbers, and do not fall below the SU(3) values until spins just below the cutoff. This is not to say that the Dy and actinide nuclei are examples of SU(5) but only that, though close to the SU(3) limit, small symmetry breaking, for example, in the direction of SU(5), might significantly alter the B(E2) predictions of the pure limit, especially in cases like these of large N values. A second point, emphasized by Gelberg,[32] is that the boson number itself is defined by reference to major shells. In the case of strongly deformed nuclei, the single particle (Nilsson) orbits from different shells intermingle and large shell gaps are less evident. It may well be that one needs to employ effective N values in such nuclei, thereby again modifying the IBA predictions. Since such an effective N would have consequences for lower lying low spin states one therefore has additional constraints on N which must be simultaneously satisfied. Finally, the simple IBA may be extended by the introduction of a g(L=4) boson which clearly will obliterate the spin cutoff and significantly affect B(E2) values.

The effects of the g boson, however, are best discussed in connection with the high spin multiple Coulomb excitation data on ^{194,196}Pt. Reanalysis of the data has resulted[29] in a revised set of B(E2) values (shown in fig. 13) that do not exhibit a falloff. These nuclei are excellent examples of the 0(6) scheme, and any perturbation thereto would be caused by a slight tendency toward the SU(3) limit, which has the same predicted B(E2) falloff as in 0(6). Further, the boson number N is rather small so that the discrepancies occur at fairly low spin ($I=8^+$ or 10^+), in particular, below the backbend at spin 10^+. Finally, these are not strongly deformed nuclei and the boson number should be well defined. Thus some of the possible sources of the discrepancies applicable to the Dy, Th and U nuclei are unlikely here. On the other hand, the neighboring Os isotopes have well known low lying K=4 bands which, though apparently having large amplitudes for excitations built on the σ=max, τ=4 state which resembles a two phonon γ vibration, also exhibit properties expected for a hexadecapole vibration. Thus one might expect the g boson to be important in this region. Clearly,

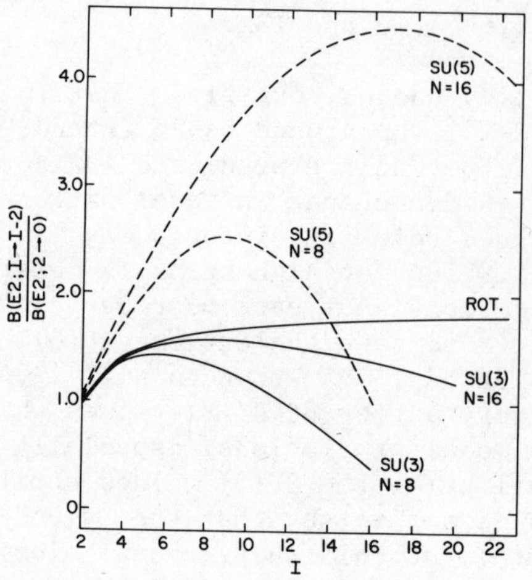

Fig. 12. B(E2) values in the ground band for the symmetric rotor and the SU(3) and SU(5) limits of the IBA for two values of the boson number N.

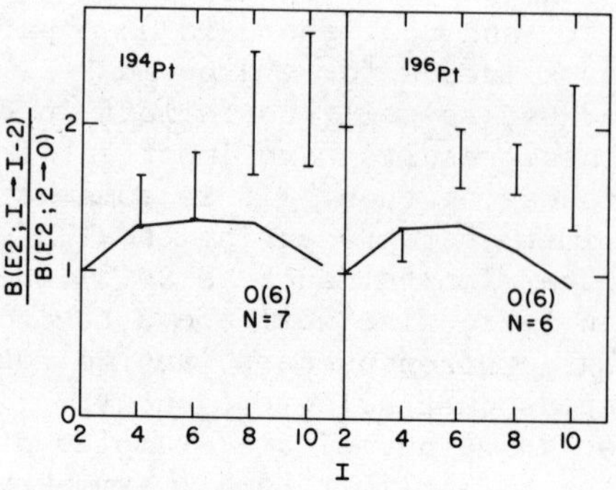

Fig. 13. B(E2) values in Pt compared with predictions of the O(6) limit. Taken from ref. 29.

since it also leads to low lying K=3 and 4 excitations, there are strong constraints on its energy and interactions with the s,d boson system: if relevant, the g boson must simultaneously account for the hexadecapole vibrations and the high spin B(E2) values in the ground band, and its properties must be consistent with evidence for K=3 or 4 bands in nearby (Er,Yb) nuclei.

Given the enormous and incontrovertable overall success of the IBA, these discrepancies for high spin states should be treated, not as disproving the IBA, but as a subtle means of probing possible extensions to it, always, though, in conjunction with the constraints imposed by the data for low spin states.

SUMMARY

Overall, the IBA has exhibited a remarkable ability to account for the character and properties of a broad variety of collective states over a major portion of the periodic table. It does this with a minimum of parameters which, furthermore, usually vary smoothly and, in many cases, have a priori predictable behaviour. Its predictions of new nuclear symmetries, its simple treatment of complex transitional regions and its ability to use the smooth behaviour of its parameters to predict as yet unstudied nuclei have been noted. With the rapid growth in the number of tests of the IBA, the first significant discrepancies have also appeared. Thus far, the most important seem to center on the calculations for ^{194}Os, in particular in relation to the expected microscopic shell structure and the consequent expected IBA-2 parameter variations, on the possibly related discrepancies for quadrupole moments in Pt, and on the expected spin cutoffs and B(E2) falloffs for high spin states. Given the overriding success of the IBA, it is suggested that these discrepancies are best treated as an opportunity to probe possible refinements and extensions of the IBA. Finally, it should be clear that the IBA is complementary to geometrical models. It is a replacement for them primarily in the sense that it provides a single unified model and framework that can subsume the realms of many individual geometrical models and thereby treat broad classes of nuclei, often highly disparate in structure, in a unified way without the need to resort to the ad hoc choice of particular schemes for particular nuclei or classes of states.

I would like to express deep appreciation to F. Iachello for innumerable discussions and for aid in keeping abreast of theoretical developments in the IBA and of experimental tests of it. Discussions along these lines with A. Arima, I. Talmi, H. Feshbach, O. Scholten and R. Broglia are gratefully acknowledged. Obviously this review would not have existed without the help of a number of others, both experimentalists and theorists, who are engaged in testing the IBA and who often provided results prior to publication. In particular I would like to thank J. A. Cizewski, A. Gelberg, P. Brentano, K. P. Lieb, B. Barrett, D. D. Warner, A. Dieperink, K. Stelzer, C. Baktash, Th. W. Elze, and J. Stachel, and E. Grosse.

REFERENCES

1. R. F. Casten, International Conference on Band Structure and Nuclear Dynamics, New Orleans, Louisiana, Feb. 1980, to be published.
2. A. Arima and F. Iachello, Phys. Rev. Lett. 35 (1975) 1069; Ann. Phys. (N.Y.) 99 (1976) 253; 111 (1978) 201; Phys. Rev. Lett. 40 (1978) 385; Phys. Rev. C16 (1977) 2085.
3. J. A. Cizewski, R. F. Casten, G. J. Smith, M. L. Stelts, W. R. Kane, H. G. Borner, and W. F. Davidson, Phys. Rev. Lett. 40 (1978) 167.
4. O. Scholten, F. Iachello and A. Arima, Ann. Phys. (N.Y.) 115 (1978) 366.
5. R. F. Casten and J. Cizewski, Nucl. Phys. A309 (1978) 477.
6. A. Arima, T. Otsuka, F. Iachello and I. Talmi, Phys. Lett. 66B (1977) 205.
7. T. Otsuka, A. Arima, F. Iachello and I. Talmi, Phys. Lett. 76B (1978) 139.
8. T. Otsuka, A. Arima and F. Iachello, Nucl. Phys. A309 (1978) 1.

9. F. Iachello, in Interacting Bosons in Nuclear Physics, ed. by F. Iachello (Plenum Press, New York, 1979), p. 1.
10. O. Scholten, in Interacting Bosons in Nuclear Physics, ed. by F. Iachello (Plenum Press, New York, 1979), p. 17 and O. Scholten, Thesis, 1980.
11. Manual for IBA-1 Code Package PHINT, written by O. Scholten.
12. R. Casten, in Interacting Bosons in Nuclear Physics, ed. by F. Iachello (Plenum Press, New York, 1979), p. 37.
13. R. F. Casten and J. A. Cizewski, to be published.
14. H. H. Bolotin, I. Katayama, H. Sakai, Y. Fujita, M. Fujiwara, K. Hosono, T. Itahashi, T. Saito, S. H. Sie, P. Conley, D. L. Kenedy and A. E. Stuchbery, J. Phys. Soc. Jap. 47 (1979) 1397.
15. K. Kumar and M. Baranger, Nucl. Phys. A122 (1968) 273; K. Kumar, Nucl. Phys. A92 (1967) 653; and Nucl. Phys. A231 (1974) 189.
16. R. Bijker, A. E. L. Dieperink, O. Scholten and R. Spanhoff, Preprint, KV1-224 and this conference.
17. P. D. Duval and B. R. Barrett, preprint and this conference.
18. R. F. Casten, A. I. Namenson, W. F. Davidson, D. D. Warner and H. G. Borner, Phys. Lett. 76B (1978) 280.
19. C. Baktash, J. X. Saladin, J. J. O'Brien and J. G. Alessi, preprint.
20. J. A. Cizewski, E. R. Flynn, R. E. Brown and J. W. Sunier, Phys. Lett. 88B (1979) 207.
21. T. Tamura, K. Weeks and T. Kishimoto, Phys. Rev. C20 (1979) 307.
22. A. Saha, O. Scholten, D.C.J.M. Hageman and H. T. Fortune, Phys. Lett. 85B (1979) 215.
23. U. Kaup and A. Gelberg, Z. Phys. A293 (1979) 311.
24. H. P. Hellmeister, U. Kaup, J. Keinonen, K. P. Lieb, R. Rascher, R. Ballini, J. Delaunay and H. Dumont, Phys. Lett. 85B (1979) 34 and Nucl. Phys. A332 (1979) 241 and P. Lieb, private communication, 1979.
25. P. Van Isacker and G. Puddu, preprint, 1980.
26. G. Puddu, O. Scholten and T. Otsuka, preprint, KV1-226.
27. D. D. Warner, R. F. Casten and W. F. Davidson, to be published.
28. H. Ower et al., International Conference on Nuclear Behaviour at High Angular Momentum, Strasbourg, France, April 22-24, 1980.
29. J. Idzko et al., International Conference on Nuclear Behaviour at High Angular Momentum, Strasbourg, April 22-24, 1980.
30. H. Emling et al., International Conference on Nuclear Behaviour at High Angular Momentum, Strasbourg, April 22-24, 1980.
31. J. Panqueva, K. P. Hellmeister, F. J. Bergmeister and K. P. Lieb, preprint and K. P. Lieb, this conference.
32. A. Gelberg and A. Zemel, preprint and this conference.

Research has been performed under contract DE-AC02-76CH00016 with the U.S. Department of Energy.

STRUCTURE OF THE IBA-1 EFFECTIVE HAMILTONIAN IN THE Z=50-82 PROTON SHELL*

O. Castaños†, P. Federman†† and A. Frank†

†Centro de Estudios Nucleares and ††IFUNAM
Universidad Nacional Autónoma de México
A.P. 20-364, México 20, D.F. MEXICO

The kind of questions that motivate this work are: Can detailed agreement with experiment be obtained in the framework of I.B.A. models[1)] using the same parameters -or "effective" interaction - for whole series of isotopes, or is the hamiltonian a strongly varying function of the nucleon numbers? In the former case, do the experimental data constrain the "effective" hamiltonians to forms exhibiting the dynamical symmetries[2-4)], or do they converge into forms with more complicated interpretations? Which terms of the complete hamiltonian are effectively important in different regions?

In order to answer these questions, we have performed some rather detailed energy levels calculations for chains of isotopes in the Z = 50-82 shell starting with the complete, except for binding energies, IBA-1 hamiltonian[5)]:

$$H = k_1\hat{n}_d + k_2\hat{n}_d^2 + k_3\hat{n}_d\hat{N} + k_4\hat{L}^2 + k_5\hat{\Lambda}^2 + k_6\hat{P}^2 + k_7\hat{Q}^2 \quad (1)$$

where $\hat{n}_d$ and $\hat{N}$ are the d-boson and total-boson number operators respectively, $\hat{L}$ is the total angular momentum, $\hat{\Lambda}^2$ is the quadratic Casimir operator of O(5), $\hat{P}^2$ is related to the quadratic Casimir operator of O(6) and counts the number of boson pairs coupled to L = 0, and $\hat{Q}^2$ is the quadrupole-quadrupole interaction between the bosons, and is related to the Casimir operator of SU(3).

*Work supported by CONACYT-PNCB 1510

The above parameters k_i can be expressed in term of the single-boson energy splitting ε and the two-boson interaction matrix elements:

$$
\begin{aligned}
\varepsilon &= k_1 + k_2 + k_3 + 6\,k_4 + 4\,k_5 - \frac{9}{4}\,k_7 \\
\langle d^2 0 | V | d^2 0 \rangle &\; 2(k_2 + k_3) - 12\,k_4 - 8\,k_5 + \frac{5}{2}\,k_6 + \frac{7}{2}\,k_7 \\
\langle d^2 2 | V | d^2 2 \rangle &= 2(k_2 + k_3) - 6\,k_4 + 2\,k_5 - \frac{3}{4}\,k_7 \\
\langle d^2 4 | V | d^2 4 \rangle &= 2(k_2 + k_3) + 8\,k_4 + 2\,k_5 + k_7 \qquad (2) \\
\langle ds2 | V | ds2 \rangle &= 2\,k_3 + 4\,k_7 \\
\langle s^2 0 | V | s^2 0 \rangle &= \frac{1}{2}\,k_6 \\
\langle d^2 0 | V | s^2 0 \rangle &= \frac{\sqrt{5}}{2}\,k_6 + 2\sqrt{5}\,k_7 \\
\langle d^2 2 | V | ds2 \rangle &= 2\sqrt{7}\,k_7
\end{aligned}
$$

If binding energies are also considered, two more terms should be included in the hamiltonian (1): $k_8\,\hat{N}$ and $k_9\hat{N}^2$. Of course they would also contribute to ε and the two-body matrix elements in (2), but are ignored here since we deal only with excitation energies.

The relations (2) can be used to interpret the effect of the different terms in the general hamiltonian on the effective matrix elements.

The matrices of the hamiltonian (1) are constructed using analytic solutions obtained in the $U(6) \supset U(5) \supset O(5) \supset O(3)$ chain[5]. For practical reasons the calculations are limited to $N = 17$ bosons and $L = 10$.

A least square fitting procedure is used to obtain the parameters k_i, much the same way as done for Shell Model effective-interaction calculations, although here the initial values are more ambiguous. The procedure followed was to obtain the best fits for succesively 2,3, etc. parameters, and choose the best among them. For values of Z near closed shells (50 or 82), this procedure seems to converge to a unique answer, but for values of Z near the middle of the shell convergence seems more difficult. This may be related to the fact that the IBA-1 model includes only the completely symmetric representations of IBA-2, as pointed out recently by Scholten[6,15].

We have performed preliminary calculations for the Xe, Ba, Ce and Nd isotopes in the region near Z = 50; for the Pt, Os and Dy in the region near Z = 82; and for Nd, Sm and Gd in the intermediate region.

The results for Xe and Ba isotopes are compared with experiment[7] in Tables 1 and 2 respectively. The rather detailed fits are obtained with only three parameters: k_1, k_4 and k_6, which are listed in Table 3.

The number of experimental levels included in each case is also listed in Table 3. For the Xe isotopes all experimentally known energy levels are reproduced by the effective IBA hamiltonian. In other cases single levels seem to fall outside the model space and thus require the introduction of additional degrees of freedom. For instance, in the Ba isotopes there is one single level (out of 50), namely the third excited 0^+ state in ^{134}Ba, that is calculated with a deviation of about 2 Mev or some 20 times the r.m.s. and is therefore left out in the fitting procedure.

This region has been described in terms of a transition between O(6) and U(5) dynamical symmetries[2]. The present results support such a description. The values of k_1, k_4 and k_6 listed in Table 3 give the admixtures of U(5) and O(6) operators necessary to describe with the same "effective" hamiltonian the whole chain of isotopes. There is one difference though with IBA-2 results[8]: no important SU(3) influence is obtained in the present treatement. Rather the influence of the k_7 term starts to be important in the next region: Nd, Sm, Gd, etc....

Results for Sm and Gd are compared with experiment in Tables 4 and 5 respectively. The fits are obtained for the parameters listed in Table 3. In this region four parameters seem to be enough to describe the "effective" hamiltonian, representing an admixture of U(5) and SU(3) operators, as expected[9]. If less energy levels are included in the analysis, only three parameters are enough for the Sm isotopes[10]. As seen from Table 3, the influence of SU(3) appears in the present calculations starting with the Nd isotopes. Also in this case (as well as in the last) some levels had to be left out the least square search. Typically the correspond to second excited 0^+ and levels above it, again indicating the need for additional degrees of freedom.

Table 1. Experimental and calculated energy levels for the Xe isotopes.

A	J	E_{exp} (MeV)	$E_{cal}-E_{exp}$ (MeV)	A	J	E_{exp} (MeV)	$E_{cal}-E_{exp}$ (MeV)
116	2	.39	-.03	122	5	1.77	.14
	4	.92	-.08		6	1.46	.12
	6	1.53	-.12		6	2.06	-.03
	8	2.21	-.11		8	2.22	.11
	10	2.96	-.08		10	3.03	.12
118	0	.83	.00	124	2	.35	.09
	2	.34	.05		2	.85	.01
	2	.93	-.19		3	1.25	.10
	2	1.23	.08		4	.88	.12
	3	1.37	-.19		4	1.44	-.01
	4	.81	.07		5	1.84	.15
	4	1.44	-.18		6	1.55	.10
	5	1.52	.24		8	2.33	.07
	6	1.40	.08		10	3.23	.01
	6	2.00	-.12	126	2	.39	.08
	8	2.07	.10		2	.88	.02
	10	2.81	.16		3	1.32	.09
120	0	.91	-.04		4	.94	.09
	2	.32	.08		4	1.49	-.00
	2	.88	-.10		5	1.90	.15
	2	1.27	.09		6	1.64	.07
	3	1.27	-.04		6	2.21	-.04
	4	.80	.12		8	2.44	.03
	4	1.40	-.08		10	3.32	.01
	4	1.71	.02	128	2	.44	.04
	5	1.82	.02		2	.97	-.03
	6	1.40	.14		4	1.03	.04
	6	1.99	-.03	130	2	.54	-.03
	8	2.05	.20		2	1.12	-.15
	10	2.87	.19		4	1.21	-.09
122	2	.33	.09		6	1.94	-.13
	2	.84	-.02	132	2	.67	-.14
	3	1.21	.08		2	1.30	-.29
	4	.83	.13		3	1.80	-.23
	4	1.40	-0.03		4	1.44	-.29

Finally, some results for the region near Z=82 are presented in Tables 6 and 7. There we compare calculated and experimental[7] energies for the Pt and Os isotopes respectively. The fits are obtained with three or four parameters, listed in Table 3. The parameters represent admixture of O(6) and U(5) operators, quite symmetrically with the Xe-Ba region. The expected strong O(6)

Table 2. Experimental and calculated energy levels for the Ba isotopes.

A	J	E_{exp} (MeV)	$E_{cal}-E_{exp}$ (MeV)	A	J	E_{exp} (MeV)	$E_{cal}-E_{exp}$ (MeV)
120	2	.18	.03	128	4	1.37	-.06
	4	.54	.02		10	3.08	.08
	6	1.04	.01		5	1.93	.01
122	2	.20	.04		6	1.41	.06
	4	.57	.04		8	2.19	.06
	6	1.08	.05	130	2	.36	.02
	8	1.70	.07		2	.91	-.08
	10	2.40	.14		4	.90	.03
124	2	.23	.03		6	1.59	.03
	4	.65	.02	132	0	1.50	-.01
	6	1.23	.00		2	.47	-.02
	8	1.92	-.01		2	1.03	-.08
	10	2.69	-.01		3	1.51	.07
126	2	.26	+.03		4	1.13	-.08
	2	.59	.04		4	1.73	-.08
	4	.71	.03		6	1.93	-.11
	6	1.33	.00	134	0	1.76	-.06
	8	2.09	-.02		0	2.16	-.20
	10	2.94	-.01		2	.61	-.09
128	2	.28	.05		2	1.17	-.08
	2	.89	-.14		3	1.64	.12
	3	1.33	-.08		4	1.40	-.19
	4	.76	.06		4	1.97	-.12

influence[11]) is quite apparent from the values of k_6. But again no important k_7 term seems to be required in this treatement, as assumed in ref. 12.

The trends of the parameters in Table 3 seem to indicate that the O(6) dynamical symmetry is important for few proton (proton-holes) outside closed shells. Towards the middle of the shell SU(3) takes over, as expected if the $\hat{Q}^2$ term reflects mainly the neutron-proton interaction.

The present preliminary results indicate that it is possible to define effective IBA-1 hamiltonians that describe the energy levels of groups of isotopes. The effective hamiltonians reduce to simple forms including the admixture of symmetries required to describe the group of isotopes with the same parameters. In the Shell Model case the contributions to the effective operatores arise from neglected configurations. In the IBA case, in addition the effective hamiltonians may include contributions due to modifications in the boson structure[13]).

Table 3. Number of levels included in the fits, r.m.s. deviation and values of the parameters in the hamiltonian (1).

Isotope	#levels	σ	k_1	k_3	k_4	k_5	k_6	k_7
Xe	69	95	.519	-	.00985	-	.0382	-
Ba	46	80	.793	-	.00745	-	.232	-
Ce	25	80	.875	-.0896	.0136	.0281	-	-
Nd1	15	70	1.058	-.115	-	.0736	-	-
Nd2	18	75	1.574	-.118	-	-	-	-.0199
Sm	42	125	2.678	-.246	-	.0211	-	-.0139
Gd	62	115	3.085	-.266	-	.0248	-	-.0113
Dy	61	150	1.497	-.108	-	.0262	-	-.00347
Os	76	130	.286	.0369	.00280	-	.176	-
Pt	82	120	.565	-	-.00823	.00905	.0850	-

Table 4. Exp. and calculated energy levels for the Sm isotopes.

A	J	E_{exp} (MeV)	$E_{cal}-E_{exp}$ (MeV)	A	J	E_{exp} (MeV)	$E_{cal}-E_{exp}$ (MeV)
148	2	.550	.042	152	4	.367	-.022
	2	1.455	-.076		4	1.023	.037
	2	1.664	.148		4	1.372	.005
	4	1.180	.096		6	.707	-.008
	6	1.906	.156		8	1.125	.033
150	0	.740	-.195		10	1.615	.099
	2	.334	-.056	154	0	1.100	-.238
	2	1.046	-.179		2	.082	-.001
	2	1.194	-.076		2	1.178	-.233
	3	1.505	-.011		2	1.440	-.253
	4	.773	-.091		3	1.539	-.254
	4	1.449	-.069		4	.267	.000
	4	1.643	.068		4	1.372	-.240
	6	1.279	-.072		4	1.661	-.257
	8	1.837	.006		6	.547	.005
	10	2.432	.153		8	.903	.025
152	0	.685	-.041	156	0	1.068	-.081
	2	.122	-.014		2	.076	-.005
	2	.810	-.021	158	2	.073	-.008
	2	1.084	-.039		4	.240	-.022
	3	1.234	-.007		6	.499	-.047

Table 5. Experimental and calculated energy levels for the Gd isotopes.

A	J	E_{exp} (MeV)	$E_{cal}-E_{exp}$ (MeV)	A	J	E_{exp} (MeV)	$E_{cal}-E_{exp}$ (MeV)
150	0	1.21	-.18	154	8	1.15	-.04
	2	.64	-.04		8	1.76	.03
	2	1.43	-.07		10	1.64	.00
	2	1.52	.24	156	0	1.05	-.27
152	0	.62	-.14		2	.09	-.01
	2	.34	.08		2	1.13	-.27
	2	.93	-.14		2	1.15	.02
	2	1.11	-.08		3	1.25	-.02
	3	1.43	-.02		4	.29	-.02
	4	.76	.10		4	1.30	-.25
	4	1.28	-.01		4	1.36	-.23
	4	1.55	.04		5	1.51	-.00
	5	1.82	.20		6	.59	-.04
	6	1.23	-.07		8	.97	.05
	6	1.67	.19	158	0	1.20	-.30
	6	2.00	.22		2	.08	-.01
	8	1.75	.02		2	1.19	-.26
	10	2.30	.18		2	1.26	-.14
154	0	.68	-.07		3	1.27	-.07
	2	.12	-.02		4	.26	-.02
	2	.82	-.09		4	1.36	.23
	2	1.00	.02		4	1.41	-.09
	3	1.13	.05	160	2	.08	.00
	4	.37	.04		2	.99	.01
	4	1.05	-.07		3	1.06	.15
	4	1.26	.06		4	.25	-.02
	5	1.43	.11		4	1.15	-.05
	6	.72	-.05		5	1.48	-.02
	6	1.37	-.04		6	.51	-.03
	6	1.61	.11		6	.54	-.04
					8	.87	-.07

Table 6. Experimental and calculated energy levels for the Os isotopes.

A	J	E_{exp} (MeV)	$E_{cal}-E_{exp}$ (MeV)	A	J	E_{exp} (MeV)	$E_{cal}-E_{exp}$ (MeV)
182	0	.50	.05	190	2	.30	-.02
	2	.16	.02		2	.60	-.04
	2	.67	-.22		2	1.20	.05
	2	.86	.01		3	.92	.02
	4	.42	.05		4	.74	.06
	6	.77	.10		4	1.13	-.13
	8	1.20	.14		5	1.45	-.01
	10	1.70	.24		6	1.29	-.12
184	0	.49	.11		6	1.73	-.16
	2	.16	.03		8	1.92	-.12
	2	.65	-.21	192	10	2.54	-.07
	2	.84	.10		0	1.20	-.21
	3	.94	-.19		2	.32	-.01
	4	.44	.07		2	.61	.01
	4	1.03	-.22		2	1.44	-.06
	4	1.24	-.15		3	.92	.09
	5	1.31	-.14		4	.79	-.05
	6	.80	.11		4	1.20	-.10
	6	1.46	-.20		5	1.48	.08
	8	1.23	.19		6	1.37	-.09
	10	1.71	.33		8	2.02	-.13
186	0	.47	.19		10	2.52	.08
	2	.19	.03	194	0	1.27	-.23
	2	.61	-.13		2	.33	.02
	2	.80	.21		2	.62	.07
	4	.49	.07		2	1.51	-.01
	6	.88	.12		3	.92	.18
	8	1.34	.18		4	.81	-.01
	10	1.86	.29		4	1.23	-.04
188	0	.80	-.04		6	1.41	-.05
	2	.27	-.02		8	2.10	-.08
	2	.61	-.09		10	2.44	.27
	2	1.12	.03	196	0	1.14	-.08
	3	.94	-.08		2	.36	.02
	4	.67	-.06		2	.69	.07
	4	1.08	-.15		2	1.36	.19
	5	1.44	.13		3	1.02	.20
	6	1.19	-.09		4	.88	-.01
	8	1.78	-.11	198	2	.41	.01
	10	2.43	-.09		2	.78	.05
190	0	.92	-.07		4	.99	-.05

Table 7. Experimental and calculated energy levels for the Pt isotopes.

A	J	E_{exp} (MeV)	$E_{cal}-E_{exp}$ (MeV)	A	J	E_{exp} (MeV)	$E_{cal}-E_{exp}$ (MeV)
178	2	.13	.06	186	5	1.28	.04
	3	.90	-.09		6	.87	.03
	4	.40	.08		6	1.49	-.16
	5	1.21	.05		8	1.42	-.04
	6	.76	.09		10	2.07	-.13
	8	1.19	.13	188	0	1.09	-.22
	10	1.68	.19		2	.16	.05
180	2	.13	.06		2	.63	-.14
	3	1.02	-.20		2	1.31	-.02
	4	.41	.07		3	.79	.07
	5	1.38	-.10		4	.48	.04
	6	.80	.07		4	.97	-.09
	8	1.26	.08		5	1.18	.15
	10	1.77	.12		6	.94	-.03
182	2	.13	.07		8	1.51	-.12
	3	1.04	-.19		10	2.17	-.22
	4	.40	.09	190	0	.91	-.05
	5	1.40	-.11		2	.19	.03
	6	.79	.08		2	.56	-.05
	8	1.28	.07		2	1.12	.19
	10	1.81	.09		3	.76	.12
184	0	1.04	-.22		4	.55	-.02
	2	.12	.08		4	.96	-.06
	3	1.08	-.21		6	1.05	-.12
	4	.38	.11		8	1.67	-.25
	5	1.43	-.13	192	0	.96	-.08
	6	.77	.11		2	.21	.01
	8	1.28	.09		2	.49	.02
	10	1.87	.05		3	.69	.20
186	0	1.06	-.23		4	.58	-.04
	2	.14	.07		4	.91	-.00
	2	.77	-.26		5	1.14	.22
	2	1.21	.07		6	1.09	-.15
	3	.91	-.06		6	1.47	-.08
	4	.43	.07	194	0	.70	.19
	4	1.07	-.20		2	.22	.01
	4	1.46	-.16		2	.66	.13
					4	.60	.05

We wish to thank J. Zylicz for calling our attention to the correct ^{116}Xe experiment levels[14]) and Guadalupe Estrada for her careful typing.

REFERENCES

1. A. Arima and F. Iachello, Phys. Rev. Lett. 35 (1975) 1069.
2. A. Arima and F. Iachello, Ann. Phys. (N.Y.) 99 (1976) 253.
3. A. Arima and F. Iachello, Ann. Phys. (N.Y.) 111 (1978) 201.
4. A. Arima and F. Iachello, Phys. Rev. Lett. 40 (1978) 385.
5. O. Castaños, E. Chacón, A. Frank and M. Moshinsky. Journ. Math. Phys. 20 (1979) 35.
6. O. Sholten, Thesis, 1980.
7. Table of Isotopes, Ed. by C.M. Lederer and V.S. Shirley, John Wiley & Sons, 1978; M. Sakai and Y. Gono, preprint, July 1979.
8. G. Puddu, O. Scholten and T. Otsuka, preprint, December 1979.
9. O. Scholten, F. Iachello and A. Arima, Ann. Phys. (N.Y.) 115 (1978) 325.
10. O. Castaños, A. Frank and P. Federman, Phys. Lett. 88B (1979) 203.
11. A. Arima and F. Iachello, Ann. Phys. (N.Y.) 123 (1979) 468.
12. R.F. Casten and J. A. Cizewski, Phys. Lett. 79B (1978) 5.
13. P. Federman and S. Pittel, Phys. Rev. C 20 (1979), 820.
14. A. Plochocki et al., Proc. of the Canberra Conference on Nuclear Interactions, p. 409, Sfringler-Verlaz, 1978.
15. P. Van Isacker and G. Puddu, preprint 3074-622 Yale University.

COLLECTIVE FEATURES IN LIGHT NUCLEI *

Heinz Clement

Sektion Physik der Universität München
8046 Garching, West-Germany

ABSTRACT

Collective features in the low energy excitation spectrum of light even-even nuclei are discussed. It is attempted to describe their mainly transitional character in the formalism of IBA. In comparing to experimental results the main emphasis is put onto transition amplitudes, which give a detailed information on the collective character of nuclear states.

INTRODUCTION

The excitation spectrum of light nuclei has been the topic of extensive studies since many years. Therefore, the decay properties of most of the low-lying states are known. In addition the static quadrupole moments for the 2_1^+-states have been determined from Coulomb excitation experiments. Further insight into the dynamical properties has been achieved most recently by polarized deuteron scattering experiments[1)], which allow the extraction of transition amplitudes not only in their magnitude but also in their phase supplying thus a further stringent test of nuclear models.

The collective behaviour of the nucleons common in heavy elements is known even in nuclei as light as the ones in the fp- and sd-shell. Fig. 1 sketches typical energy levels of even parity for α-type nuclei, which reveal evidently collective features. Whereas in the lower part of the shell the rotational behaviour clearly dominates, occurs in the upper part a transition to a pronounced vibrational structure. It is these systematic changes in the collectivity of these nuclei on the one hand and the ability of the IBA[2)] to handle rotational, vibrational and transitional nuclei in the same

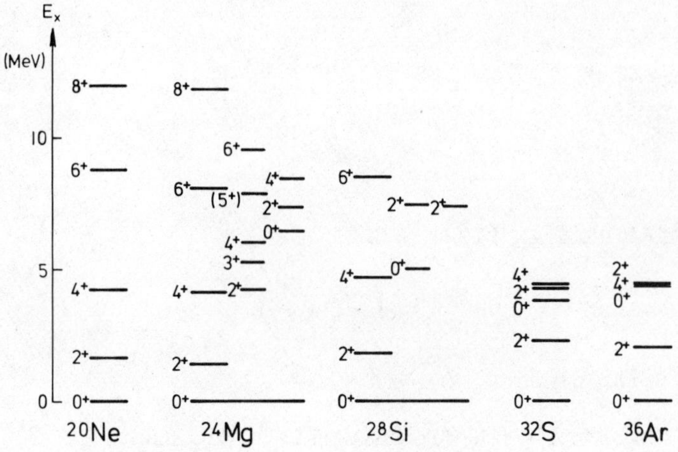

Fig. 1. Sketch of collective even-parity levels in sd-shell nuclei.

framework on the other hand, which gives the motivation to describe these nuclei in the formalism of IBA. Since these nuclei are also light enough for a microscopic shell model description, the sd-shell gives the unique opportunity to study the many-particle and collective nature of the states at the same time. The observation of pronounced rotational features in such nuclei as ^{24}Mg, e.g. was the impetus for the development of the SU3-model by Elliott[3], his classification of shell model states according to the group SU3 lead not only to a simplification in microscopic calculations, but also to a better understanding of the results of such complicated and elaborate calculations. One should note that, although the SU(3) group introduced by Elliott is formal identical to the SU(3)-limit of IBA, it refers to fermions rather than bosons.

Very sensitive measures of the collectivity are transition matrix elements and combinations of them like the quantity $P_4=Q_{2_1^+}\cdot\langle 0_1^+\|E2\|2_1^+\rangle\langle 2_1^+\|E2\|2_2^+\rangle.\langle 2_2^+\|E2\|0_1^+\rangle$, which connects the reduced matrix elements of the lowest 2^+-states. According to Kumar[4] P_4 should be negative, if these states are either of vibrational or rotational (ground and γ-band) character. For all the nuclei discussed in the following the measurements[1] show, that this rule is fulfilled.

IBA-CALCULATIONS

To induce both the rotational and the vibrational features in the sd-shell in a semi-microscopic approach we have utilized the

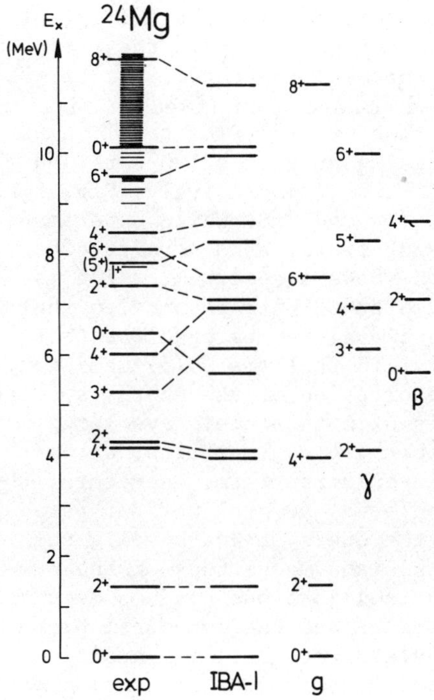

Fig. 2. Comparison of the experimental[6] and the calculated energy spectrum of even-parity levels in ^{24}Mg. The parameters of the IBA-1 hamiltonian are (in MeV): ε=0, κ=+.053, κ'=0.090, κ"=-1.0, OCT=+0.09, HEX=-0.10. For explanation of notation see refs. 2 and 5.

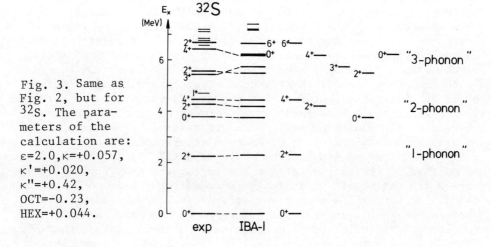

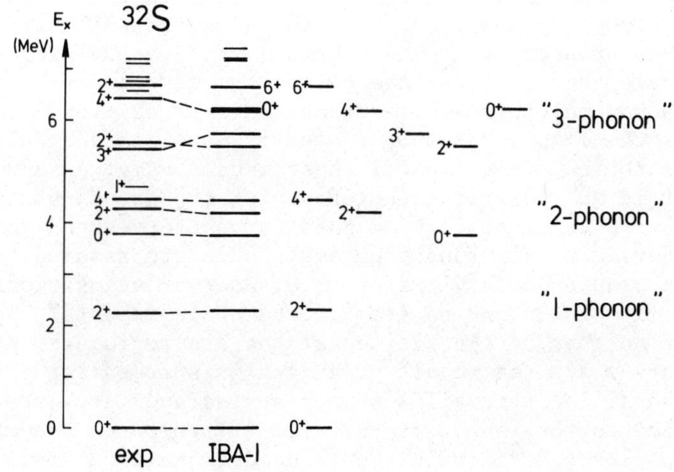

Fig. 3. Same as Fig. 2, but for ^{32}S. The parameters of the calculation are: ε=2.0, κ=+0.057, κ'=+0.020, κ"=+0.42, OCT=-0.23, HEX=+0.044.

SU(6)-formalism of the IBA. The aim of this undertaking is that the parameters involved in the description of IBA may depend in some characteristic way on the mass-number, i.e. the filling of the sd-shell. So far calculations have been performed in the framework of IBA-1[2] the one-kind-of-boson approach. The parameters of this hamiltonian in the multipole expansion have been fixed by fits to the energy spectra utilizing the code PHINT[5]. The total number N is taken as the number of nucleon pairs above the ^{16}O-core and the number of hole pairs below the ^{40}Ca-core, respectively. Fig. 2 and 3 show the results for the nuclei ^{24}Mg and ^{32}S. While experimentally the first one exhibits a clear rotational band structure corresponding to a triaxial nucleus, ^{32}S shows a spectrum, which is characteristic for an anharmonic vibrator. With respect to their energy spectra these nuclei can be classified being close to the limits O(6) and SU(5), respectively. In fact the SU(5)-hamiltonian gives already a very reasonable description of the energy spectrum of ^{32}S. However, the excited levels of both nuclei have large static quadrupole moments and exhibit decay properties, which require the inclusion of the Q.Q-interaction term. Therefore, the full SU(6) hamiltonian has been employed. Several sets of parameters are found, which reproduce the energy spectra well, but differ in the reproduction of phases and magnitudes of the transition amplitudes. For each nucleus just one set has been found, which describes both the energy spectra and the dynamical properties. This points to the extreme selectivity of the transition amplitudes to the SU(6)-hamiltonian. In this way both nuclei can be described very well up to about 6-8 MeV excitation energy (i.e. up to and above the pairing gap), with one-to-one correspondence between the experimental and the calculated even-parity levels (with exception of 1^+-states which are excluded in IBA-1).

According to its dynamical properties the calculated levels can be grouped into bands or multiplets well-known from geometrical pictures as shown on the right hand sides in Figs. 1 and 3. For ^{24}Mg pronounced γ- and β-bands exist in addition to the ground state band, for ^{32}S the calculated levels can be divided into n-phonon multiplets with strong $\Delta n=2$ transitions. The actual mixing of the 1- and 2-phonon states is unveiled very clearly in the hadronic excitation of these levels. Fig. 4 shows the scattering[7] of 18 MeV polarized deuterons for the ground-state and the $2_1^+, 0_2^+$, 2_2^+ and 4_1^+-states. The dashed curves correspond to a description, where all the transition amplitudes are assumed to be those of a pure quadrupole vibrator, i.e. no $\Delta n=2$ transitions as in the SU(5)-limit of IBA and no static quadrupole moments. The scattering data of neither of the excited states are reproduced properly. The full curves are the result of an analysis revealing strong $\Delta n=2$ transitions for all 2-phonon states and large static quadrupole moments for the 2^+-states. Table 1 compares the experimental results for ^{32}S with the reduced transition amplitudes of the IBA-calculation employing the full SU(6)-hamiltonian (see caption

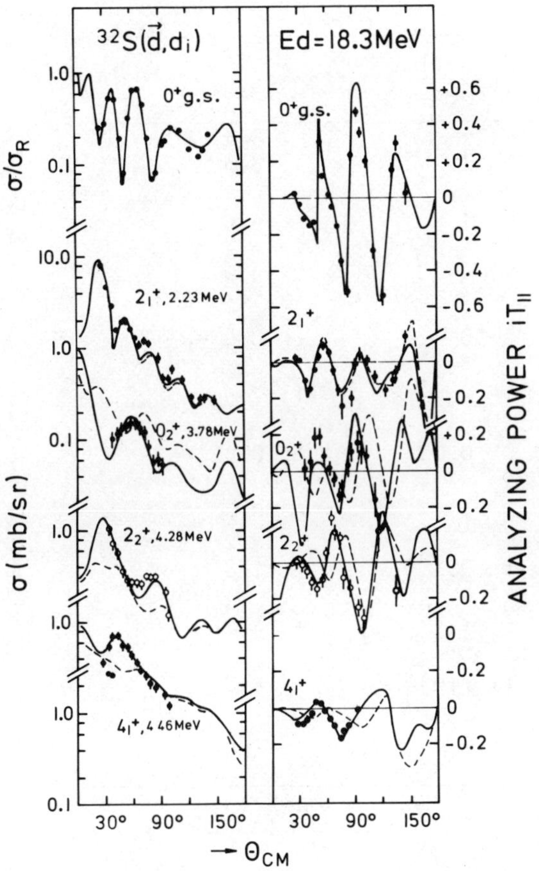

Fig. 4. Cross section and analyzing power of the scattering of polarized deuterons on ^{32}S.

of Fig. 3). Choosing the E2-transition operator in its simplest form $T(E2)=q_2(s^\dagger d+d^\dagger s)$ and adjusting the effective E2-strength parameter to fit the decay of the 2_1^+-state leads to full agreement of the calculated values with all measured E2-transitions and quadrupole moments, both in magnitude and in phase. In addition also the phases for the E4 and E0 transitions coincide with the experimental results. Table 2 gives a similar comparison for the transition properties of ^{24}Mg. Here, too, the agreement is remarkably good with the exception of the E4 transitions. The cut-off of the Yrast-band at spin 8^+ predicted by the model agrees with the experimental finding, though the $8^+ \rightarrow 6^+$ transition has still a large experimental uncertainty.

Table 1. Reduced matrix elements and static quadrupole moments Q_{2^+} for ^{32}S. For the IBA-results q_2=4.5 efm^2 has been used

I_i	I_f	<f‖ Eλ‖ i> in efm^λ exp.[a)]	IBA-1
2_1^+	$\to 0_1^+$	+ 18 ± 1	+ 16.4
2_2^+	$\to 0_1^+$	− 5 ± 1	− 6.0
	$\to 2_1^+$	− 19 ± 3	− 18.2
4_1^+	$\to 0_1^+$	− 300 ± 50	− 0.2 $*q_4$
	$\to 2_1^+$	+ 26 ± 2	+ 27.6
0_2^+	$\to 0_1^+$	+ 0.11 ± 0.03	+ 0.43 $*q_0$
	$\to 2_1^+$	− 8 ± 1	− 10.7
2_3^+	$\to 0_1^+$	? 1.7 ± 0.3[b)]	+ 1.8
	$\to 2_1^+$	? 3.9 ± 1.5[b)]	+ 4.2
3_1^+	$\to 2_1^+$	? 10.6 ± 1.0[c)]	+ 8.8
4_2^+	$\to 2_1^+$	? 12.7 ± 1.5[c)]	− 11.6
$Q_{2_1^+}$		− 14 ± 5	− 10.3
$Q_{2_2^+}$		+ 14 ± 10	+ 3.2

a) Ref. 1 b) Ref. 9 c) Ref. 8

For ^{24}Mg and ^{32}S the formalism of IBA has been shown to be appropriate to describe satisfactorily both the static and the dynamic collective properties and to correlate the characteristics of the transition amplitudes to the characteristics of the excitation spectrum of these nuclei. Other nuclei in the sd-shell are presently under investigation with the aim to find a systematic trend in the IBA-parameters across the sd-shell. For ^{28}Si and ^{36}Ar promising results have been obtained so far, for ^{20}Ne we face the well-known problem, that there exists a pronounced rotational band up to 8^+, whereas the IBA-model as used here predicts a cut-off at

Table 2. Reduced matrix elements and static quadrupole moments Q_{2^+} for ^{24}Mg. For the IBA-results q_2 = 4.0 efm^2 has been used.

I_i	I_f	<f‖ Eλ‖ i> in efm^2 exp.[a]	IBA-1
2_1^+	→ 0_1^+	− 20.4 ± 0.5	− 20.4
2_2^+	→ 0_1^+	+ 4.5 ± 0.7	+ 7.1
	→ 2_1^+	− 14 ± 0.2	− 14.7
4_1^+	→ 0_1^+	0 ± 40	− 0.9 $*q_4$
	→ 2_1^+	+ 31 ± 2	+ 30.4
3_1^+	→ 2_1^+	+ 7 ± 2	+ 8.3
	→ 2_2^+	+ 32 ± 6	+ 23.5
4_2^+	→ 0_1^+	+345 ± 50	+ 0.15 $*q_4$
	→ 2_1^+	− 3 ± 2	− 3.1
	→ 2_2^+	+ 19 ± 4	+ 20.4
	→ 4_1^+	? 6.1 ±$^{2.5}_{6.1}$ [c]	− 15.4
0_2^+	→ 2_1^+	? 1.5 ± 0.2[b]	− 0.2
	→ 2_1^+	? 6.0 ± 0.7[b]	− 10.0
2_3^+	→ 0_1^+	? 2.3 ± 0.3[b]	+ 0.6
6_1^+	→ 4_1^+	? 43 ±$^{18}_{8}$ [c]	+ 33.2
4_3^+	→ 2_1^+	? 4.7 ± 0.9[b]	− 0.5
	→ 2_2^+	? 5.4 ± 1.1[b]	− 9.8
6_2^+	→ 4_1^+	? 6.5 ±$^{3}_{1.5}$ [c]	+ 6.8
	→ 4_2^+	? 34 ±$^{15}_{7}$ [c]	− 24.1
8_1^+	→ 6_1^+	? 33 ±$^{20}_{7}$ [c]	+ 29.0
	$Q_{2_1^+}$	− 16 ± 4	− 14.2
	$Q_{2_2^+}$	+ 16 ± 10	+ 14.0

a) Ref. 1 b) Ref. 8 c) Ref. 10

spin 4^+. A solution of this situation could be supplied, e.g. by the introduction of g bosons, which have been proposed due to the fact that in many nuclei large E4 moments are observed, which are not explained in the current IBA-model (e.g.) the Pt-isotopes show large $4_2^+ \rightarrow 0_1^+$ transitions[11]), as is the case also in ^{24}Mg).

For the fp-shell Fig. 5 shows ^{54}Cr as an example. The transition amplitudes and static quadrupole moments for the levels $0_1^+ - 2_1^+ - 4_1^+ - 2_2^+$ obtained from polarized deuteron scattering[1]), agree very well with the triaxial rotor picture of Davydov[12]). In the IBA-1 description, where ^{40}Ca is assumed as core, the low-lying spectrum is described satisfactorily by adjustment of 3 parameters in the hamiltonian. In addition to the ground and γ-band, which are fitted similarily also by the Davydov model, it predicts moreover ß- and higher bands, whose members correspond in energy well to experimentally known levels. For the lowest states the dynamical properties are sketched on the right hand side of fig. 5. Again the B(E2)-values, quadrupole moments and phases agree reasonably with the experimental results. However, as in the other cases the parameters of the SU(6)-hamiltonian could be fixed uniquely only by comparison to the experimentally determined transition amplitudes. Especially the amplitudes connected in the quantity P_4 turned out to show a very large sensitivity to the IBA-parameters for all nuclei investigated.

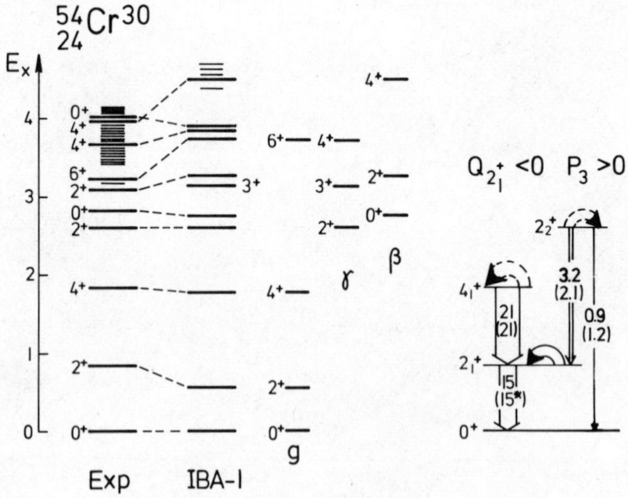

Fig. 5. Same as Fig. 2, but for ^{54}Cr. The parameters of the calculation are: κ=+0.027, κ'=+0.072, κ"=-0.047, the others being zero. On the right hand side the dynamical properties of the low-lying levels are sketched. The numbers refer to B(E2)-values[1]), in brackets are the IBA-results.

CONCLUSIONS

Attention has been paid to distinctive light nuclei, which show in their excitation spectrum as well as in their dynamical behaviour pronounced collective features comparable to those known in heavy nuclei. It has been shown that it is possible to describe these features in the mathematical framework of IBA. With the model a connection between the dynamical properties and the excitation spectrum can be achieved. It has to be emphasized, however, that several sets of parameters can be found for each nucleus supplying equally good descriptions of the excitation spectrum but very different sets of transition ampltides. This points to the extreme selectivity of the dynamical properties to nuclear models and to the need of detailed experimental information.

It is a pleasure to thank Dr. O. Scholten for making his computer codes available.

* supported in part by the BMFT

References:

1) H. Clement, R. Frick, G. Graw, F. Merz, P. Schiemenz, N. Seichert and Sun Tsu Hsun, Phys.Rev.Lett. 45 (1980) 599, and to be published
2) A.Arima, F. Iachello, Ann.Phys. 99 (1976) 253 and 111 (1978) 202 O. Scholten, F. Iachello, A. Arima, Ann.Phys. 115 (1978) 325 and quotations herein
3) J.P. Elliott, Proc.Roy.Soc.Ser. A245 (1958) 128 and 562 M. Harvey, Adv.Nucl.Phys.Vol.1, eds. M. Baranger and E.Vogt, Plenum Press, N.Y. 1968
4) K. Kumar, Phys.Lett. 29B (1969) 25
5) O. Scholten, Groningen, code PHINT, 1980
6) C.M. Lederer and V.S. Shirley, Table of Isotopes (Wiley, N.Y., 1978) 7th ed.
7) H. Clement, D. Ehrlich, R. Frick, G. Graw, P. Schiemenz, N. Seichert, Proc. AIP Conf. on Clustering Aspects, Winnipeg 1978, 742 and to be published
8) P.M. Endt, Nucl. Data Tables 23 (1979) 3
9) W.F. Coetzee, M.A. Meyer, D. Reitmann, Nucl.Phys. A185 (1972) 644
10) D. Branford, A.C. Mc Gough, I.F. Wright, Nucl.Phys. A241 (1975)349
11) P.T. Deason, C.H. King, R.M. Ronningen, T.L. Khoo, F.M. Bernthal, J.A. Nolen, preprint Michigan State Univ. (1980) to be published
12) A.S. Davydov, G.F. Filippov, Nucl.Phys. 8 (1958) 237

THE LEVEL STRUCTURE OF ^{156}Gd

Anders Bäcklin

Tandem Accelerator Laboratory, Box 533

S-751 21 Uppsala, Sweden

INTRODUCTION

The spectra of even-even deformed nuclei are well known up to the region including the first vibrational bands. Above that region, however, only partial information exist on these nuclei. For instance, no complete information on the states in the region where two-phonon vibrational states are expected is available in any deformed nucleus.

The extremely large thermal neutron capture cross sections of ^{155}Gd (61 000 b) and ^{157}Gd (255 000 b) offer a unique possibility to obtain a complete information for states with spin up to about five units up to an energy of about 2 MeV in the deformed even nuclei ^{156}Gd and ^{158}Gd.

We here report some results obtained from a study [1)] of the reaction ^{155}Gd(n,γ)^{156}Gd using high-resolution spectroscopy. Gamma-rays from the reaction mentioned were measured with the Risø bent crystal spectrometer [2)] and several Ge(Li) spectrometer systems at Idaho Falls [3)]. The corresponding conversion electron spectra were studied with magnetic spectrometers at Studsvik [4)] and Munich [5)]. About 650 transitions were observed. These data and information from radioactive decay [6)] and transfer reactions [7)] were used for the construction of the level scheme [1)].

The experimental data are compared with predictions of the IBA model, at first hand IBA-1. This model, which so successfully reproduces the systematic trends of the lowest states in even-even nuclei over vast mass regions [8)], also predicts numerous collective states at higher energies [9,10)].

POSITIVE PARITY STATES

Fig 1 shows the experimentally observed $K^{\pi}=0^+$, 2^+ and 4^+ bands to the left and those calculated in IBA-1 to the right. Also the strongest inter-band transition modes are indicated with arrows representing the average of the reduced transition probabilities between the bands.

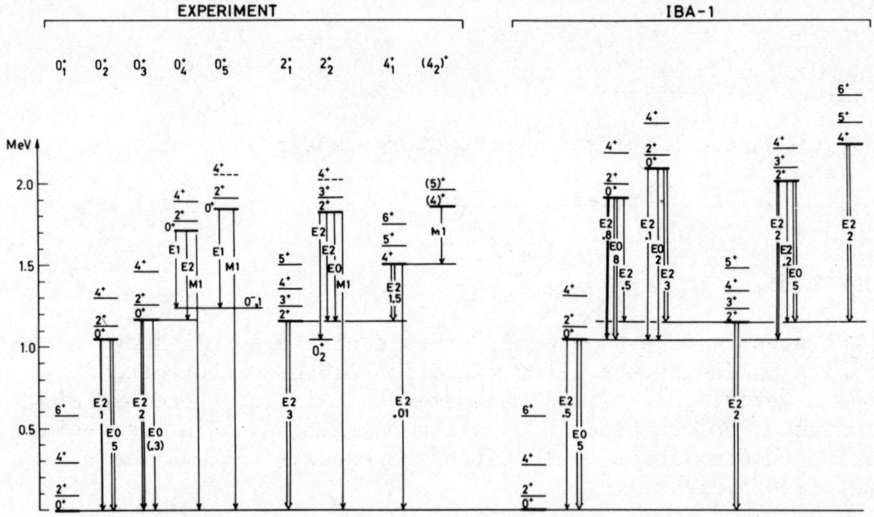

Fig 1. $K^{\pi}=0^+$, 2^+ and 4^+ bands in ^{156}Gd observed in the (n,γ) reaction (left) and calculated in the IBA-1 model (right). The vertical arrows indicate the main mode of decay of the bands. In cases where the transition probabilities are known the number below the multipolarity symbol indicates the average reduced strength of the transitions between the bands connected by the arrow. The units employed are: For E0: $10^{-2}\rho^2$; for E1: $10^{-8}e^2b$; for E2: $10^{-2}e^2b^2$; for M1: $10^{-3}\mu_N^2$. The corresponding Weisskopf units are: For E1: $1.9\cdot10^{-2}e^2b$; for E2; $0.50\cdot10^{-2}e^2b^2$; for M1: 1.8 μ_N^2. The widths of the arrows have been drawn approximately proportional to the reduced strengths. The experimentally observed bands have been labelled in order of raising energy.

It is known from other work [11] that the level energies of the ground band (0_1^+), the first excited 0^+ band (0_2^+) and the first 2^+ band (2_1^+) up to spin 10 to 12 are well reproduced by IBA-1. As seen in Fig 1 the 0_2^+ band decays with on the average enhanced E2 transitions [12] and strong E0 transitions to the 0_1^+ band, which (in the phenomenologic terminology) is characteristic for a β band.

In the present experiment three additional 0^+ bands are observed. The lowest of these, the 0_3^+ band, decays to the 0_1^+ band with E2 transitions on the average twice as strong as those from the 0_2^+ band, while the corresponding E0 transition is much weaker. For the 0_4^+ and 0_5^+ bands no transition rates have been measured and Fig 1 only indicates the strongest decay modes of these bands.

In IBA-1 no low-lying band corresponding to the 0_3^+ band is obtained. It has been suggested [9,13] that this band may arise through the promotion of particles to the next major shell. The strong E2 transitions to the 0_1^+ band may be explained as essentially due to mixing [1].

IBA-1 predicts two additional 0^+ bands in the energy region where the 0_4^+ and 0_5^+ bands are observed. Their E2 and E0 decay is compatible with that expected for bands that may be characterized as two-phonon 2β and 2γ bands. Experimentally bands with these properties are not observed. The 0_4^+ and 0_5^+ states decay preferentially with E1 transitions to the octupole states and do not seem to correspond to the states predicted by IBA.

In addition to the well known 2_1^+ band (γ-band) we observe a 2^+ band at higher energy. Its decay mode with strong E2 and E0 transitions to the 0_2^+ band and E2 transitions to the 2_1^+ band is similar to that of the second 2^+ band predicted by the IBA. The characteristics of this corresponds to what is expected for the β+γ band.

The first 4^+ band occurs at about 1.5 MeV, while the IBA predicts it at about 2.2 MeV. This discrepancy is not possible to resolve within the IBA-1 [10], but the inclusion of a 4^+ boson in the model may bring the state down [13]. One should however note that the 4_1^+ state and the IBA-1 state show a similar strong E2 decay to the 2_1^+ state and may therefore both be classified at least to a considerable extent as 2γ bands.

Another possible 4^+ band is observed at 1.86 MeV. The nature of this state is not clear.

Several 1^+ bands are observed around 2 MeV, cf Fig 2. In the IBA, collective 1^+ states cannot be formed in the IBA-1, but can be obtained if neutron bosons and proton bosons are considered separately, which is done in the IBA-2 [9,14]. In a limited calculation without f-bosons the first 1^+ state is obtained as indicated in Fig 2 with the main decay with M1 transitions to the ground state band.

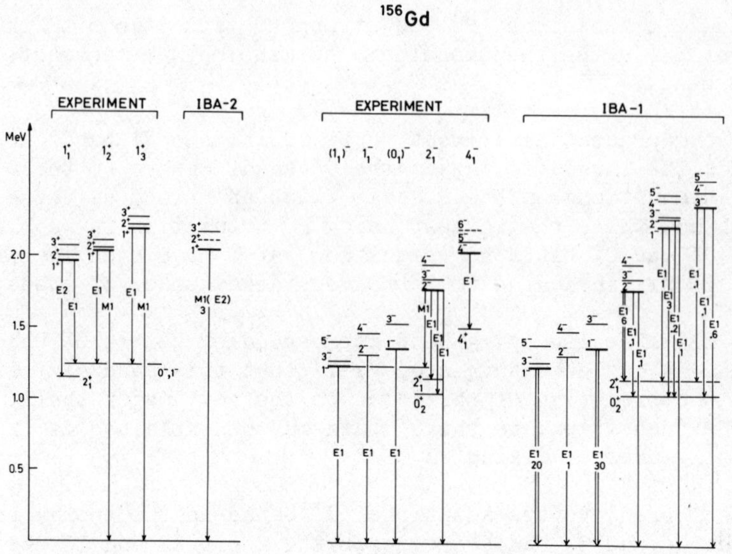

Fig 2. Experimentally observed and theoretically calculated 1^+ bands (left) and negative parity bands (right) in ^{156}Gd with their main decay modes indicated. Further explanations are given in the caption of Fig 1.

NEGATIVE PARITY STATES

Experimentally low-lying $K^\pi=0^-$ and $K=1^-$ bands with strong mixing of the odd spin levels are observed. About 0.5 MeV higher there is a $K^\pi=2^-$ band and above that a $K^\pi=4^-$ band. The expected $K^\pi=3^-$ band, that would complete the first octupole multiplet, has bot been identified, but it is likely to be one of the four 3^- levels that have been observed in the region 1.9-2.1 MeV and which cannot be definitely ordered into regular band structures.

The IBA-1, including f-bosons [10,11], predicts the $K^\pi=0^-$, 1^- and 2^- bands close to the observed positions and a strong mixing of the 0^- and 1^- bands, as found experimentally. The main decay of the 2^- band should be through E1 transitions to the 2_1^+ band, which is also observed experimentally. At higher energy a 1^- and a 3^- band are predicted, which have not yet been identified.

CONCLUSIONS

The (n,γ) experiment has revealed all or almost all low-spin levels in ^{156}Gd up to about 2 MeV. The IBA-1 gives a good description of the states up to and including the one phonon quadrupole and octupole vibrational states. Above this region the agreement between model and experiment seems to decrease.

Special problems are the nature of the 0_3^+ band and the low energy of the 4_1^+ band, which may be solved by including further degrees of freedom into the IBA.

The experimental data of this work are the result of a common effort of the former groups at Risø, Idaho Falls, Munich and Studsvik and this cooperation is gratefully acknowledged. I would like to express my gratitude to Professor F Iachello and Dr O Scholten for enlightening discussions and for communication of programs and unpublished results.

REFERENCES

1. A Bäcklin, G Hedin, B Fogelberg, M Saraceno, R C Greenwood, C W Reich, H R Hoch, H A Baader, H D Breitig, O W B Schult, K Schreckenbach, T von Egidy and W Mampe, to be published
2. U Gruber, B P Maier and O W B Schult, Kerntechnik 5 (1963) 17
M P Maier, U Gruber and O W B Schult, Kerntechnik 5 (1963) 19
3. R C Greenwood and C W Reich, Nucl Phys A223 (1974) 66
4. A Bäcklin, Nucl Instr Meth 57 (1967) 261
5. T von Egidy, E Beiber and Th W Elze, Z Phys 195 (1966) 489
6. A F Kluk, N R Johnson and J H Hamilton, Phys Rev C10 (1974) 1451
J E Cline, unpublished
D J McMillan, J H Hamilton and J J Pinajian, Phys Rev C4 (1971) 542
7. P O Tjøm and B Elbeck, private communication
8. Interacting Bosons in Nuclear Physics, ed F Iachello, Plenum Press, New York 1979
9. F Iachello, Structure of Medium-Heavy Nuclei 1979, The Institute of Physics, London, Conference Series 49, p 161
10. O Scholten, private communication
11. J Konijn, W N de Boer, A van Poelgeest, W H A Hesselink, M J A de Voigt, H Verheul and O Scholten, to be published
12. Part of the data have been taken from F K McGowan, W T Milner and P H Stelson, Int Conf on Band Structure and Nuclear Dynamics, New Orleans 1980, contributed papers p 130
13. F Iachello and P van Isacker, private communication
14. A Arima, T Ohtsuka, F Iachello and I Talmi, Phys Lett 66B (1977) 205
O Scholten, Interacting Bosons in Nuclear Physics, ed F Iachello, Plenum Press, New York 1979, p 17 and references therein

RECENT IBA-2 CALCULATIONS: A THEORETICAL STUDY OF THE TUNGSTEN ISOTOPES*

Bruce R. Barrett and Philip D. Duval

Department of Physics, University of Arizona, Tucson, AZ 85721
and
Theoretical Physics Division, PAS 81, University of Arizona, Tucson, AZ 85721

INTRODUCTION

Much work has been carried out recently in the application of the so-called IBA-2 model[1,2] (the Interacting Boson Approximation (IBA) formulation of the model for both protons and neutrons interacting with each other) to nuclei throughout the medium to heavy mass region. We will report in detail only on our calculations for the tungsten isotopes and refer those interested in other such investigations to the literature.[2,3]

THE INTERACTING BOSON APPROXIMATION FOR PROTONS AND NEUTRONS (IBA-2)

We consider a system of N_π proton bosons and N_ν neutron bosons outside a closed shell and assume that these bosons do not interact with the bosons inside the closed shell (the core bosons). So we write the Hamiltonian in terms of the valence bosons only:

$$H = H_\pi + H_\nu + V_{\pi\nu} \quad (1)$$

where $H_{\pi(\nu)}$ represents the single-boson energies and boson-boson interactions for the proton (neutron) bosons and $V_{\pi\nu}$ is the interaction between the proton and neutron bosons. These energies and interactions can, in principle, be derived from a microscopic theory[1-4], but here we treat the problem phenomenologically. The form of $H_{\pi(\nu)}$ is for only one kind of boson and corresponds to what is called the IBA-1 Hamiltonian.[2,5]

*Research supported in part by the NSF, Grant No. PHY-7902654.

The IBA-2 Hamiltonian incorporates the essential features of the underlying microscopic picture in the fermion space. These features are a strong pairing force between identical particles and a strong quadrupole interaction between non-identical particles. If two identical nucleons are completely paired (i.e. J = 0), they form an s-boson. They can also align their spins to give J = 2, forming a d-boson. Higher even values of J are also possible but for realistic nuclear interactions the pair is either unbound or only weakly bound for $J \geq 4$. So the nature of the IBA is to work in a model space made up of only s- and d-bosons.

The form taken for the IBA-2 Hamiltonian[1-3] is

$$H = \varepsilon_\pi \hat{n}_{d_\pi} + \varepsilon_\nu \hat{n}_{d_\nu} + \kappa Q_\pi \cdot Q_\nu + V_{\pi\pi} + V_{\nu\nu} + M_{\pi\nu} \qquad (2)$$

where $\varepsilon_{\pi(\nu)}$ is the single-boson energy for protons (neutrons), $\hat{n}_{d_{\pi(\nu)}}$ is the number operator for d-proton (neutron) bosons, $Q_{\pi(\nu)}$ is the quadrupole interaction for protons (neutrons):

$$Q_\rho = (d^\dagger \times s + s^\dagger \times \tilde{d})^{(2)}_\rho + \chi_\rho (d^\dagger \times \tilde{d})^{(2)}_\rho \; ; \; \rho = \pi,\nu \qquad (3)$$

$V_{\rho\rho}$ is the form of the residual interaction between like bosons[5];

$$V_{\rho\rho} = \sum_{L=0,2,4} \tfrac{1}{2} (2L+1)^{\frac{1}{2}} C^\rho_L [(d^\dagger_\rho \times d^\dagger_\rho)^{(L)} \cdot (\tilde{d}_\rho \times \tilde{d}_\rho)^{(L)}]^{(0)} \qquad (4)$$

where $\rho = \pi,\nu$; and $M_{\pi\nu}$ is a Majorana term which fixes the location of states with mixed proton-neutron symmetry with respect to the totally symmetric states, which lie lower in energy.[2,3]

APPLICATION TO THE TUNGSTEN ISOTOPES

We now apply this model to the calculation of the energy spectra of the tungsten isotopes ($Z = 74$, $N_\pi = 4$ and $82 < N < 126$, $0 < N_\nu \leq 11$). To reduce the number of free parameters, we make the following simplifications.

First, we set $\varepsilon_\pi = \varepsilon_\nu \equiv \varepsilon$, which is the usual assumption[2,3]. Second, we include only the C_0 and C_2 terms in the $V_{\nu\nu}$ interaction and do not include at all the $V_{\pi\pi}$ interaction, since $N_\nu > N_\pi$ for most of region fitted. Last we chose the coefficients in the Majorana terms so that this term primarily pushes up in energy those states with large antisymmetric parts.[3]

With these simplifications, the Hamiltonian used in our fit to the tungsten isotopes becomes:

$$H = \varepsilon \hat{n}_d + \kappa Q_\pi \cdot Q_\nu + V_{\nu\nu} + M_{\pi\nu} \quad (5)$$

where $\hat{n}_d = \hat{n}_{d_\pi} + \hat{n}_{d_\nu}$, and we now have only six free parameters: ε, κ, χ_π, χ_ν, C_0 and C_2. After one isotope is fitted, we establish χ_π, which is kept constant for the remaining isotope fits, leaving only five free parameters.

The experimentally determined energy levels for the even-even tungsten isotopes span the range in neutron number from N = 96 to 114. We can make predictions beyond this region by a smooth extrapolation of the above parameters. In Fig. 1 we give a detailed comparison with the experimental data according to the quasi-ground state rotational band and the quasi-γ and β-vibrational bands. Figure 2 contains graphs of the parameters used.

In general, the agreement with experiment for the ground-state-band and γ-band energy levels is extremely good. The agreement with the β-band energies, however, is not so successful. Perhaps the most striking feature of the energy spectra is the sharp rise in the γ-and β-bands at neutron number N = 108, which is thought to originate from a partial shell closure at the $i_{13/2}$ level. This same rise also occurs in the γ-band of the neighboring osmium isotopes (Z = 76).[3] Fitting this feature has led to a dip in the value of χ_ν at N = 108.

Having obtained the wave functions for the energy states in the tungsten isotopes by fitting the experimental energy levels, we can determine the electromagnetic transition rates between these states.

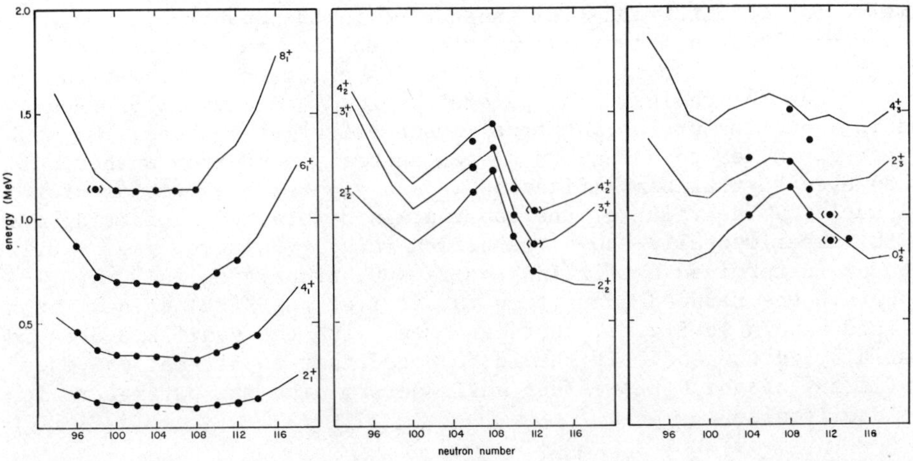

Fig. 1. Comparison between calculated (lines) and experimental[6] (points) energy levels of the tungsten isotopes in the ground state, quasi-γ and quasi-β bands, respectively.

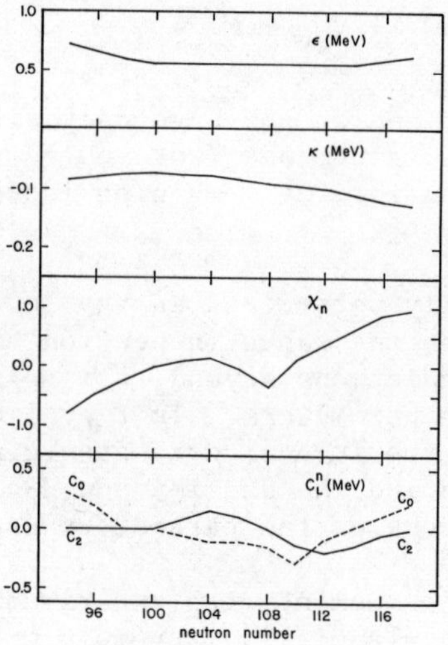

Fig. 2. The parameter set used for the tungsten isotopes. The value used for χ_π was -1.6.

We obtained reduced transition rates for E4, E2 and E0 transitions, but will present only the E2 results here.

The form of the E2 transition operator is given by

$$T^{(E2)} = e_\pi Q_\pi + e_\nu Q_\nu \qquad (6)$$

In principle, the parameters χ_π and χ_ν in Q_π and Q_ν, respectively, may be different from those used in the quadrupole operators in the Hamiltonian; however, we have taken them to be the same in our calculations, so as to reduce the number of free parameters; this seems a natural choice. The parameters $e_{\pi(\nu)}$ have units of e-barns and indicate the proton (neutron)-boson effective charges. We might expect e_π and e_ν to depend on proton number and neutron number but as an even further simplification we use $e_\pi = e_\nu$ = a constant for all nuclei. The value of the constant is determined by fitting one of the experimentally known transition rates. Using $e_\pi = e_\nu = 0.126$ e-barns (determined by fitting the $2_1^+ \to 0_1^+$ transition in $^{182}_{74}W_{108}$) we obtain the reduced transition rates, i.e. the B(E2) values, such as those shown in Fig. 3. Our agreement with the available data is generally quite good. It should be noted that no attempt was made to fit any of the B(E2) values while determining the parameters in the Hamiltonian.

We also calculated the quadrupole moments, two-neutron separation energies, isotope shifts and isomer shifts, but present only the results for the quadrupole moments. The E2 transition operator is a quadrupole operator and can be directly related to the quadru-

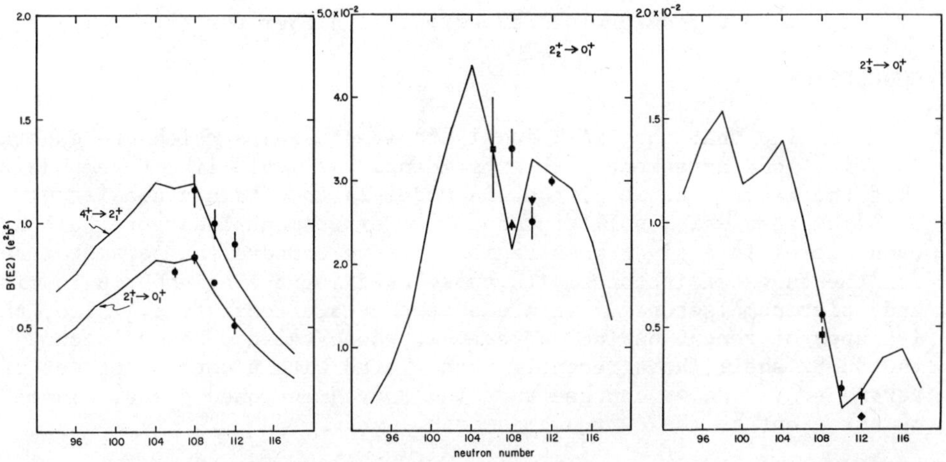

Fig. 3. Comparison between calculated (lines) and experimental[7] (points) B(E2) values for transitions noted on the figure.

pole moment for a nucleus. Using the IBA wavefunctions and the E2 transition operator given by Eq. (6), we obtain the results shown in Fig. 4 for $J^{\pi} = 2_1^+$ and 2_2^+. Note that the parameters e_{π}, e_{ν}, χ_{π} and χ_{ν} have already been determined, so that we fit no new free parameters in determining the quadrupole moments. The IBA predicts the correct sign in both of the above cases, and the agreement with experiment is very good for $Q_{2_1^+}$. But in the case of $Q_{2_2^+}$ the IBA value differs

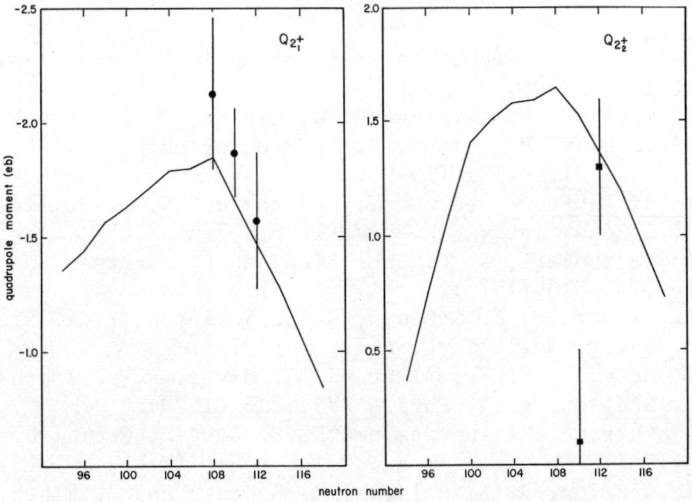

Fig. 4. Comparison between calculated (lines) and experimental[8] (points) quadrupole moments for the 2_1^+ and 2_2^+ states.

from the recently determined experimental number for ${}^{184}_{74}W_{110}$.

CONCLUSIONS

We find that the IBA-2 model provides results which are generally in quite good agreement with experiment. We would like to emphasize that the global scheme of the IBA model is to obtain a consistent set of parameters which will reproduce the experimental data for all even-even nuclei in a given mass region. If we compare the parameter set for the tungsten isotopes with those used for the neighboring osmium and platinum isotopes,[3] we see that they are consistent. Also, the isotopes of xenon, barium and cerium, whose valence bosons occupy the 50-82 shell, have recently been fitted with a consistent set of parameters.[3] So we can see that the above-mentioned global scheme is borne out by the current research.

REFERENCES

1. T. Otsuka, A. Arima, F. Iachello and I. Talmi, Phys. Letters 76B, 139 (1978).
2. Iachello, F., ed., Interacting Bosons in Nuclear Physics, Plenum Press, New York (1979).
3. O. Scholten, Ph.D. Thesis, Drukkerij Dijkstra Niemeyer bv, Groningen (1980).
   G. Puddu, O. Scholten and T. Otsuka, to be published in Nucl. Phys. A (1980).
   R. Bijker, A. E. L. Dieperink, O. Scholten and R. Spanoff, Kernfysisch Versneller Instituut preprint 224 (1980).
4. T. Otsuka, A. Arima and F. Iachello, Nucl. Phys. A309, 1 (1978).
5. A. Arima and F. Iachello, Ann. Phys. (N.Y.) 99, 253 (1976); 111, 201 (1978).
6. M. Sakai and Y. Gono, Institute of Nuclear Study, Japan, preprint 160 (1979).
   L. G. Mann, J. B. Carlson, R. G. Lanier, G. L. Struble, W. M. Buckley, D. W. Heikkinen, I. D. Proctor and R. K. Sheline, Phys. Rev. C19, 1191 (1979).
   Table of Isotopes, ed. C. M. Lederer and V. S. Shirley, (John Wiley & Sons, Inc., New York, 1978), 7th ed.
7. W. Andrejtscheff, K. D. Schilling and P. Manfrass, Nuclear Data Tables 16, 515 (1975).
   W. T. Milner, F. K. McGowan, R. L. Robinson, P. H. Stelson and R. O. Sayer, Nucl. Phys. A177, 1 (1971).
   F. K. McGowan, W. T. Milner, R. O. Sayer, R. L. Robinson, and P. H. Stelson, Nucl. Phys. A289, 253 (1979).
   C. Günther, P. Kleinheinz and R. F. Casten, Nucl. Phys. A172, 273 (1971).
   J. J. O'Brien, J. X. Saladin, C. Baktash and B. Elbek, Nucl. Phys. A291, 510 (1977).
8. J. J. O'Brien, J. X. Saladin, C. Baktash and J. G. Alessi, Phys. Rev. Letters 38, 324 (1977).

DO WE REALLY UNDERSTAND THE Pt ISOTOPES ?

M. Vergnes

Institut de Physique Nucléaire, 91406 Orsay, France

1. INTRODUCTION

The even Pt nuclei, lying in a well known transitional region between the deformed rare earth nuclei and the doubly closed shell ^{208}Pb, have been one of the testing grounds of the IBA. Although many features have indeed been succesfully explained[1] by the O(6) limit of this model, with a tendency towards SU(3) when going to the lightest isotopes, there are still several unsolved problems in these nuclei and it is important to better understand where the model is successfull and where it is not (or not yet). The goal of this paper is to outline some of the remaining difficulties and to review some experimental data concerning the particle transfer reactions. The results shall be compared, not only to the IBA but also to other microscopic and geometric models. More details may be found in reference 2.

2. ENERGIES AND ELECTROMAGNETIC PROPERTIES

If one looks at the level schemes of the even Pt nuclei, two facts are apparent at once : i) There is a 0^+ level going down in energy when A decreases, with a very low minimum at A = 186; ii) There is a crossing of the 4_1^+ and 2_2^+ levels between A = 188 and A = 186. These two facts have been considered as an experimental indication for a shape transition around A = 188-186. This has been confirmed by the independant observation[3] of a similar transition in odd-A nuclei (change from a decoupled to a strongly coupled band) between A = 187 and A = 185. Theoretically, Kumar and Baranger[3] (K.B) have calculated long ago that an oblate to prolate shape transition occurs between A = 192 and A = 186 and most of the other microscopic models : Götz et al[5], BET[6], IBA-2[7], predict it between A = 192 and

A = 190. There appears to be a discrepancy between the place where theories predict the shape transition and the place where it seems to be observed experimentally. The values of Q_{2^+} are all positive between A = 198 and 192 and are not known for the lightest stable isotopes.

There has been a lot of discussions on the subject : are the Pt nuclei soft or rigid ? The geometrical analogue of the O(6) limit of the IBA is the γ-unstable model and all the microscopic calculations (K.B, BET and many others[8]) predict shallow potential energy surfaces, particularly soft in the γ direction. However, in recent years, the assymetric rigid rotor model (AROT) has been used quite successfully in core plus particle calculations[9] to describe the properties of odd-A Pt nuclei. The branching ratio $B_4 = B(E2, 4_2^+ \rightarrow 4_1^+)/B(E2, 4_2^+ \rightarrow 2_2^+)$ is model dependant and it has been argued that it was a way to make a choice between soft and rigid models. Indeed, the comparison of the experimental results[10] in the case of ^{194}Pt (see Table 1) is clearly in favour of γ-soft models. However, ^{192}Pt has been Coulomb excited, up to spin 12^+ for the g.s.b and up to spin 10^+ for the quasi γ - band, using 1 GeV ^{208}Pb projectiles at G.S.I., by an Orsay, Bordeaux, Stockholm, G.S.I, collaboration[11] : Preliminary B(E2) extracted from this experiment[12] are in much better agreement with AROT than with the IBA-1 O(6) limit or with the few BET or K.B values available. In particular the cut-off, due to the finite number of bosons, does not seem to be observed.

Table 1. Comparison of the B(E2) ratio B_4, with different model predictions in ^{194}Pt.

	Exp.[10]	O(6) γ.unst.	BET[6]	γ-soft[8]	AROTO2[13] (γ=30°)	AROT (γ=30°)
$\dfrac{B(E2, 4_2^+ \rightarrow 4_1^+)}{B(E2, 4_2^+ \rightarrow 2_2^+)}$	1.26	0.9	0.79	0.72	0.58	0.45

The experimental energies of relatively low-lying levels of the even Pt nuclei $(2_1^+, 2_2^+, 4_1^+, 4_2^+)$ compare quite well with IBA-2[7], the agreement being a little worse with BET[6]. For the 0_2^+ level the agreement with IBA-2 is much better than with BET, the trend being opposite for the 0_3^+ (see ref.2 for details). It should be remarked that in ^{194}Pt two very close 0^+ levels are known at about the energy predicted for the 0_3^+. The level at 1546 keV has about the same energy and two neutron transfer strength[14] as the levels at 1617 and 1670 keV in ^{192}Pt and ^{190}Pt respectively, and is identified as the 0_3^+ level of IBA. The 1479 keV level, which is clearly 0^+, is not predicted in the IBA scheme. The g.s.b. has been followed up to high spins in many Pt nuclei and an accute backbending is observed around I = 10^+, several 10^+ levels being observed in a narrow energy interval. This phenomenon, unexplained by the purely collective models, has been interpreted as due to the crossing of the g.s.b by 2 quasi-particle (q.p) bands (rotation aligned) of $(\nu i13/2)^{-2}$ and $(\pi h11/2)^{-2}$ character. Yadav et al. have been able[13] to get a satisfactory agreement by adding to AROT the 2 q-p excitations proposed above and taking into account the coupling between 0 q-p and 2 q-p (AROTO2 model). For the quasi γ - band their results are better than AROT but still disagree appreciably from experiment.

3. TWO NUCLEON TRANSFER REACTIONS

The (p,t) reaction on the Pt isotopes has been studied in Orsay[14], M.S.U[15] and Tohoku[16] and the (t,p) reaction was recently performed at L.A.S.L[17]. The results for the g.s → g.s transitions are shown together in Fig.1 and compared to the 3 limits of IBA-1. The Tohoku results[16] for W, Os and Ir targets are also shown in the same figure. It is clear that the two neutrons strengths for the Pt nuclei are in quite acceptable agreement with the O(6) limit, with perhaps a slight tendency towards SU(3) for the lightest isotopes. The most striking fact is, however, the sudden and important change between the strengths observed for the W, Os, Ir and those observed for the Pt nuclei. This change could possibly correspond to an SU(3) → O(6) transition between Ir and Pt (see Fig. 1). The (t,p) cross sections for the Ir and Pt nuclei have however been found of the same order[18] and a carefull experimental confirmation of the discontinuity shown in Fig.1 seems necessary.

In our Orsay study (targets between ^{198}Pt and ^{190}Pt) several 0^+ excited states have been populated : the ratios of the populations of these states to that of the g.s are shown in Fig.2 and compared to ratios of two neutrons strengths obtained in an IBA-2 calculation[7]. A similar comparison is shown for the ratio of the populations of the 2_1^+ and 2_2^+ levels. The IBA-2 calculation, as already indicated, predicts a change of sign of Q_{2^+} between A = 192 and A = 190 corresponding, roughly speaking, to an O(6) → SU(3) transition. Accordingly, besides the g.s, the 0^+ level which is predicted to be populated in (p,t) is the 0_3^+ for A ⩾ 192, but the 0_2^+ for A < 190. Although the agreement is not perfect, the general experimental features are in qualitative agreement with the IBA calculation. Particularly, it is certain experimentally that the 0_3^+ strength, about constant between ^{194}Pt and ^{190}Pt, decreases very appreciably in ^{188}Pt. Our (p,t) results are therefore in qualitative agreement with the beginning of a transition O(6) → SU(3) when A decreases, proposed already on the basis of energies and electromagnetic rates by Casten et al.[1]. Caution is however necessary in such comparisons of two neutrons transfer strengths, particularly when the experimental population is small, and rigorous theoretical estimates of the cross sections (usual coherent summation of products of structure factors by kinematic factors) would be necessary for a detailed test. For example, in the case of ^{196}Pt - a typical O(6) nucleus - the 0_3^+ level at 1403 keV should (due to a selection rule)

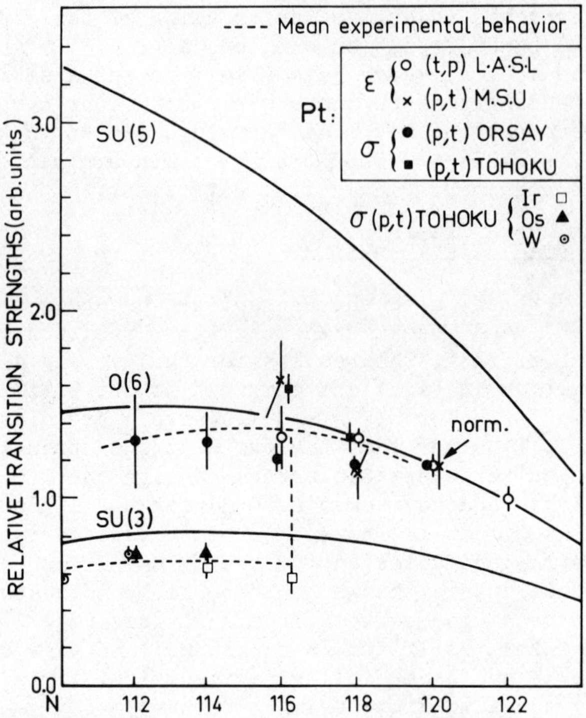

Fig. 1 : Comparison of the g.s → g.s transition strengths in the two neutron transfer reactions, with the 3 limits of IBA-1.

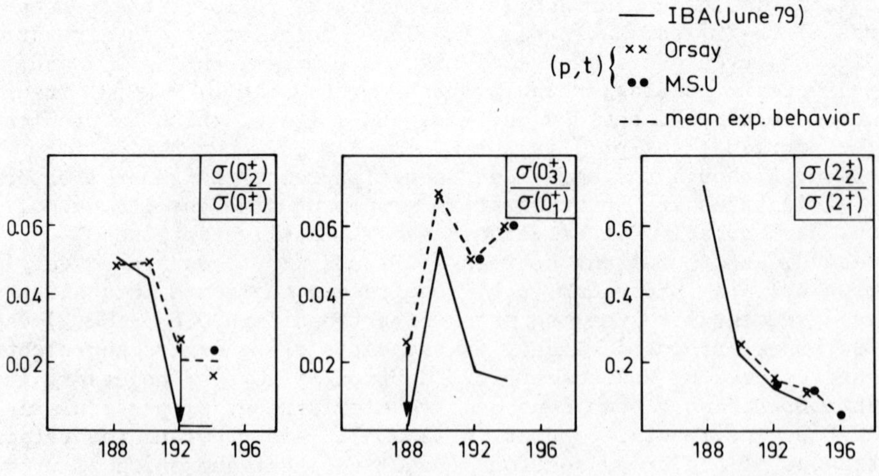

Fig. 2 : Comparison of ratios of (p,t) experimental cross sections for 0^+ and 2^+ levels in Pt, with strengths ratios computed in an IBA-2 calculation[7].

be populated in (p,t) but not in (t,p)... indeed the population is about the same (≈ 2 % of the g.s) in both reactions[15,18].

4. ONE NUCLEON TRANSFER REACTIONS

These reactions are very good spectroscopic tools and should be used to test the IBA as well as the other microscopic theories (up to now no theoretical predictions have been available for the Pt region). The strong population of a level in these reactions is generally an indication of particle (rather than collective) character. Particularly striking in ^{194}Pt is the strong population of the 0_2^+ level in the ^{193}Ir(^{3}He,d)^{194}Pt reaction[19] and of the 4_2^+ level in the ^{195}Pt(p,d)^{194}Pt reaction[19]. It will be very interesting to see what are the predictions of the supersymetries model[20] - which works reasonably well for the ^{194}Pt(t,α)^{193}Ir reaction[18] - for the relative populations of the different 0^+ levels in the ^{193}Ir(^{3}He,d)^{194}Pt reaction.

5. CONCLUSION

Although the IBA permits to explain reasonably well a lot of results in the Pt region, there are still many unsolved problems (for example the question of the finite number of bosons, of the shape transitions, of the softness or rigidity) and these nuclei are certainly among the most difficult ones to understand in detail ... therefore very fascinating.

REFERENCES

1. J.A. Cizewski, R.F. Casten, G.J. Smith, M.L. Stelts, W.R. Kane, H.G. Boerner and W.F. Davidson, Phys. Rev. Letters 40 (1978) 167. R.F. Casten and J.A. Cisewski, Phys. Lett. 79B (1978)5.
2. M. Vergnes, Orsay Internal Report, 1980, IPNO - PhN-80-17.
3. C. Bourgeois, M.G. Desthuilliers, P. Kilcher, J. Letessier, J.P. Husson, the Isocele collaboration, V. Berg, A. Höglund, A. Huck, A. Knipper, C. Richard-Serre, C. Sebille-Schück, the Isolde collaboration, M.A. Deleplanque, C. Gerschel, M. Ishihara, N. Perrin, B. Ader, Proceedings of the 3rd Int. Conf. on nuclei far from stability, Cargèse (1976), CERN report 76-13, 456.
4. K. Kumar and M. Baranger, Nucl. Phys. A 122 (1968) 273. K. Kumar, Proceedings of Conf. on Properties of Nuclei far from stability, Leysin, CERN report 70-30 (1970) 779.
5. U. Goetz, H.C. Pauli and K. Alder, Nucl. Phys. A 192 (1972) 1.
6. K.J. Weeks and T. Tamura, Phys. Rev. Letters 44 (1980) 533 and private communication (to be published).
7. F. Iachello, O. Schölten and R. Bijker, private communication, June 1979.
8. K. Kishimoto and T. Tamura, Nucl. Phys. A 270 (1976) 317.
9. T.L. Khoo, F.M. Bernthal, C.L. Dors, M. Piiparinen, S. Saha,

P.J. Daly and J. Meyer-Ter-Vehn, Phys. Lett. 60B (1976) 341.
10. C. Baktash, J.X. Saladin, J.J. O'Brien and J.G. Alessi, Phys. Rev. C18 (1978) 131.
11. C. Roulet, H. Sergolle, P. Hubert, T. Lindblad, E. Grosse, D. Schwalm, P. Fuch, H. Wollersheim, J. Idzko, H. Emling, R. Simon, to be published.
12. C. Roulet, private communication.
13. H.L. Yadav, H. Toki and A. Faessler, Phys. Lett. 76B (1978) 144 and references therein.
14. M. Vergnes, G. Rotbard, J. Kalifa, J. Vernotte, G. Berrier, R. Seltz, H.L. Sharma and N.M. Hintz, B.A.P.S. II, 21, 8 (1976) 959 and to be published, M. Vergnes, C. Rotbard, J. Kalifa, G. Berrier, J. Vernotte, Y. Deschamps and R. Seltz, J. de Physique Lettres 17 (1978) L192.
15. P.T. Deason, C.H. King, T.L. Khoo, J.A. Nolen Jr and F.M. Bernthal, Phys. Rev. C20 (1979) 927.
16. K. Miura, T. Suehiro, Y. Hiratate, T. Shoji, H. Yamaguchi and Y. Ishizaki, Proceedings of Int. Conf. on nuclear structure, Tokyo (1977) 552, and private communication.
17. J.A. Cizewski, E.R. Flynn, R.E. Brown and J.W. Sunier, Phys. Lett. 88B (1979) 207.
18. J.A. Cizewski, Contribution to this Conference.
19. M. Vergnes, G. Rotbard, G. Ronsin, J. Kalifa, J. Vernotte and R. Seltz, to be published.
20. F. Iachello, Contribution to this Conference.

E2 MATRIX ELEMENTS AND SHAPE COEXISTENCE IN ^{110}Pd

Lennart Hasselgren* and Douglas Cline

Nuclear Structure Research Laboratory**
University of Rochester
Rochester, N.Y. 14627

The Ru, Pd and Cd nuclei lie in a well known shape transition region. The low-lying level spectra exhibit features characteristic of an anharmonic vibrator while the B(E2) values are enhanced by around 50 single particle units.[1] Calculations have been made for some of these nuclei using both the Interacting Boson Model[2] and the Boson Expansion Theory.[3] The purpose of the present work is to measure the E2 properties, which are especially sensitive to the quadrupole collective shape parameters in order to investigate the collectivity and to differentiate between the competing collective models.

Coulomb excitation of ^{110}Pd has been performed using 48 MeV ^{16}O, 170 MeV and 190 MeV ^{58}Ni and 954 MeV ^{208}Pb ions from the accelerators at Uppsala, Rochester, Brookhaven and Berkeley. The many collaborators involved in the work are listed in reference 4. A high efficiency γ-ray detection efficiency system[5] was used with the ^{16}O beam to study all known states up to 1.7 MeV in excitation energy. The experimental and analysis techniques for the heavier ion Coulomb excitation work are described in reference 6. The γ-ray spectra resulting from Coulomb excitation with ^{208}Pb ions were quite complicated and insufficient knowledge is available to extract an unambiguous level scheme for the high spin states. Known level systematics[1,7] in the Pd isotopes were used to assist in extracting the high spin part of the level spectrum from the γ-ray spectra. The resultant level spectrum is shown in figure 1. Two 0^+ bands are observed in addition to the ground band, the γ-band and the

* Permanent address, Institute of Physics, University of Uppsala, Box 530, S-5121, Uppsala, Sweden.
**Supported by the National Science Foundation.

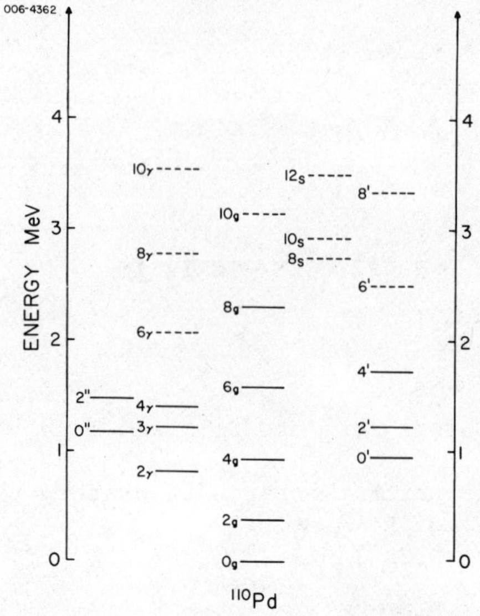

Fig. 1 The level spectra of ^{110}Pd derived from the present Coulomb excitation data. The assignment is not conclusive for the levels shown dashed.

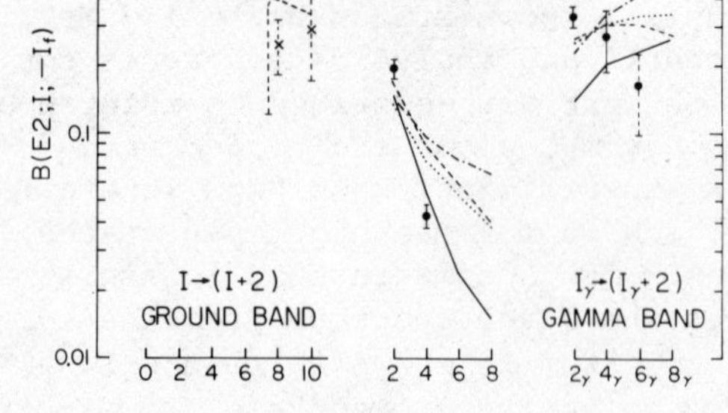

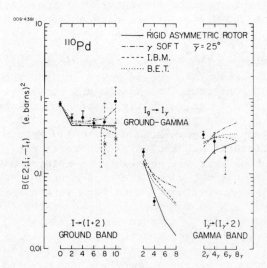

Fig. 2 The enhanced B(E2) values for the ground and gamma bands in ^{110}Pd. The results marked by x correspond to the $8_g^+ \rightarrow 10_s^+$ and $10_s^+ \rightarrow 12_s^+$ transitions.

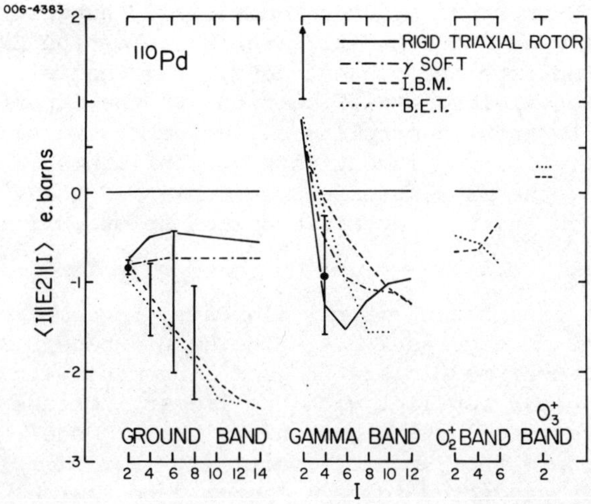

Figure 3 The diagonal E2 matrix element for ^{110}Pd.

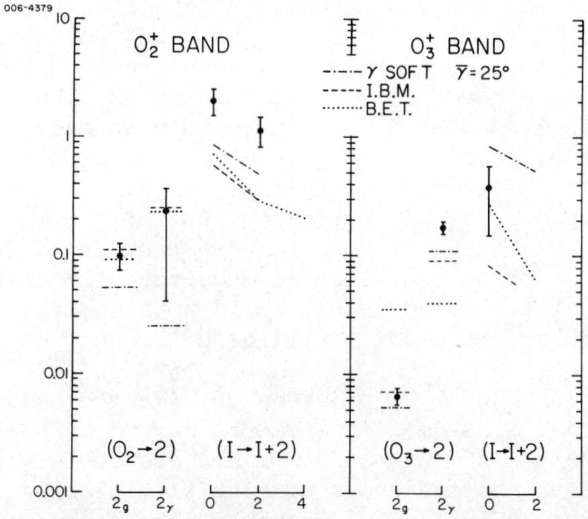

Figure 4 The in-band B(E2) values and inter-band B(E2) values for the 0_2^+ and 0_3^+ bands in ^{110}Pd.

expected two-quasiparticle superband.[7,8] Some γ-rays involving the levels shown dashed in figure 1 had similar energies and Coulomb excitation yields and could equally well be exchanged within the decay scheme. However, this produces only small changes in the absolute level energies and has little influence on the E2 matrix elements extracted from the present data. The analysis of the Coulomb excitation yields is difficult due to the importance of the static moments and the coupling between the several bands. It took one-year of effort to investigate the influence of the signs and magnitudes of the many interference terms and the static quadrupole moments in order to obtain a unique solution for all these unknowns.

Some of the enhanced E2 matrix elements are compared with model predictions in figures 2,3,4. The ground-band B(E2) values predicted by the four models are in good agreement with the data. The data are in clear conflict with the linear increase with the phonon number predicted by the harmonic vibrator model. None of the models reproduce the γ-band transitions. For example, the $B(E2; 4^+_\gamma \rightarrow 4^+_g)$ is predicted to be a factor 2 too large by the I.B.M. calculation.[2] The experimental ground band diagonal E2 matrix elements, shown in figure 3, appear to exhibit the behaviour predicted by the boson models.

The in-band B(E2) values for the rotational band based on the 0^+_2 state at 946.7 keV are 2.5 larger than the corresponding ground band values. The $B(E2; 4^+ \rightarrow 6^+)$ in this band, not shown, is equally strong.These in-band B(E2) values correspond to the quadrupole deformation parameter $\beta_2 = 0.4$ for the 0^+_2 band compared with $\beta_2 = 0.25$ for the ground band. The 0^+_2 band also has a larger effective moment of inertia. The I.B.M. predicts only one low-lying 0^+ state in addition to the ground state. The measured in-band B(E2) values shows that the IBM 0^+_2 band must correspond to the experimental 0^+_3 band. If this is true then the I.B.M. predicts the $B(E2; 0^+_3 \rightarrow 2^+_g)$ to be one order of magnitude too large while the other B(E2) values are reasonable. It is notable that the I.B.M.[2] and B.E.T.[3] calculations predict similar E2 properties.

This experiment on ^{110}Pd provides the first case where a relatively complete set of E2 matrix elements have been measured, i.e. B(E2) values, excited state static moments and the signs of the many E2 interference terms involving several collective bands. Additional data now being analyzed should give a more accurate and complete set of E2 matrix elements. The present data show that the IBM,[2] and BET,[3] are similar and moderately successful in reproducing the data with the notable exception of the strongly deformed 0^+ rotational band we have found to coexist at low excitation energies in ^{110}Pd. Similar experiments are in progress to study such shape coexistence in neighboring nuclei in order to try to determine the structure of these interesting low-lying deformed bands.

References

1. L. Hasselgren, Interacting Bosons in Nuclear Physics, ed. F. Iachello (Plenum Press, New York 1979) p. 67.
2. P. Van Isacker and G. Puddu, Interacting Bose-Fermi Systems, ed. F. Iachello (Plenum Press, New York 1980).
3. T. Kishimoto and T. Tamura, Nucl. Phys. A270 (1976) 317.
4. L. Hasselgren, J. Srebrny, D. Cline, T. Czosnyka, C. Y. Wu, R. M. Diamond, F. S. Stephens, H. Körner, D. Habs, C. Fahlander, L. Westerberg, H. Bäcklin, C. Baktash, G. R. Young, and H. Sang, Bull. Amer. Phys. Soc. 25 (1980) 596.
5. N. G. Jonsson, J. Kantele and A. Bäcklin, Nucl. Inst. and Methods 152 (1978) 485.
6. D. Cline, Interacting Bose-Fermi Systems, ed. F. Iachello (Plenum Press, New York 1980).
7. A. Graue, L.E.Samuelson, F. A. Rickey, P. C. Simms and G. J. Smith, Phys. Rev. C14 (1976) 2297.
8. C. Flaum, and D. Cline, Phys. Rev. C14 (1976) 1224.

EVIDENCE OF THE FINITE BOSON NUMBER

FROM LIFETIME MEASUREMENTS IN ^{78}Kr AND ^{79}Rb*

H.P.Hellmeister, K.P.Lieb and J.Panqueva

II. Physikalisches Institut der Universität,
3400 Göttingen, and Institut für Kernphysik
der Universität zu Köln, 5000 Köln,
Federal Republic of Germany

INTRODUCTION

At the first Erice Meeting on Interacting Bosons, Gelberg and Kaup[1] presented an interpretation of the ground state and γ bands of the even Kr isotopes in terms of IBA-2 [2]. They recognized that the spectra change from that of the rather strongly deformed ^{74,76}Kr to the γ-unstable shapes of the ^{80,82}Kr isotopes. These fits required remarkably few force constants; furthermore, it was found that all interaction parameters remained essentially constant, except for the quadrupole term χ_ν and the boson number N_ν which varied linearly with the neutron number.

At a about the same time, an experiment on the lightest stable isotope ^{78}Kr had been started together with the group of Dr. J. Delaunay at Saclay. This nucleus has eight proton particles and eight neutron holes in the (1f 2p $1g_{9/2}$) shell. The motivation of this study was

(i) to extend the positive parity bands to higher spins;
(ii) to study systematically E2 transition probabilities in particular between states which may be affected by the finite boson number $N = N_\nu + N_\pi = 8$; and
(iii) to clarify the role of the negative parity states.

The reactions used were ^{65}Cu(^{16}O,p2n) at 42 - 58 MeV and ^{68}Zn(^{12}C,2n) at 30 - 36 MeV bombarding energy; the heavy ion beams were supplied by the Köln and Saclay FN tandem accelerators. As a result of this work[3], the gsb was ex-

*Supported by the German Bundesministerium für Forschung und Technologie

tended up to 16^+, the γ-band up to 10^+ and the octupole band up to 13^-. The 11^+ candidate of the γ-band has been identified at 5443 keV[4]. Recent linear polarization measurements by Robinson et al.[5] support our spin-parity assignments for the states below 4 MeV.

E2 AND E1 TRANSITION PROBABILITIES IN ^{78}Kr

Lifetime measurements have been performed for some 20 states with the recoil distance and Doppler shift attenuation methods. We have considerably improved the mechanical and thermal stability of our plunger apparatus so that the flight distance can now be adjusted with a precision of 0.2 μm and the thermal drift is less than 0.5 μm over several hours. This allows us to measure lifetimes as short as 1 ps with an accuracy of 10%; this high quality of data is required, since the majority of lifetimes lies between 0.3 ps and a few ps.

The shorter lifetimes were determined by the DSA technique. As the singles spectrum is very complex, DSA line shapes at 0° were measured in coincidence with gates set in the 90° detector. In order to minimize the effect of the unknown stopping power of the recoiling nuclei, spectra with five different backings and at several beam energies were taken. Furthermore, the precision of the plunger data provided a cross check for the lifetime of the 1978 keV 6^+ state. Fig. 1 illustrates the consistency of the results.

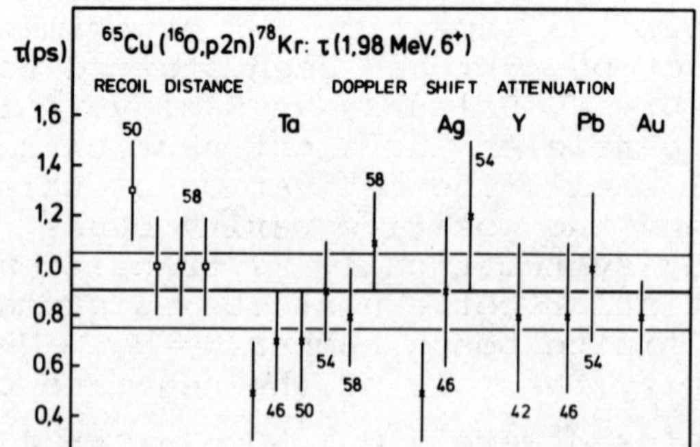

Fig. 1: DSA and recoil distance measurements for the 1978 keV 6^+ state in ^{78}Kr. The DSA data have been labelled by the backing and beam energy used.

The most serious drawback of lifetime measurements employing heavy ion fusion reactions is the simultaneous feeding of many discrete states with comparable lifetimes, and the continuum feeding time associated with the unobserved γ-ray cascade connecting the entry states with the discrete states. For instance, due to the time delay of discrete feeder states, the effective time constant of the 1978 keV state increases by a factor of 2 between 42 - 58 MeV beam energy. From our experience, it is essential to account for the intensities and time constants of all feeder states, to take measurements at different bombarding energies, e.g. for different angular momentum windows in the final nucleus, and to have clean DSA line shapes with good statistics, preferentially in coincidence with a preceding feeder transition. (We attribute the deviations of our results with respect to a more recent DSA experiment[5] in part to different assumptions on the feeding as well as to several other shortcomings of these data[5]: fewer spectra with much less counting statistics, less good detector resolution and incomplete separation of doublets, no cross check against recoil distance method,..)

The problem of the continuum feeding time is more difficult to attack. From a recent survey on the shortest lifetimes in medium nuclei measured with heavy ion reactions, $\tau_f \leqq 0.15$ ps was found[6]. We have reproduced these short feeding times by a computer simulation in which the flux and time evolution of the particle evaporation and subsequent γ-decay are calculated in terms of the statistical model formalism. Fig. 2 shows the distribution of feeding times for the reaction ^{68}Zn(^{12}C,2n) at 36 MeV. The area under each curve is proportional to the side feeding intensity. Although the average feeding time $\langle\tau_f\rangle$ increases for decreasing spin, e.g. increasing length of the cascade, it is in all cases shorter than the measured lifetime.[11]

The interpretation of the B(E2) values is given in Fig. 3. The IBA-1 and IBA-2 fits for the gsb are close to the SU(3) and O(6) limits[7], but rule out the SU(5) limit. The band structure is further supported by the strong in-band and weak inter-band transitions of the γ-band; here, the IBA-2 fit (one effective charge, same interaction terms as in ref. 1) is somewhat better than the IBA-1 fit (two E2 charges). One additional argument in favor for this interpretation are the E1 transitions connecting the octupole and ground state band illustrated in Fig. 4. If the effective E1 charge is fitted to the experimental $7^- \rightarrow 6^+$ strength, the other weak inter-band E1 transitions from the high spin states as well as the weak in-band E2

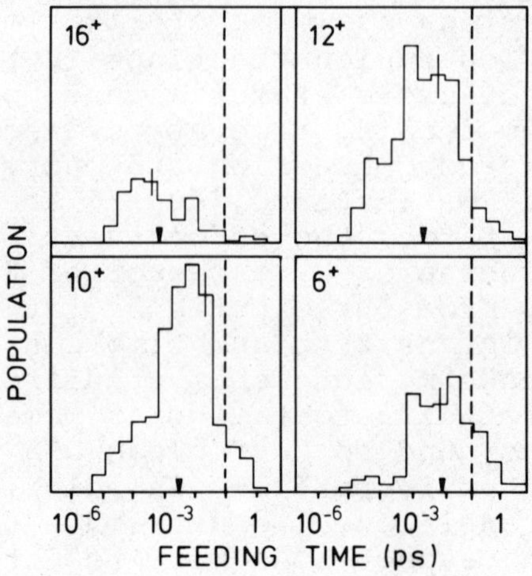

Fig. 2: Calculated distribution of feeding times of the reaction $^{68}Zn(^{12}C,2n)^{78}Kr$ at 36 MeV leading to various states of the ground state band

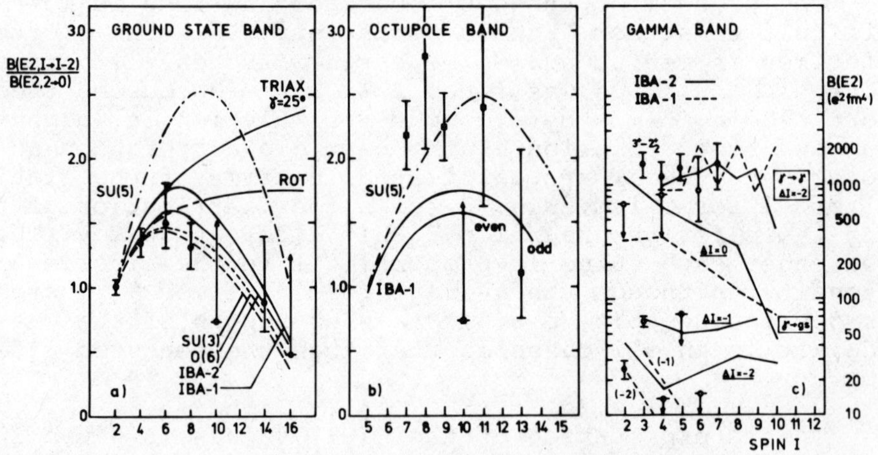

Fig. 3: Fits to the B(E2) values in ^{78}Kr in terms of IBA-1 and IBA-2 as well as the limits O(6), SU(3) and SU(5)

transitions (labelled b(E2) in Fig. 4) between the low spin members of the octupole band are well reproduced. Therefore, in spite of the missing $5^- \rightarrow 3^-$ transition, the negative parity bands are indeed associated with an octupole vibration rather than with a $(g_{9/2}p_{1/2})$ configuration.[5]

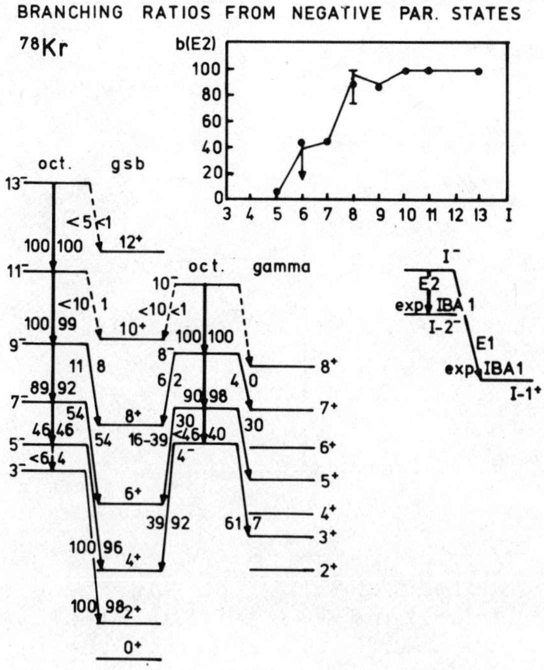

Fig. 4: Experimental and IBA-1 branching ratios normalized to the decay of the 3288 keV 7^- state.

One remarkable feature of the E2 strengths is the reduction of the $8^+ \rightarrow 6^+$, $14^+ \rightarrow 12^+$ and $13^- \rightarrow 11^-$ B(E2) values with respect to the classical symmetric and triaxial rotor limits ($\beta = 0.33$, $\gamma = 25^o$). This reduction has been attributed[3] to the finite boson number N = 8. Alternatively, the decrease of B(E2, $14^+ \rightarrow 12^+$) has been related to the crossing of the gsb with the aligned $\pi^2(g_{9/2})$ band which leads to strong backbending in ^{80}Kr . One possibility to differentiate between the two explanations is to study the core plus-proton system ^{79}Rb.

E2 TRANSITION STRENGTHS IN THE DECOUPLED BAND OF ^{79}Rb

High spin states in ^{79}Rb have been investigated via the reactions $^{70}Ge(^{12}C, p2n)$ and $^{63}Cu(^{19}F,p2n)$. From γ-γ-coincidences, excitation functions, angular distributions and n-γ-multiplicity measurements, we extended the $g_{9/2}$ decoupled band up to $29/2^+$, possibly up to $33/2^+$. In Fig. 5a, the transition energies $\hbar\omega$ are plotted versus $I(\omega)$ and compared with the bands in the core ^{78}Kr. It is evident that the decoupled band shows only a very slight backbending effect between I = 13/2 and I = 17/2 and then

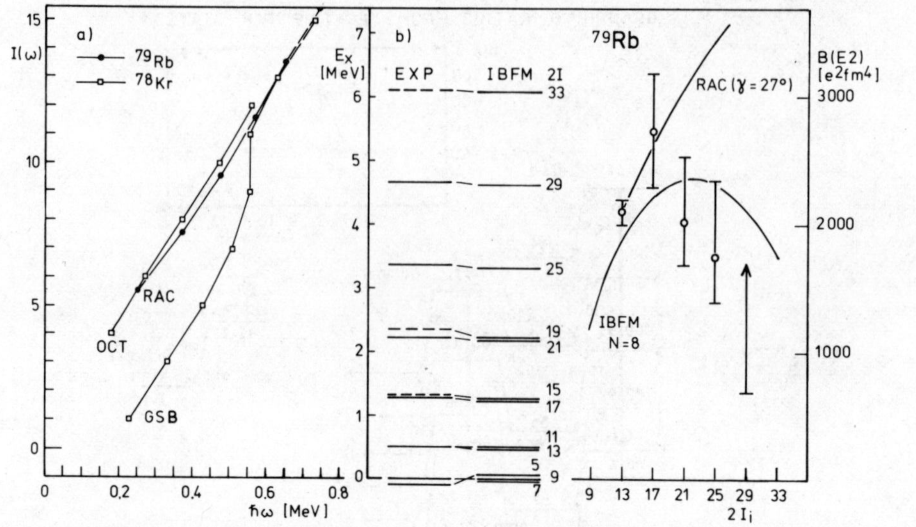

Fig. 5: a) Angular frequency ħω versus spin I(ω) for the decoupled band in ^{79}Rb and the ground state and octupole band in ^{78}Kr. b) B(E2) values of the decoupled band compared with the predictions of RAC ($\gamma = 0^{o}$ and 27°) and IBFM (see text).

follows the octupole band and the continuation of the gsb in the core. As to the E2 transition strengths given in Fig. 5b, the $13/2^{+} \rightarrow 9/2^{+}$ and $17/2^{+} \rightarrow 13/2^{+}$ transitions confirm nicely the rotational alignment coupling scheme (RAC). Indeed, the experimental ratios $R_1 \equiv B(E2, 13/2 \rightarrow 9/2)/B(E2, 2 \rightarrow 0) = 1.62 \pm 0.14$, and $R_2 \equiv B(E2, 17/2 \rightarrow 13/2)/B(E2, 13/2 \rightarrow 9/2) = 1.30 \pm 0.23$ agree well with the RAC predictions $R_1 = 1.5$ and $R_2 = 1.23$, respectively[8], corresponding to a triaxial core with $\beta = 0.33$, $\gamma = 27^{o}$. This excellent agreement supports the idea that the $g_{9/2}$ proton serves as spectator and does not polarize the core. The E2 strength of the $21/2 \rightarrow 17/2$ transition, however, already starts decreasing and that of the $25/2 \rightarrow 21/2$ transition confirms this trend. The theoretical values in Fig. 5b represent a fit in terms of the Interacting Boson Fermion model[9] with core-particle interaction parameters of Γ = 1.9 MeV, $\Lambda = -1.9$ MeV and $A = -0.78$ MeV; for the ^{78}Kr the IBA-1 force constants and effective E2 charges of ref. 3 have been adopted.[10]

From the comparison of the band structure in ^{78}Kr and ^{79}Rb and the falloff of the B(E2) values in both nuclei we suggest that, in spite of the upbending effect due to the aligned $g_{9/2}$ proton pair in ^{78}Kr, the reduction of the B(E2) values in both nuclei can be attributed to the finite boson space.

The authors gratefully acknowledge stimulating discussions with F. Iachello, U. Kaup, A. Gelberg, P. von Brentano and O. Scholten; the help of C. F. Casten during some of the calculations; and the help of F. J. Bergmeister, J. Delaunay, H. Dumont, J. Keinonen and R. Rascher during some of the measurements.

References

1. U.Kaup and A.Gelberg, Z.Phys. A293:311 (1979).
2. T.Otsuka, A.Arima, F.Iachello, and I.Talmi, Phys. Lett. 76B:139 (1979).
3. H.P.Hellmeister, U.Kaup, J.Keinonen, K.P.Lieb, R.Rascher, R.Ballini, J.Delaunay, and H.Dumont, Phys. Lett. 85B:34 (1979) and Nucl. Phys. A332:241 (1979).
4. L.Funke, priv. communication.
5. R.L.Robinson, et al., Phys. Rev. C21:607 (1980).
6. H.P.Hellmeister, K.P.Lieb, and W.F.J.Müller, Nucl. Phys. A307:515 (1978) and to be publ.
7. A.Arima and F.Iachello, Ann. Phys. (N.Y.) 99:253 (1976); 111:201 (1978); 123:468 (1979).
8. F.Stephens, Rev. Mod. Phys. 47:43 (1975). J. Meyer-ter-Vehn, Nucl. Phys. A249:111, 141 (1975). H.Toki and A.Faessler, Phys. Lett. 63B:121 (1976).
9. F.Iachello and O.Scholten, Phys.Rev.Lett. 43:679(1979); O.Scholten, thesis University of Groningen, unpubl.
10. An improved fit of the positive parity states in ^{78}Kr in terms of IBA-1 has been achieved for the parameters: $\varepsilon = 0.41$, $C_0 = 0.26$, $C_2 = -0.14$, $C_4 = 0.15$, $v_0 = -0.24$, $v_2 = 0.10$ MeV; effective E2 charges $q_2 = 8.6$ efm^2, $q_2' = -16.0$ efm^2.
11. In the calculations of the continuum feeding times, E1 transitions of $5 \cdot 10^{-4}$ Wu, statistical E2 transitions of 1 Wu as well as collective E2 transitions of 200 Wu were assumed. When changing the latter quantity to 50 Wu the feeding times increased somewhat, but still are considerably smaller than the lifetimes measured.

A SHELL MODEL STUDY OF THE PROTON-NEUTRON INTERACTING BOSON MODEL

Takaharu Otsuka

Physics Division,
Japan Atomic Energy Research Institute
Tokai, Ibaraki, 319-11, Japan

1. Introduction

The proton - neutron interacting boson model (P-N IBM ; also called IBM-2) has been recently proposed to describe quadrupole collective states in medium-heavy nuclei.[1,2] The P-N IBM consists of proton bosons of L=0 (called s_π) and L=2 (called d_π), and neutron bosons of L=0 (called s_ν) and L=2 (called d_ν). A two-body interaction between them is assumed. In terms of this model, numerous phenomenological analyses of the collective states have been performed over a wide range of the periodic table.[3]

I shall discuss in this talk the microscopic basis of the P-N IBM, taking a degenerate many j-orbit shell in order to simplify discussions. A prescription to derive the P-N IBM Hamiltonian was proposed by Otsuka, Arima and Iachello, and the method is called hereafter the OAI method.[1,2,4] The OAI method is based on an important assumption that the quadrupole collective states can be described in terms of the 0^+ and 2^+ collective nucleon pairs. This assumption has not been examined especially for systems with active protons and neutrons, although it holds exactly in some special schematic cases, for instance, the Ginocchio model.[5] I first show how and where this assumption holds. Note that other approaches to the IBM are also based on this assumption.[5,6]

It will be next shown that nucleon pairs other than the 0^+ and 2^+ collective ones have non-negligible effects on energies of low-lying states. Although this is an important problem in the study of the collective states, no definite discussion has been presented. It will be demonstrated that the P-N IBM is a good approximation to the exact shell model calculation if the effects are included by renormalization of the P-N IBM Hamiltonian.

2. The OAI method

In the OAI method[2] the collective nucleon pairs of $J=0^+$ (S) and $J=2^+$ (D) are first introduced. Proton pairs are denoted as S_π and D_π, while neutron pairs S_ν and D_ν. States constructed by the S and D pairs are called S-D states, which span the S-D subspace. The importance of the S-D subspace is mentioned above. An S-D state is mapped onto the corresponding s-d boson state. The P-N IBM Hamiltonian is constructed so that matrix elements between S-D states should be reproduced by the corresponding s-d boson matrix elements. In the zeroth order OAI method[2] the P-N IBM Hamiltonian contains up to two-body terms for the two-body nucleon Hamiltonian.

3. S and D pairs

Degenerate $0g_{7/2}$, $1d_{5/2,3/2}$, $2s_{1/2}$ orbits are taken in this talk for both protons and neutrons for the sake of simplicity, expecting that detailed properties of individual nucleon orbits are not so relevant to the present purpose. Systems with identical nucleons are first considered. The S pair is created in this case by the generalized seniority operator, S_+, of Kerman[7]

$$S_+ = \Sigma_j \sqrt{(j + \frac{1}{2}) \cdot \frac{1}{2}} \; [a_j^\dagger a_j^\dagger]^{(0)} . \tag{1}$$

The S pair thus defined is nothing but the Cooper pair. The surface delta interaction (SDI) is assumed between identical nucleons[8] The SDI is written as

$$- g \sum_{(i,j)\; K} (2K+1) \, (C^{(K)}(\theta_i\phi_i) \, C^{(K)}(\theta_j\phi_j)), \tag{2}$$

where g is the strength, i and j are nucleon suffixes, and the surface multipole operator $C^{(K)}$ is[9]

$$C^{(K)}(\theta,\phi) = \sqrt{\frac{4\,\pi}{2K+1}} \cdot Y^{(K)}(\theta,\phi). \tag{3}$$

I take the same value of g for protons and neutrons, with g=0.5 (MeV). This value is determined so that the excitation energy of the 2_1^+ state is approximately equal to that in real single closed shell nuclei ($\sim$1.4 MeV)[1]

The SDI between identical nucleons conserves the generalized seniority mentioned above.[7,10] In Fig. 1, spectra of the identical nucleon system are shown for (generalized) seniority v=0, 2, 4, 6. Note that the excitation energies are independent of the nucleon number for the SDI in degenerate orbits. In Fig. 1 one finds lower-lying 2^+, 4^+, 6^+ states of v=2, which consist of one nucleon pair of the corresponding J^π and the S pairs. These pairs, including the S pair, are called the favoured pairs by Hecht et al.[9], and creation operators of them are denoted as $F^{\dagger(J)}$ (J=0, 2, 4, 6). The favoured pairs satisfy commutation relations

$$F^{\dagger(J)} \propto [\, S_+ \, , \, C^{(J)} \,] \, , \qquad (4)$$

where $C^{(J)}$ is defined in eq. (3). This suggests that the favoured pair possesses the particle-hole excitation character.

The creation operator of a D pair, $D^\dagger$, is defined uniquely in this case as

$$D^\dagger \;=\; P \cdot F^\dagger \, , \qquad (5)$$

where P is an operator for the seniority projection introduced in Ref. 2. I emphasize that the D pair thus determined has both the particle-hole excitation character and the quadrupole pairing character.

The S-D states are constructed by using operators S_+ and $D^\dagger$ as[2]

$$|\, S^{n_s} \; D^{n_d} > \, = \frac{1}{\mathcal{N}} \, (\, S_+ \,)^{n_s} \, (\, D^\dagger \,)^{n_d} \, |0> \, , \qquad (6)$$

where n_s and n_d denote respectively the numbers of S and D pairs, and $\mathcal{N}$ is a normalization constant. Although an S-D state is in principle distributed in eigenstates of the SDI, as shown in Table I, the S-D component is concentrated in the lowest eigenstate of each v and J for v=0$\sim$6 except the case of v=6 and J=2^+. Namely, for a given v and J, the S-D state is already the lowest eigenstate of the SDI in a good approximation.

The SDI between identical nucleons is mapped onto the boson Hamiltonian, H^B, with the single d-boson energy, ε_d, and the interaction between two d-bosons, c_L (L=0, 2, 4) ;

$$H^B = \varepsilon_d \, n_d + \frac{1}{2} \sum_{L=0,2,4} c_L \, ([d^\dagger d^\dagger]^{(L)} [\tilde{d}\, \tilde{d}]^{(L)}) \, , \qquad (7)$$

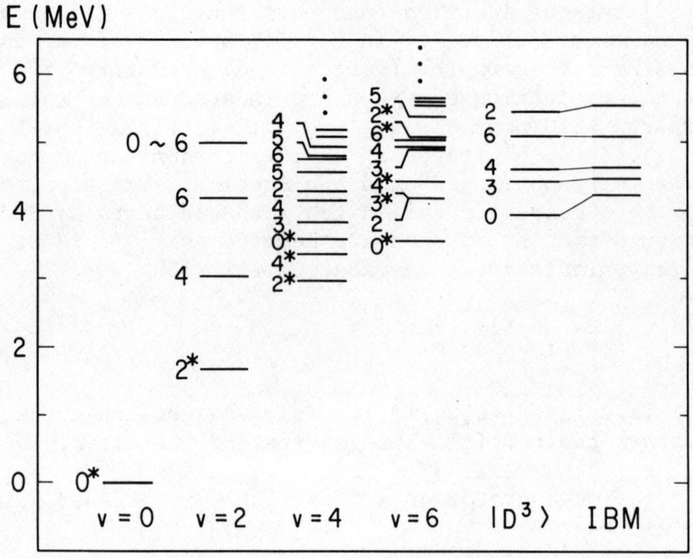

Fig. 1. Spectra of identical nucleon system classified by seniority v. The numbers beside energy levels are spins, while all states have positive parity. Asterisks mean states in which the S-D components are concentrated. Excitation energies of states with three D pairs ($|D^3>$) are compared with their boson predictions (IBM).

where n_d is the number operator of d bosons. In the OAI[2] ε_d and c_L's are determined from the excitation energies of S-D states $|S^{N-1} D>$ and $|S^{N-2} D^2>$. Once boson Hamiltonian (7) is fixed, one can predict excitation energies of states $|S^{N-M} D^M>$ for $M \geq 3$. Figure 1 also shows such boson predictions for excitation energies of states $|S^{N-3} D^3>$, and these predictions are compared with the exact fermion values. One finds a good agreement as in the single j-orbit cases[2] Similar tests of the OAI were made also for the quadrupole operator $C^{(2)}$ in eq. (3)[11]

4. S and D pairs in quadrupole collective states

Systems with active protons and neutrons are discussed here. Since the quadrupole-quadrupole interaction is considered to be a good effective interaction for the description of the quadrupole collective state[1,12] the proton-neutron interaction $V_{\pi\nu}$ is assumed to be the (surface) quadrupole-quadrupole interaction

Table I. Occupation probabilities in the S-D subspace for eigenstates of identical nucleon system. The surface delta interaction is assumed. The eigenstates are specified by seniority (v), total angular momentum (J), and excitation energy (E_x). The numbers "i" indicate sequential state numbers in ascending order with excitation energy for a fixed set of v and J. Some eigenstates with smaller occupation probabilities are omitted.

v	J	i	E_x(MeV)	Prob. (%)
0	0	1	0.0	100.0
2	2	1	1.67	100.0
4	4	1	3.35	99.1
		2	4.15	0.8
	2	1	2.96	95.1
		2	4.26	2.9
	0	1	3.59	91.3
		5	6.05	8.4
6	6	1	5.06	96.0
		2	5.64	3.2
	4	1	4.42	89.2
		2	4.92	3.0
		3	5.46	2.8
	3	1	4.17	90.0
		4	5.72	4.4
	2	1	4.16	10.8
		2	5.08	60.0
		4	5.60	14.1
	0	1	3.55	88.5
		6	6.66	9.1

$$V_{\pi\nu} = - g_{\pi\nu} \times 5 \times (C_{\pi}^{(2)} \; C_{\nu}^{(2)}) , \qquad (8)$$

where $g_{\pi\nu}$ is the strength, and $C_{\pi}^{(2)}$ and $C_{\nu}^{(2)}$ are $C^{(2)}$ in (3) of proton and neutron, respectively. I adopted $g_{\pi\nu} = 0.3$ (MeV) following consequences of effective interaction studies.[13,14]

Shell model diagonalizations[18] are performed for g and $g_{\pi\nu}$ fixed, while the number of active protons (n_{π}) and that of active neutrons (n_{ν}) are varied as $0 \leq n_{\pi},\ n_{\nu} \leq 6$ and $-6 \leq n_{\pi},\ n_{\nu} \leq 0$. The negative n stands for holes. Spectra for various combinations of n_{π} and n_{ν} are shown in Fig. 2. The whole spectra come down from the both sides towards the middle of the shell as seen in real nuclei.

In the left hand side of Fig. 2, spectra seem to be the intermediate situation between the SU(5) and SU(3) limits of the IBM,[15,16] and become more rotational (or closer to the SU(3) limit[16]) as the number of particles increases. In fact, the excitation energies of 2_1^+, 4_1^+, 6_1^+, 8_1^+ at $n_{\pi}=n_{\nu}=6$ follow the I(I+1) rule[12] within 5 % deviations.

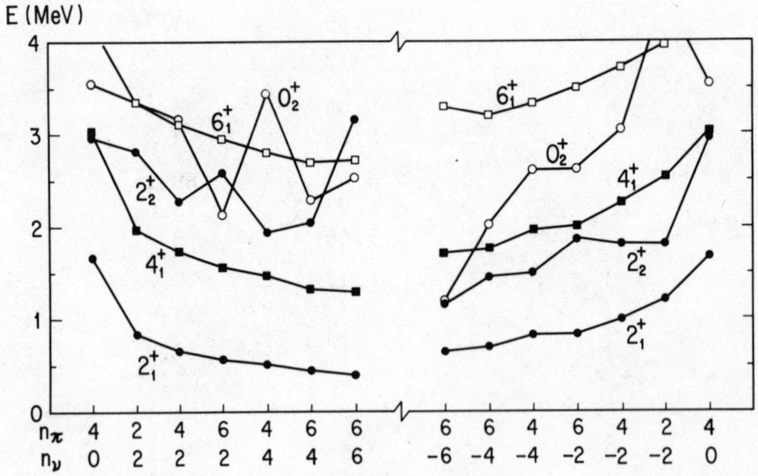

Fig. 2. Excitation energies of $0_{1,2}^+$ (open circles), $2_{1,2}^+$ (solid circles), 4_1^+ (solid squares), and 6_1^+ (open squares) in systems with n_{π} active protons and n_{ν} active neutrons. The negative n stands for holes. Solid lines are drawn to guide the eye.

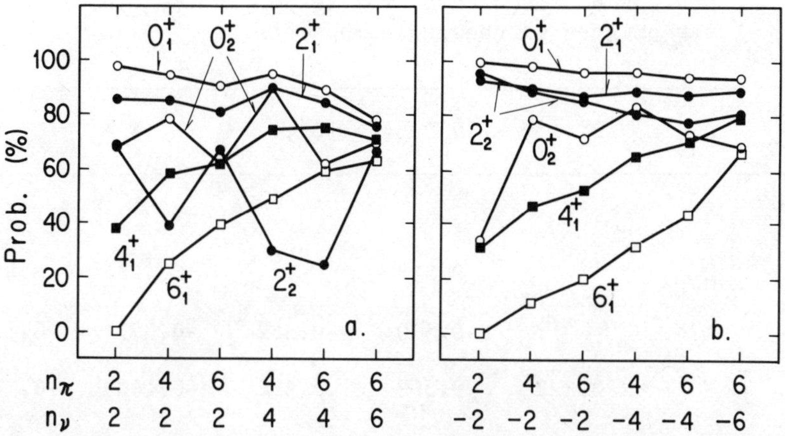

Fig. 3. Occupation probabilities in the S-D subspace for eigenstates shown in Fig. 2. See the caption of Fig. 2. Neutrons are particles in (a), while holes in (b). Protons are particles in both cases.

Spectra in the right hand side of Fig. 2 resemble those of the O(6) limit of the IBM,[1,7] especially at $n_\pi = -n_\nu = 4$. In this side of Fig. 2, the 2_2^+ state comes down below 4_1^+ as seen in the O(6) region of real nuclei[1,17] and low-lying 3_1^+ states are also obtained as shown later.

Occupation probabilities that an eigenstate of the proton-neutron system is found in the S-D subspace are shown in Fig. 3 for several low-lying states. Here by the S-D subspace I mean a subspace spanned by the products of the proton S-D states and the neutron S-D states. In general, the occupation probabilities increase first as the number of particles (or holes) increases. The occupation probabilities are saturated around 70 ∿ 90 %, and they decrease gradually in the middle region.

The occupation probability of the 4_1^+ state, for instance, increases around $|n_\pi| = |n_\nu| = 4$ in both cases. In order to see this point, components of the 4_1^+ state are investigated. The 4^+ favoured pairs are denoted as G_π and G_ν hereafter. The S pairs are omitted here in showing states. In Table II, amplitudes of components $|D_\pi^2; 4^+\rangle$, $|D_\pi D_\nu; 4^+\rangle$, $|D_\nu^2; 4^+\rangle$, $|G_\pi\rangle$ and $|G_\nu\rangle$ in the 4_1^+ state are shown for several combinations of n_π and n_ν.

Table II. Amplitudes of some components in the 4_1^+ state with n_π active protons and n_ν active neutrons. The negative n stands for holes, and G_π and G_ν mean the proton and the neutron 4^+ favoured pairs, respectively. The S pairs are omitted in showing components.

n_π	n_ν	$\|D_\pi^2;4^+>$	$\|D_\pi D_\nu;4^+>$	$\|D_\nu^2;4^+>$	$\|G_\pi;4^+>$	$\|G_\nu;4^+>$
0	4					1.0
2	2		0.618		0.438	0.438
2	4		-0.570	-0.382	-0.377	0.174
4	4	0.322	0.502	0.322	-0.140	-0.140
4	6	0.249	0.404	0.298	-0.103	-0.034
0	-4					1.0
2	-2		0.569		-0.564	-0.564
2	-4		0.595	-0.312	-0.535	0.321
4	-4	0.360	-0.581	0.360	-0.263	-0.263
4	-6	0.375	-0.530	0.310	-0.205	-0.167

For $|n_\pi|$ or $|n_\nu|=2$, amplitudes of components $|G_\pi>$ and $|G_\nu>$ are large, while they are small for $|n_\pi|$, $|n_\nu| \geq 4$. Three components $|D_\pi^2;4^+>$, $|D_\pi D_\nu;4^+>$ and $|D_\nu^2;4^+>$ are strongly connected by $V_{\pi\nu}$ in (8), and these matrix elements become larger as the number of particles or holes increases. If these three are present (i.e. $|n_\pi|$, $|n_\nu| \geq 4$), a coherent linear combination of them comes down, and the occupation probability in the S-D subspace becomes large.

I would like to point out that the proton-neutron qaudrupole-quadrupole interaction in (8) connects the S pair only to the D pair because of the relation,

$$D^\dagger \propto P \cdot [\ C^{(2)},\ S_+\ ], \tag{9}$$

which is obtained by eqs. (4) and (5). Such matrix elements become larger as the number of particles or holes increases. This situation can be seen in the following relation ;

$$C^{(2)} \,|\tilde{j}^n\,(S^{n/2});0^+\rangle \tag{10}$$

$$= \sqrt{\frac{n\,(2\Omega - n)}{4\,(\Omega - 1)}}\cdot\langle\tilde{j}^2(D)\,\|\,C^{(2)}\,\|\,\tilde{j}^2(S)\rangle\cdot|\tilde{j}^n\,(S^{(n/2)-1}\,D);2^+\rangle.$$

where $\Omega = \Sigma\,(2j+1)/2$. The matrix element of $V_{\pi\nu}$ between $|D_\pi^2\rangle$ and $|D_\pi D_\nu\rangle$ mentioned above is an example of those matrix elements ;

$$\langle D_\pi^2 S_\pi^{N-2}\;S_\nu^{N'};L^+\,|\,(C_\pi^{(2)}C_\nu^{(2)})\,|\,D_\pi S_\pi^{N-1}\;D_\nu S_\nu^{N'-1};L^+\rangle$$

$$= W(L\,0\,2\,2\,;\,L\,2)\times\langle D_\pi^2 S_\pi^{N-2}\,\|\,C_\pi^{(2)}\,\|\,D_\pi S_\pi^{N-1}\rangle \tag{11}$$

$$\times\,\langle S_\nu^{N'}\,\|\,C_\nu^{(2)}\,\|\,D_\nu S_\nu^{N'-1}\rangle\,,$$

where one D_π pair is changed to S_π, while an S_ν pair is altered to D_ν. Note that the importance of the S pair is based on the pairing nature in the interaction between identical nucleons.

The above argument is a general one, and may be extended to more realistic cases with non-degenerate orbits by utilizing the approximate quasi-spin scheme[19] for the generalized seniority operator of Talmi.[20] One thus see that the dominant roles of the S and D pairs originate in the basic features of the effective nucleon-nucleon interaction, while large configuration spaces of valence nucleons are also needed for coherence properties of the S and D pairs. It is mentioned here that, because of competition against pairs other than S and D, higher spin states and most of side band states need more particles or holes to have large occupation probabilities in the S-D subspace.

5. Shell Model and P-N IBM Spectra

The shell model spectra are compared in Fig. 4 with spectra of the P-N IBM Hamiltonian obtained in the zeroth order OAI method. In Fig. 4(a), $n_\pi = n_\nu = 4$, while $n_\pi = -n_\nu = 4$ in Fig. 4(b). Similarities between the shell model and the P-N IBM spectra are found only in relative locations of states, and there are notable descrepancies especially in the absolute scale.

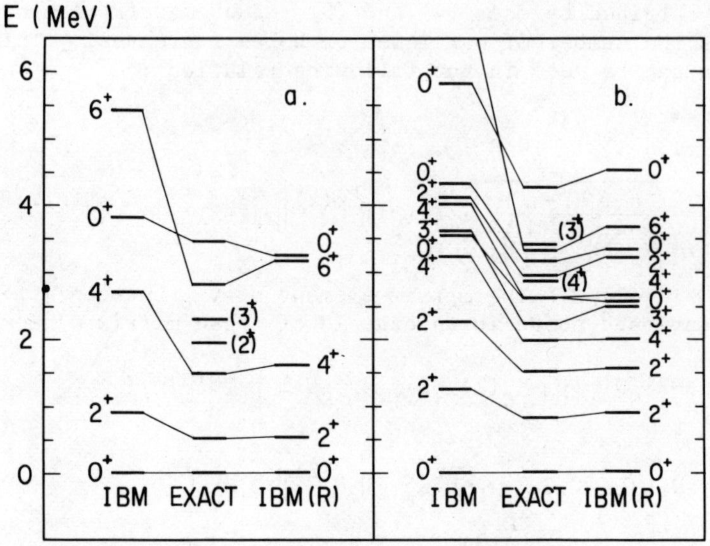

Fig. 4. Spectra of $n_\pi = |n_\nu| = 4$ in Fig. 2 (EXACT) compared with those of the P-N IBM (IBM) and the P-N IBM with renormalization (IBM(R)). In (a) $n_\nu = 4$, while $n_\nu = -4$ in (b). The boson Hamiltonians are derived microscopically. The shell model states with spins in parentheses are "intruders".

The descrepancies are due to non-S-D components admixed in low-lying eigenstates. One can see their mixing probabilities in Fig. 3. As far as low-lying eigenstates are concerned, most of admixed non-S-D components consist of S, D, and G(4^+) nucleon pairs. This is because matrix elements of $V_{\pi\nu}$ in which a D pair is changed to a G pair are relatively large. For instance, $V_{\pi\nu}$ couples $|D_\pi S_\pi S_\nu^2;2^+\rangle$ with $|G_\pi S_\pi D_\nu S_\nu;2^+\rangle$ as

$$\langle D_\pi S_\pi S_\nu^2;\ 2^+ \,|\, (C_\pi^{(2)} C_\nu^{(2)}) \,|\, G_\pi S_\pi D_\nu S_\nu;\ 2^+\rangle$$
$$= W(2\,0\,4\,2\,;\,2\,2)\ \langle D_\pi S_\pi \| C_\pi^{(2)} \| G_\pi S_\pi\rangle \langle S_\nu^2 \| C_\nu^{(2)} \| D_\nu S_\nu\rangle. \quad (12)$$

The first matrix element in the right hand side of this equation is not negligibly small compared to the second one.

Neither the proton Hamiltonian nor the neutron Hamiltonian connects S-D states to non-S-D states strongly, since S-D states are their eigenstates in a good approximation (see Table I).

Effects of the 6^+ nucleon (favoured) pair are much smaller than those of the G pair. Pairs other than the favoured pairs, i.e. S', D', G', etc. have almost no effects on low-lying states as pointed out by Hecht et al.[9]

In the original OAI method,[2] the boson Hamiltonian is determined by equating its matrix elements with those of the nucleon Hamiltonian between S-D states. Effects of non-S-D states are therefore not included. The OAI has to be extended. The extension is such that one should use nucleon matrix elements between states where non-S-D states are already admixed properly, instead of matrix elements between pure S-D states. Those "correlated" matrix elements are equated to the corresponding boson ones to determine the boson Hamiltonian.

The system of $n_\pi = n_\nu = 4$ is considered in order to illustrate this extended OAI method. A state $|D_\pi S_\pi S_\nu^2\rangle$ is taken as an example. One should keep in mind that the single proton d boson energy, $\varepsilon_{d\pi}$, is determined from the energy of this state. State $|D_\pi S_\pi S_\nu^2\rangle$ is coupled by $V_{\pi\nu}$ in (8) to a non-S-D state $|G_\pi S_\pi D_\nu S_\nu; 2^+\rangle$. I introduce a state $|D_\pi S_\pi S_\nu^2\rangle\!\rangle$ where $|D_\pi S_\pi S_\nu^2\rangle$ is the main component and non-S-D state $|G_\pi S_\pi D_\nu S_\nu\rangle$ is admixed by $V_{\pi\nu}$. The admixture is evaluated by the Feshbach method.[21] In detail, states $|D_\pi S_\pi S_\nu^2\rangle$ and $|G_\pi S_\pi D_\nu S_\nu\rangle$ are treated as p and q states in the Feshbach method, respectively. Energy spacing between the former and the latter states is mainly determined by the excitation energy of G_π pair, which is determined by the proton Hamiltonian. The energy denominator which appears in the Feshbach method is approximated by the sum of the energy spacing thus fixed and the energy gain due to the coupling between states $|D_\pi S_\pi S_\nu^2\rangle$ and $|G_\pi S_\pi D_\nu S_\nu\rangle$. Once the energy of this "correlated" state $|D_\pi S_\pi S_\nu^2\rangle\!\rangle$ is obtained, one can calculate the renormalized single proton d boson energy from the energy.

The same procedure can be taken for state $|S_\pi^2 D_\nu S_\nu\rangle$ to calculate the renormalized single neutron d boson energy. In practical calculations, a slightly more sophisticated method is adopted so that matrix elements of $V_{\pi\nu}$ are also taken into account in the energy denominator.

Several parameters in the boson Hamiltonian are thus renormalized, and all major renormalization effects are included. Note that the Hamiltonian still consists of one- and two-body terms. Once the single d-boson energy, for instance, is renormalized, its effects appear in all boson states except a state with no d boson. The renormalized single d boson energy is, however, determined at $|D_\pi S_\pi S_\nu^2\rangle$ and $|S_\pi^2 D_\nu S_\nu\rangle$. Namely it is assumed that effects of non-S-D components admixed to states with many D pairs can be evaluated approximately by the renormalization of the P-N IBM Hamiltonian. One possible way to examine this assumption is to compare spectra obtained by the Hamiltonian thus renormalized with the exact shell model spectra. Figure 4 also displays such comparisons. Agreement with the shell model is considerably improved in the renormalized P-N IBM results.

Similar improvements are also found in wave functions. For instance, amplitudes of $|S_\pi^2 S_\nu^2>$ and $|D_\pi S_\pi D_\nu S_\nu;0^+>$ in the ground state of Fig. 4(a) are respectively 0.582 and 0.620, while those of $|s_\pi^2 s_\nu^2)$ and $|d_\pi s_\pi d_\nu s_\nu)$ in the renormalized P-N IBM are 0.570 and 0.638, respectively. Without renormalization, however, these amplitudes are 0.667 and 0.621, respectively. It should be emphasized that these improvements in energies and wave functions are made without playing any adjustable parameters.

Figure 4 also shows shell model states which are not connected to any boson state. These are two lowest examples of "intruder" states where non-S-D components are so dominant that their effects cannot be taken into account by the present renormalization method.

Major changes due to the renormalization are found in the following terms,

$$\varepsilon_{d_\pi} n_{d_\pi} + \varepsilon_{d_\nu} n_{d_\nu} + \kappa_0 \{ (d_\pi^\dagger \tilde{d}_\nu) s_\pi s_\nu^\dagger + \text{h.c.} \} \\ + \kappa_2 \{ (d_\pi^\dagger d_\nu^\dagger) s_\pi s_\nu + \text{h.c.} \} + g_{\pi\nu}^{(4)} ([d_\pi^\dagger d_\nu^\dagger]^{(4)} [\tilde{d}_\pi \tilde{d}_\nu]^{(4)}), \tag{13}$$

where ε_{d_π}, ε_{d_ν}, κ_0, κ_2 and $g_{\pi\nu}^{(4)}$ are coefficients. The last term is renormalized mainly due to the coupling between $|D_\pi S_\pi D_\nu S_\nu;4^+>$ and $|G_\pi S_\pi S_\nu^2>$ (or $|S_\pi^2 G_\nu S_\nu>$), which has a large effect on energy spacings between members of a band. The renormalized values (original values) of these parameters in the $n_\pi = n_\nu = 4$ case are respectively 1.13 (1.67), 1.13 (1.67), -0.60 (-0.71), -0.64 (-0.71), and -0.64 (-0.19) MeV. In the $n_\pi = -n_\nu = 4$ case, those values are respectively 1.27 (1.67), 1.27 (1.67), -0.64 (-0.71), -0.61 (-0.71), and -0.73 (0.19) MeV.

Mixing of non-S-D components is smaller in Fig. 4(b) than in Fig. 4(a) (see also Fig. 3). Since protons are particles and neutrons are holes in Fig. 4(b), various S-D components in an eigenstate are usually coupled incoherently to a given non-S-D component because of the phase change by the particle-hole transformation. However, since both of protons and neutrons are particles in Fig. 4(a), such couplings are coherent and larger mixing of non-S-D occurs. Figure 4(a) corresponds to the SU(5)-SU(3) region, while Figure 4(b) to the O(6) region.[1,4] This suggests that, for a given set of $|n_\pi|$ and $|n_\nu|$, more states are described well by the P-N IBM in the O(6) region than in the SU(5)-SU(3) region, although this difference should disappear for $|n|$ larger.

The rotational spectrum is seen at $n_\pi = n_\nu = 6$ in Fig. 2, where the occupation probabilities in the S-D subspace somewhat decrease for 0_1^+, 2_1^+, 4_1^+, though still more than 70 % (see Fig. 3). In this case, mixing between S and D pairs becomes very large as in the SU(3) limit of IBM.[16] The admixture of G pairs to low-lying states therefore becomes relatively large, since each D pair is coupled to G pairs by $V_{\pi\nu}$.

In summary, the present calculation shows that the S and D pairs play dominant roles in various situations of the qaudrupole collective motion. The truncation to the S-D subspace becomes better in general as the number of particles or holes increases. It is also suggested that the P-N IBM seems to be a good approximation to the exact shell model calculation if effects of non-S-D components are properly included by renormalization of the P-N IBM Hamiltonian.

Similar studies on the validity of the P-N IBM are being carried out in non-degenerate orbits as well as in other degenerate orbit shells. Properties of electro-magnetic transitions and nucleon transfer reactions are also under investigation.

References

1. T. Otsuka, A. Arima, F. Iachello and I. Talmi, Phys. Lett. 76B: 139 (1978).
2. T. Otsuka, A. Arima and F. Iachello, Nucl. Phys. A309: 1 (1978). (1978); T. Otsuka, page 93 in "Interacting Bosons in Nuclear Physics", F. Iachello, ed., Plenum, New York (1979).
3. For example, F. Iachello, G. Puddu, O. Scholten, A. Arima and T. Otsuka, Phys. Lett, 89B: 1 (1979): Lecture by B. Barrett in this workshop.
4. T. Otsuka, Doctor Thesis, University of Tokyo (1978).
5. J. Ginocchio, Ann. Phys. (N.Y.) 126: 234 (1980), and references therein.
6. R. A. Broglia, P. F. Bortignon, A. Vitturi and E. Maglione, lectures in this workshop.
7. A. K. Kerman, Ann. Phys. (N.Y.) 12: 300 (1961).
8. P. J. Brussard and P. W. M. Glaudemans,"Shell-model applications in nuclear spectroscopy", North-Holland, Amsterdam (1977), and references therein.
9. K. T. Hecht, J. B. McGrory and J. P. Draayer, Nucl. Phys. A197: 369 (1972); J. B. McGrory, Phys. Rev. Lett, 41: 533 (1978).
10. A. Arima and M. Ichimura, Prog. Theor. Phys. 36: 296 (1966).
11. T. Otsuka, to appear.
12. A. Bohr and B. R. Mottelson, "Nuclear Structure" vol. II, Benjamin, Reading, (1975).

13. J. P. Schiffer and W. W. True, Rev. Mod. Phys. 48: 191 (1976); and references therein.
14. G. H. Herling and T. T. S. Kuo, Nucl. Phys. A181: 113 (1972); and private communication from T. T. S. Kuo.
15. A. Arima and F. Iachello, Ann. Phys. (N.Y.) 99: 253 (1976).
16. A. Arima and F. Iachello, Ann. Phys. (N.Y.) 111: 201 (1978).
17. A. Arima and F. Iachello, Ann. Phys. (N.Y.) 123: 468 (1979).
18. The shell model space is truncated in most of diagonalizations because of the limitation of the computing facility. The dimension of the shell model space is, however, still much larger ($\sim$3000 dim.) than that of the S-D subspace ($\lesssim$60 dim.), and this truncation causes practically no effects on final conclusions. Discussions in this talk do not have to be in 0.1 % precision.
19. T. Otsuka and A. Arima, Phys. Lett. 77B: 1 (1978).
20. I. Talmi, Nucl. Phys. A172: 1 (1971).
21. H. Feshbach, Ann. Phys. (N.Y.) 5: 357 (1958); 19: 287 (1962).

THE GINOCCHIO MODEL AND THE INTERACTING BOSON APPROXIMATION

Akito Arima

Department of Physics
Faculty of Science
University of Tokyo
Hongo, Bunkyo-ku, Tokyo, Japan

INTRODUCTION

Various approaches to nuclear collective motions have been made since Bohr and Mottelson published their classic works.[1] Among these, the Interacting Boson Approximation[2] (IBA) and the boson expansion theory[3] have been successful in explaining properties of vibrational, transitional and rotational nuclei. The interacting boson approximation assumes that the collective low-lying states of heavy nuclei are composed primarily of monopole and quadrupole pairs of fermions that are approximated as monopole and quadrupole bosons.[4] The boson expansion method[5] determines an expansion for the fermion operators by demanding that the commutation relations are preserved in the boson space. In general this leads to an infinite expansion for the fermion pair operators in powers of bosons. Furthermore all the fermion pairs, not just the collective pairs, are mapped onto bosons. This theory is implemented in practice by first making a BCS transformation from particles to quasiparticles.[3] The boson expansions are then made in the quasiparticle basis. Ultimately only a collective quadrupole boson in this quasiparticle space is retained and an effective boson Hamiltonian for this collective quadrupole boson is derived.[3]

It is highly desirable to test the mapping used for these models. Recently Ginocchio invented schematic fermion shell model Hamiltonians which have pairing and quadrupole-quadrupole interactions.[6] These Hamiltonians are exactly soluble for many-fermion systems. The Ginocchio Hamiltonians thus provide us with an ideal test for any theory of nuclear collective motions.

We test two methods of deriving Hamiltonians for the Interacting Boson Model Hamiltonian. They are the Belyaev-Zelevinsky-Marshalek (BZM) method and the Otsuka-Arima-Iachello method.[6] The Boson Hamiltonians thus obtained were diagonalized. The results were found to be in good agreement with exact solutions. On the other hand, the BCS + (Quadrupole) Boson Expansion method gave rather poor results, especially in high spin states.

I will explain briefly the Ginocchio model in §2 and derive the Interacting Boson Model Hamiltonian in §3. Comparison between exact and approximate solutions will be shown in §4, where we will learn that the Interacting Boson Approximation indeed provides a very good approximation.

The Ginocchio Model

I explain the Ginocchio model briefly in this section. Its model space is spanned by four single particle orbits

$$j = k + \frac{3}{2},\ k + \frac{1}{2},\ |k - \frac{1}{2}|,\ |k - \frac{3}{2}|,$$

where k is an integer. We introduce twenty-eight operators

$$S^+ = \frac{1}{2} \Sigma (-1)^{j-m} a^+_{jm} a^+_{j-m},$$

$$D^+_\mu = \sum_{j,j'} (-1)^{k+\frac{3}{2}+j} \sqrt{(2j+1)(2j'+1)} \begin{Bmatrix} j & j' & 2 \\ \frac{3}{2} & \frac{3}{2} & k \end{Bmatrix} [a^+_j a^+_{j'}]^{(2)}_\mu,$$

$$P^{(r)}_\mu = 2 \sum_{j,j'} (-1)^{r+k+\frac{3}{2}+j} \sqrt{(2j+1)(2j'+1)} \begin{Bmatrix} j & j' & r \\ \frac{3}{2} & \frac{3}{2} & k \end{Bmatrix} [a^+_j \tilde{a}_{j'}]^{(r)}_\mu$$

$$r = 0,\ 1,\ 2,\ 3,$$

and their hermitian conjugate operators S and D_μ. They form a closed algebra, SO_8 group. Some subsets of these operators form subgroups;

SO_7; D^+, D, $P^{(0)}$, $P^{(1)}$, $P^{(3)}$

SO_6; $P^{(1)}$, $P^{(2)}$, $P^{(3)}$

SO_5; $P^{(1)}$, $P^{(2)}$

SO_4; S^+, S, $P^{(0)}$, $P^{(1)}$

SO_3; $P^{(1)}$

The most general Hamiltonian of the Ginocchio model takes the following form;

$$H = G_0\, S^+\cdot S + G_2(D^+\cdot\tilde{D}) + \frac{1}{4} b_2\, P^{(2)}P^{(2)} + \frac{1}{4}\sum_{r=1,3} b_r P^{(r)}P^{(r)} .$$

This Hamiltonian can be diagonalized in a subspace spanned by states generated by operating S^+ and D^+ on the vacuum

$$|NN_d\gamma JM\rangle = \mathcal{N}(S^+)^{N-N_d}(D^+)^{N_d}_{\gamma JM}|0\rangle$$

where $\mathcal{N}$ is a normalization constant, N the total number of pairs ($N = \frac{1}{2}n$) and N_d the total number of quadrupole pairs. J and M are total angular momentum and its projection. An additional quantum number γ is introduced to label the states completely.

There are three limits in which one finds analytical expressions of energy spectra.[6] The first in the seniority ($SO_5 \times SU_2$) limit. The level structure in this limit is that of anharmonic vibrator. The second is the SO_6 limit which corresponds to the γ-unstable nucleus. The last is the SO_7 limit. The energy spectrum in this situation is similar to that in the seniority limit. The difference is that in the SO_7 limit the energy gap varies with N whereas it is independent of N in the seniority limit.

Since the Ginocchio model is exactly solvable, it is very useful for studying approximations such as the Interacting Boson Approximation and the Boson Expansion Technique.

Boson-mapping of the Ginocchio model Hamiltonians

The mapping of the Ginocchio model Hamiltonians to boson Hamiltonians is not unique. We test three methods for this purpose. One is the BZM method. In this method the commutation relations among the generators of SO_8 are mapped onto the boson space so that the multipole operators $P^{(r)}$ are quadratic in the boson operators. In general this means that the pair operators S^+, D^+ and their hermitian conjugates involve an infinite expansion in powers of boson operators. The second approach, the boson expansion technique (BET), is similar to the BZM but the monopole boson is dropped. The third method is the Otsuka-Arima-Iachello method.[7] In this method the matrix elements of operators in a basis with a definite seniority, v, are mapped onto those between states with a definite number of quadrupole bosons, N_d, with

$$v = 2N_d. \qquad (1)$$

(1) BZM

$$S^{+} \rightarrow \mathcal{S}^{+} = s^{+}f + (s^{+}s^{+} - d^{+}d^{+})sg$$

$$D^{+}_{\mu} \rightarrow \mathcal{D}^{+}_{\mu} = d^{+}_{\mu}f + (s^{+}s^{+} - d^{+}d^{+})\tilde{d}_{\mu}g$$

$$S \rightarrow \mathcal{S}, \quad D_{\mu} \rightarrow \mathcal{D}_{\mu}$$

The functions f and g are complicated;

$$f = \frac{1}{2}(X_{-} - X_{-}) - (N+2)g$$

$$g = \frac{1}{2}Y^{-1}(X_{+} - X_{-})$$

with

$$X_{\pm} = \sqrt{\Omega - N + 2 \pm Y}$$

and

$$Y = \sqrt{(N+2)^2 - C_{\sigma}}$$

$$C_{\sigma} = (s^{+}s^{+} - d^{+}d^{+})(ss - \tilde{d}\cdot\tilde{d}).$$

$$P^{(r)}_{\mu} \rightarrow \mathcal{P}^{(r)}_{\mu} = 2\sqrt{2}\,[d^{+}d]^{(r)}_{\mu} \qquad r = 1,\ 3 \tag{2}$$

$$n \rightarrow \frac{1}{2}N = \frac{1}{2}(s^{+}s + \Sigma\, d^{+}_{\mu} f_{\mu})$$

$$P^{(2)}_{\mu} \rightarrow \mathcal{P}^{(2)}_{\mu} = 2\sqrt{2}\,[d^{+}\tilde{d}]^{(2)}_{\mu}.$$

Inserting these boson images into the Hamiltonian H, one obtains its boson image. In practice, one expands the functions f and g into an infinite series in powers of C_{σ}. In order to obtain the numerical results shown in fig. 1, we take terms up to fourth order. The resulting boson Hamiltonain is

$$H = \frac{1}{2}(G_0+G_2)[N(\Omega-2N)+N_s(N_s-1)+d^{+}\cdot d^{+}\tilde{d}\cdot\tilde{d}+d^{+}\cdot d^{+}ss+s^{+}s^{+}\tilde{d}\cdot\tilde{d}]$$

$$+\frac{1}{2}(G_0-G_2)\{(N_s-N_d)(\Omega-2N+2)+(1-\frac{5}{\Omega})N_s(N_s-1)-(1-\frac{1}{\Omega})d^{+}\cdot d^{+}\tilde{d}\cdot\tilde{d}$$

$$+b_2[N_d(N_s+1)+N_s(N_d+5)+d^{+}\cdot d^{+}ss+s^{+}s^{+}\tilde{d}\cdot\tilde{d}]$$

This Hamiltonian is indeed the one used in the Interacting Boson Approximation. Namely, the BZM method can be used to derive the Hamiltonian for the IBA.

(2) BET

In the BET, one first performs a BCS transformation and then expands the quadrupole quasiparticle pair in a power series. Here the monopole pair is dropped. Up to sixth order this Hamiltonian is

$$\tilde{H} = \{2\varepsilon + A\Omega\}N_d - A[2N_d(N_d-1) + 5(1-\frac{1}{4\Omega})C_5]$$

$$+ B[\{\Omega + \frac{3}{2} - \frac{7}{4\Omega}\}\{d^+\cdot d^+ + \tilde{d}\cdot\tilde{d}\}$$

$$- \{1 + \frac{2}{\Omega}\}\{d^+\cdot d^+ N_d + N_d\tilde{d}\cdot\tilde{d}\}$$

$$+ \frac{1}{2\Omega}\{d^+\cdot d^+[C_5 - N_d(N_d-1)] + [C_5 - N_d(N_d-1)]\tilde{d}\cdot\tilde{d}\}$$

where

$$A = 2B + G_2$$

$$B = \frac{1}{4}(b_2 - G_2)\sin^2\beta$$

$$C_5 = d^+\cdot d^+\tilde{d}\cdot\tilde{d}$$

and

$$\sin\beta = \sqrt{\tilde{n}(2\Omega-\tilde{n})}/\Omega .$$

Here $\tilde{n}$ is the average number of valence nucleons.

(3) O-A-I

In the O-A-I method[n], one maps the fermion operators in the seniority basis onto a boson basis with the number of quadrupole bosons given by (1). The odd multipole operators map the same way as in (2).

$$S^{+} = s^{+}\sqrt{\Omega - N - N_d}$$

$$D_{\mu}^{+} = [d_{\mu}^{+} h + 2d^{+}\cdot d^{+}\tilde{d}_{\mu} i\,]\sqrt{j(\Omega - N - N_d - 1)}$$

$$- s^{+}s^{+}[h\tilde{d}_{\mu} + 2i d_{\mu}^{+}\tilde{d}\cdot\tilde{d}]\sqrt{k}$$

$$P_{\mu}^{(2)} = 2[d_{\mu}^{+} sh + 2d^{+}\cdot d^{+}\tilde{d}_{\mu} si]\sqrt{j}$$

$$+ 2\sqrt{j}\,[hs^{+}\tilde{d}_{\mu} + 2is^{+}d_{\mu}^{+}\tilde{d}\cdot\tilde{d}]$$

where

$$h = \frac{1}{2}(Z_{+}+Z_{-})-(2N_d+3)i$$

$$i = \frac{1}{2}V^{-1}[Z_{+}-Z_{-}]$$

with

$$Z_{\pm} = \frac{\sqrt{k(2\Omega-N_d+5\pm V)}}{\sqrt{2}}$$

$$V = \sqrt{(2N_d+3)^2-C_5}$$

$$j = (\Omega-N-N_d)/(\Omega-2N_d-1)$$

$$k = \Omega - 2N_d + 1$$

In reference 7, it is assumed that $C_5 = 0$, which is correct only for states that have no pairs of quadrupole bosons coupled to angular momentum zero. In addition, for evaluating i and k, $(2N_d+3)/\Omega$ is assumed small. The quadrupole pair and multipole operators then become,

$$D_{\mu}^{+} = [d_{\mu}^{+} + \frac{1}{2\Omega}d^{+}\cdot d^{+}\tilde{d}_{\mu}]\sqrt{j(\Omega-N-N_d-1)} - \frac{1}{\Omega}s^{+}s^{+}\tilde{d}_{\mu} + \cdots$$

$$P_{\mu}^{(2)} = [2d_{\mu}^{+}s + \frac{1}{\Omega}d^{+}\cdot d^{+}\tilde{d}_{\mu}s\,]\sqrt{j} + \sqrt{j}\,[\,2s^{+}\tilde{d}_{\mu} + \frac{1}{\Omega}s^{+}d_{\mu}^{+}\tilde{d}\cdot\tilde{d}] + \cdots .$$

Inserting these expressions for S^+, S, D^+, D, and $P^{(r)}$, one obtains the Interacting Boson Model Hamiltonian.

In Figs. 1 and 2 and Table 1 we show the energy levels calculated by the three approximations, the BZM method, the BET method (6th order) and the OAI method (4th order).

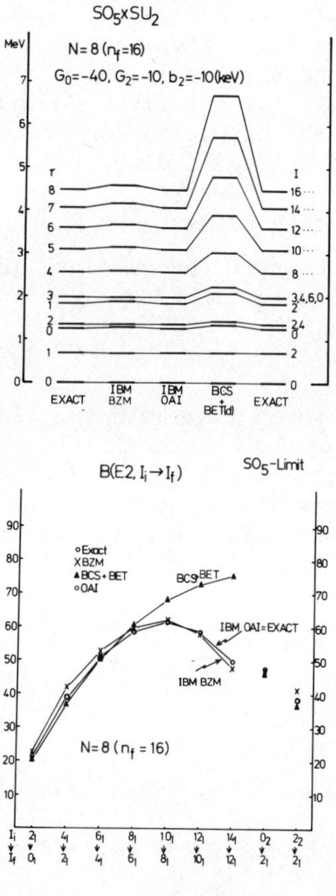

Fig. 1. SO_5 limit

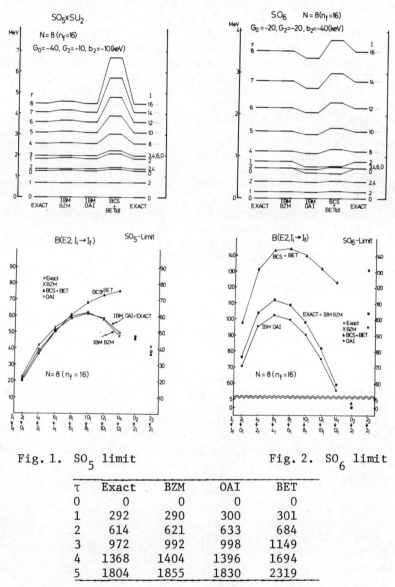

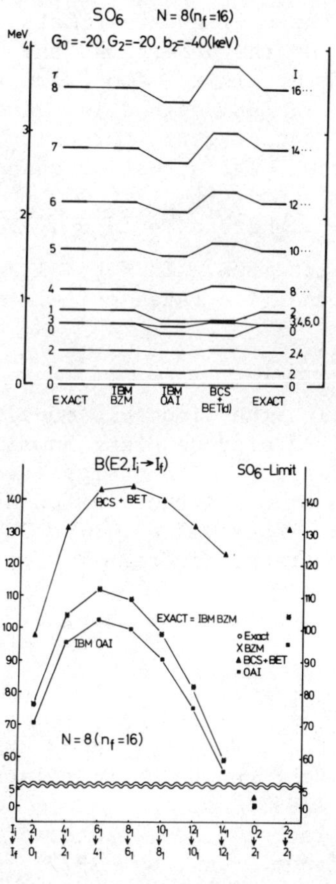

Fig. 2. SO_6 limit

τ	Exact	BZM	OAI	BET
0	0	0	0	0
1	292	290	300	301
2	614	621	633	684
3	972	992	998	1149
4	1368	1404	1396	1694
5	1804	1855	1830	2319

Table 1 The energies of yrast states in an intermediate situation between SO_5 and SO_6; $G_0 = -30$, $G_2 = -15$, $b_2 = -25$, $N = 8$, $\Omega = 22$.

These levels are compared in Figs. 1 and 2 with the exact levels for two limits; the case where $G_2 = b_2$, which gives a vibrational spectrum, and the case where $G_0 = G_2$, which gives the spectrum of a γ-unstable nuclei. Intermediate cases are also studied in Table 1. One sees that the OAI method and the BZM method give reasonable agreement in all cases. The Hamiltonian given by the OAI method becomes finite for $G_2 = b_2$ and hence gives the exact result in this limit, while it underestimates the excitation energy in the $G_0 = G_2$ limit. On the other hand the BZM method gives the exact result in the $G_0 = G_2$ limit, and overestimates the spectra in the $G_2 = b_2$ limit. The BET method is reasonable only for the first few excited states, and it gets increasingly worse as the excitation energy increases. The situation cannot be improved by adding a correction $\frac{n_d(n_d-1)}{2}$ where $n_d = \Sigma d^+_\mu d_\mu$. The correction brings down high spin states too much. In Figs. 1 and 2, we compare the absolute quadrupole transition rates calculated by all three methods.

Conclusion

Using the Ginocchio model, I have shown that both the BZM and OAI method produce boson Hamiltonians for the Interacting Boson Approximation. The energy levels and quadrupole transitions calculated by using these boson Hamiltonians are in good agreement with the exact values. The BET does poorly for both the spectra and the transition rates.

I would like to thank J. N. Ginocchio and Ń. Yoshida for their cooperation.

References

1) A. Bohr, Mat. Fys. Medd. Dan. Vid. Selsk. 26 (1952) no. 14
A. Bohr and B. R. Mottelson, Mat. Fys. Medd. Dan, Vid. Selsk. 27 (1953) no. 16.
2) F. Iachello, ed., Interacting Bosons in Nuclear Physics (Plenum Press, NY. 1979)
For example, O. Scholten, F. Iachello and A. Arima, Ann. of Phys. 115 (1978) 325
3) T. Kishimoto and T. Tamura, Nucl. Phys. A192 (1972) 246, A270 (1976) 317.
T. Tamura, K. Weeks and T. Kishimoto, Phys. Rev. C20 (1979) 307.
4) T. Otsuka, A. Arima, F. Iachello and T. Talmi, Phys. Lett. 66B (1977) 205; 76B (1978) 139.
5) S. T. Belyaev and V. G. Zelevinsky, Nucl. Phys. 39 (1962) 582.
E. R. Marshalek, Nucl. Phys. A161 (1971) 401; A224 (1974) 221, 245.
6) J. N. Ginocchio, Phys. Lett. 85B (1979) 9; Ann. of Phys. 126 (1980) 234.
7) T. Otsuka, A. Arima and F. Iachello, Nucl. Phys. A309 (1978) 1.

MICROSCOPIC STRUCTURE OF THE INTRINSIC STATE OF A DEFORMED NUCLEUS

Ricardo A. Broglia

The Niels Bohr Institute, University of Copenhagen
DK-2100 Copenhagen Ø, Denmark, and
Oak Ridge National Laboratory, Nuclear Division
Oak Ridge, Tenn. 37380, U.S.A. [a)]

INTRODUCTION

A remarkably accurate microscopic description of the properties of strongly deformed nuclei has been achieved in terms of the alignment of the orbitals of the individual nucleons in the average deformed Nilsson potential[1,2]. The main evidence for the validity of this picture has been provided by the systematic identification (cf. e.g. Ref. 3) of the Nilsson levels throughout the mass table, making use of one-particle stripping and pick-up reactions. Strong support for the model has been also found in the measurements of two-nucleon transfer reactions in which the multipole deformation associated with the Nilsson levels close to the Fermi surface is probed (cf. e.g. Refs. 4 and 5).

Recently, it has been suggested (cf. e.g. Ref. 6 and references therein) that the properties of the low-energy nuclear spectrum can be described in terms of the alignment of pairs of fermions coupled to angular momentum 0 and 2 .

In the present paper we compare, for the case of strongly deformed nuclei, the predictions of the pair aligned model with those of the fermion aligned model (Nilsson plus BCS model). For simplicity the calculations are carried out in a single j-shell and for like particles. The techniques presented here can be used to extend the calculations to the case of many j-shells and to the case of protons and neutrons. In any case we feel that the main features of the results obtained in the present paper are general properties of the two coupling schemes, and do not depend on the limitations of the calculational scheme utilized.

The present contribution is a numerical application of the work of Ref. 7, and we refer to it for further theoretical details.

1. DIAGONALIZATION OF THE PAIRING PLUS-QUADRUPOLE INTERACTION IN A SINGLE j-SHELL IN THE MEAN FIELD APPROXIMATION

Nuclear deformations arise from a competition between the shell effects and the pairing and quadrupole correlations. The average field produced by an attractive quadrupole-quadrupole residual interactions is[1,2)]

$$\delta V_Q \sim \beta M \omega_0^2 r^2 Y_{20}(\theta,\phi), \quad (1)$$

the deformation parameter β being related to the nuclear radius by

$$R = R_0 (1 + \beta Y_{20}(\theta,\phi)). \quad (2)$$

The total Hamiltonian (Nilsson Hamiltonian[2)])

$$H_0 = \mathring{H}_0 + \delta V_Q, \quad (3)$$

$$\mathring{H}_0 = \frac{1}{2} M \omega_0^2 r^2 - \frac{\hbar^2}{2M} \nabla^2, \quad (4)$$

is diagonalized by aligning the orbits of the individual particles with respect to the intrinsic symmetry axis* (strong coupling). Note that the harmonic oscillator potential is only used for convenience.

For particles moving in a single j-shell the first order correction to the single-particle energy is given by** (cf. e.g. Ref.8)

$$\Delta\varepsilon\left(\frac{m}{j}\right) = \langle jm|\delta V_Q|jm\rangle \sim M\omega_0^2 \langle r^2\rangle \sqrt{\frac{5}{16\pi}}\,\beta \times \frac{1}{2}\left(3\left(\frac{m}{j}\right)^2 - 1\right) \sim 10\beta\left(3\left(\frac{m}{j}\right)^2 - 1\right) \text{MeV}, \quad (5)$$

(expression valid for $j \to \infty$).

---

* Approximate solution of a constrained Hartree-Fock problem with constraint $Q_0 = \frac{3}{\sqrt{5\pi}} Z e R_0^2 \beta$.

** We have utilized the following approximate expressions (cf. Refs. 9 and 10): $\langle r^2\rangle = (N+3/2)\,\hbar/M\omega_0$, $\hbar\omega_0 \sim 41\, A^{-1/3}$ MeV , $\frac{2}{3}(N_{max}+2)^3 \sim A$; $N_{max} \sim (\frac{3}{2}A)^{1/3} \sim A^{1/3}$; $\Omega = \frac{1}{2}(N+1)(N+2) \sim 0.5\, A^{2/3}$.

The splitting between two successive levels is, for m/j and for large values of j , given by

$$\delta\varepsilon = \Delta\varepsilon\left(\frac{m+1}{j}\right) - \Delta\varepsilon\left(\frac{m}{j}\right) \sim 60\beta\,\frac{m}{j^2} \sim \frac{60}{\Omega}\,\beta \text{ MeV}$$

$$\sim \frac{120}{A^{2/3}}\,\beta \text{ MeV} \sim 1\text{ MeV} \quad (\beta \sim 0.3,\ A \sim 150). \tag{6}$$

The total splitting is

$$\Delta\varepsilon = \Delta\varepsilon(1) - \Delta\varepsilon(0) \sim 30\beta \text{ MeV} \sim 10 \text{ MeV} \quad (\beta \sim 0.3) \tag{7}$$

The associated Nilsson levels for $j = 41/2$ are shown in Fig. 1 as a function of the deformation parameter β . The quadrupole moment of two particles moving in these levels is also displayed in the figure. It is the maximum quadrupole moment a two nucleon system can have. This result will be modified because of the presence of pairing correlations. However, because the pairing gap $\Delta \sim 1$ MeV is much smaller than the shell splitting $\Delta\varepsilon$, the modification is expected to be small.

The average field produced by an attractive monopole pairing residual interaction is

$$\delta V_p \sim -\Delta \sum_{m>0} \left\{a^+(m)\,a^+(\tilde{m}) + a(\tilde{m})\,a(m)\right\} \tag{8}$$

$$\Delta = G \sum_{m>0} U(m)\,V(m). \tag{9}$$

The total Hamiltonian (BCS Hamiltonian)

$$H_o = H_{sp} + \delta V_p , \tag{10}$$

$$H_{sp} = \sum_{m>0} (\varepsilon(m) - \lambda)\, a^+(m)\, a(m), \tag{11}$$

is diagonalized through the linear transformation

$$\alpha^+(m) = U(m)\,a^+(m) - V(m)\,a(\tilde{m}), \tag{12}$$

$$\alpha(m) = U(m)\,a^+(\tilde{m}) + V(m)\,a(m), \tag{13}$$

provided the U , V factors have the following functional dependence on Δ and λ :

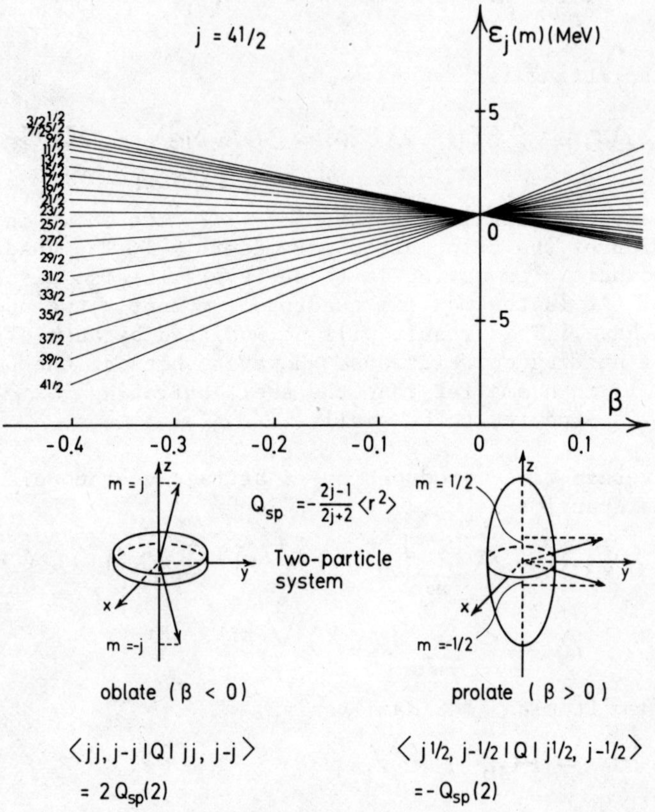

Fig. 1. Nilsson levels associated with a single j-shell as a function of the deformation. A schematic representation of the oblate and of the prolate systems corresponding to two particles moving in the lowest orbitals for $\beta < 0$ and $\beta > 0$ are given. The corresponding analytic expressions of the static quadrupole moments are also shown.

$$U(m) = \frac{1}{\sqrt{2}}\left(1 + \frac{\varepsilon(m)-\lambda}{\sqrt{(\varepsilon(m)-\lambda)^2+\Delta^2}}\right)^{1/2}, \tag{14}$$

$$V(m) = \frac{1}{\sqrt{2}}\left(1 - \frac{\varepsilon(m)-\lambda}{\sqrt{(\varepsilon(m)-\lambda)^2+\Delta^2}}\right)^{1/2}. \tag{15}$$

Utilizing the equations

$$2\sum_{m>0} V^2(m) = N\,, \qquad \text{(number equation)} \tag{16}$$

$$\sum_{m>0} \frac{1}{\sqrt{(\varepsilon(m)-\lambda)^2+\Delta^2}} = \frac{2}{G} \qquad \text{(gap equation)}, \tag{17}$$

one can determine the equilibrium values λ and Δ. Examples of occupation and probability amplitudes are displayed in Figs. 2 and 3. Note that these curves are rather universal, depending but weakly on the actual values of β and of j used. Because the occupation parameters completely characterize the system under study, the same statement above also applies to the transition matrix elements shown in Figs. 4 and 5.

The knowledge of the occupation parameters $U(m)$, $V(m)$ gives the complete solution of

$$H_{sp} + \delta V_Q + \delta V_p\,, \tag{18}$$

for any number of particles. All the properties of the system can be calculated in terms of these parameters. In fact, the wavefunction describing the intrinsic state is

$$|BCS\rangle_{\text{Nilsson}} = \prod_{m>0}\left(U(m) + V(m)\,a^{+}(m)\,a^{+}(\tilde{m})\right)|0\rangle. \tag{19}$$

This wavefunction does not conserve the number of particles. Its projection on a fixed number is given by (cf. e.g. Ref. 11)

$$|2n\rangle = \mathcal{N}\left(\sum_{m>0}\frac{V(m)}{U(m)}\,a^{+}(m)\,a^{+}(\tilde{m})\right)^{n}|0\rangle\,, \tag{20}$$

where $N = 2n$ is the number of particles, n being the number of pairs.

The fluctuation in the number of particles associated with this system is

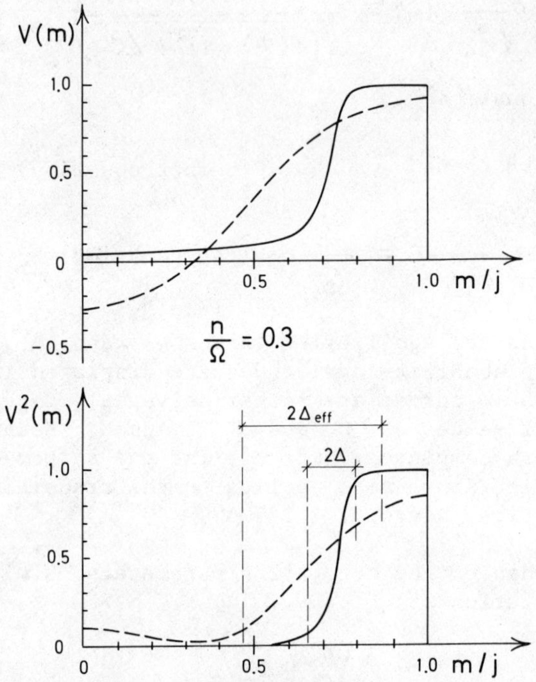

Fig. 2. Occupation amplitudes $V(m)$ and occupation probabilities $V^2(m)$ for the two models under discussion (cf. Eqs. (15) and (40)). The results associated with the alignment of particles are shown as continuous lines and those of the alignment of pairs of particles as dotted lines. The calculation was carried out in the case of the fermion alignment for a $j = 41/2$ shell, a deformation parameter $\beta = -0.4$ and a shell occupation $n/\Omega = 0.3$. The quantity $\Omega = j + \frac{1}{2}$ is the total degeneracy of the shell while n is the number of pairs. The Fermi surface and the gap parameter are obtained by solving (16) and (17) for the given number of particles. The parameter β' appearing in (40) was obtained by solving Eq. (42) for the chosen value of n/Ω.

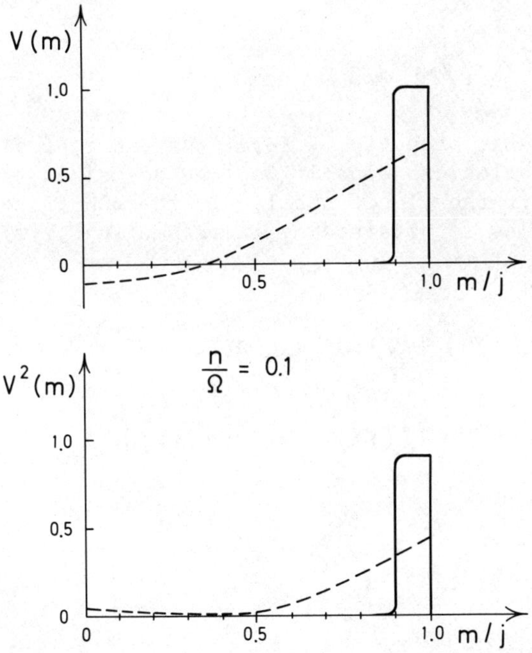

Fig. 3. Same as Fig. 2 but for $n/\Omega = 0.1$.

$$\Delta N^2 = \langle N^2\rangle - \langle N\rangle^2 = 4\sum_{m>0} U^2(m)\,V^2(m)\,. \tag{21}$$

Denoting by ρ the number of levels per unit energy, the quantity (21) is proportional to $\rho\Delta$, which measures the number of momentum states lying within the difuse Fermi surface. On the other hand, the total number of particles is porportional to $\rho\varepsilon_F$, where $\varepsilon_F = \lambda$ is the Fermi energy. Therefore, if the gap is small compared to the Fermi energy we have (cf. e.g. Ref. 12)

$$\Delta N^2 \ll N \ll N^2\,, \tag{22}$$

or

$$\Delta N/N \ll 1\,. \tag{23}$$

It is a basic property of Fermi systems that a configuration with $2n$ particles can equally well be described as a system of $2(\Omega-n)$ holes in the closed shell. In the BCS state, the description in terms of holes is obtained by interchanging $V(m)$ and $U(m)$

$$\prod_{m>0}\left(U(m)+V(m)\,a^+(m)\,a^+(\tilde{m})\right)|0\rangle \tag{24}$$

$$= \prod_{m>0}\left(V(m)+U(m)\,a(\tilde{m})\,a(m)\right)|\tilde{0}\rangle$$

where

$$|\tilde{0}\rangle = \prod_{m>0} a^+(m)\,a^+(\tilde{m})\,|0\rangle\,. \tag{25}$$

Projecting from this $|BCS\rangle$ function the states with $2(\Omega-n)$ holes one obtains

$$|2(\Omega-n)\rangle = \mathcal{N}'\left(\sum_{m>0}\frac{U(m)}{V(m)}\,a(\tilde{m})\,a(m)\right)^{(\Omega-n)}|\tilde{0}\rangle \tag{26}$$

a. Quadrupole Moment

The multipole single-particle operator can be written as

$$
\begin{aligned}
q(\lambda) &= \sum_m \langle jm | \sqrt{\frac{16\pi}{2\lambda+1}}\, r^\lambda Y_{\lambda 0} | jm \rangle \, a^+(m)\, a(m) \\
&= Q_{sp}(\lambda) \sum_m \langle jm\lambda 0 | jm \rangle \, a^+(m)\, a(m) \qquad (27)\\
&\sim Q_{sp}(\lambda) \sum_{m>0} P_\lambda(m/j) \left(a^+(m)\, a(m) + a^+(\tilde{m})\, a(\tilde{m}) \right),
\end{aligned}
$$

where we have utilized the asymptotic expressions

$$
\lim_{j\to\infty} \langle jm\lambda 0 | jm \rangle \sim P_\lambda(m/j) \; ; \; Q_{sp}(\lambda) \simeq -\langle r^\lambda \rangle . \qquad (28)
$$

The multipole moment associated with the state described by the wavefunction (19) is

$$
Q(\lambda) = \langle q(\lambda) \rangle = 2\, Q_{sp}(\lambda) \sum_{m>0} V^2(m)\, P_\lambda(m/j) . \qquad (29)
$$

Utilizing the fact that $j \gg 1$, we can define a "continuous" variable

$$
\frac{m}{j} = x , \qquad (30a)
$$

and transform the sum into an integral. Thus $\Delta m \sim \Omega \Delta x$ and

$$
\sum_{m>0} \longrightarrow \Omega \int_0^1 dx . \qquad (30b)
$$

The multipole moment (29) can be written as

$$
\frac{Q(\lambda)}{\Omega} \sim 2\, Q_{sp}(\lambda) \int_0^1 dx\, V^2(x)\, P_\lambda(x) . \qquad (31)
$$

This quantity is shown in Fig. 4 for $\lambda = 2$ in units of $Q_{sp}(2)$. The calculations were carried out for negative deformations ($\beta < 0$). This is because like particles at the beginning of a j-shell display a negative quadrupole moment.

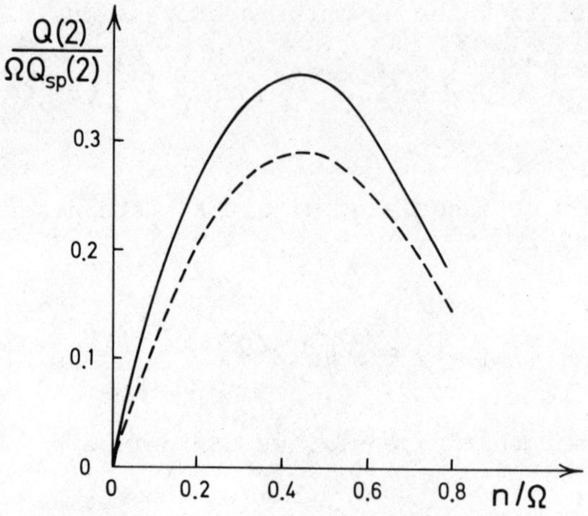

Fig. 4. Static quadrupole moment as a function of the filling of the shell parameter n/Ω . The quantity $\Omega = j+\frac{1}{2}$ is the total number of pairs that can be accommodated in a single j-shell, while n is the actual number of pairs. The calculations in the fermion aligned model were carried out utilizing Eq. (29) for $j = 41/2$ and for a deformation $\beta = -0.4$. In all cases the self-consistent values of the pairing gap and of the Fermi surface were used to calculate the occupation parameter $V(m)$. The coupling constant G was chosen such that the gap Δ reaches a value of about 1.5 MeV for $n/\Omega \sim 0.5$. The results are shown as a continuous line. The results of the pair aligned model correspond to the solution (47) ($\alpha = 0.59$ and $\beta = 0.81$). The occupation parameters of the pair aligned model were calculated utilizing equation (40), the quantity β' being determined from equation (42).

b. Two-Nucleon Transfer Spectroscopic Amplitudes

The two-nucleon transfer operator is defined as[5)]

$$T(J) = \frac{[a^+a^+]_{J0}}{\sqrt{2}} = \sqrt{2} \sum_{m>0} \left(\langle j m j -m | J0 \rangle (-1)^{j-m} \right) a^+(m)\, a^+(\tilde{m}) \tag{32}$$

Utilizing the classical limit of the coupling coefficient

$$\lim_{j \to \infty} \langle j m j -m | J0 \rangle = (-1)^{j-m} \sqrt{\frac{2J+1}{2j+1}}\, P_J(m/j) , \tag{33}$$

and replacing the sum over m by an integration according to (30) one can write

$$\begin{aligned} M_J = \langle J0 | T(J) | 00 \rangle &= \sqrt{\frac{2J+1}{\Omega}} \sum_{m>0} P_J(m/j)\, U(m) V(m) \\ &\sim \sqrt{\frac{2J+1}{\Omega}}\, \Omega \int_0^1 dx\, P_J(x)\, U(x)\, V(x) , \end{aligned} \tag{34}$$

which yields

$$\frac{M_J}{\sqrt{\Omega}} = \sqrt{2J+1} \int_0^1 dx\, P_J(x)\, U(x)\, V(x) . \tag{35}$$

Examples of this quantity are shown in Fig. 5. Note that the two-particle "cross section" is obtained by squaring (35).

2. DIAGONALIZATION OF A PAIRING PLUS QUADRUPOLE RESIDUAL INTERACTION IN A SINGLE j-SHELL AND IN THE MEAN FIELD APPROXIMATION CONSTRAINING PAIRS OF FERMIONS TO COUPLE TO ANGULAR MOMENTUM ZERO AND TWO

The basis states of the pair aligned model are

$$\begin{aligned} |(j)^2_I\rangle &= \sum_{m>0} \langle j m j -m | J0 \rangle \frac{a^+(m)\, a^+(-m)}{\sqrt{2}} |0\rangle \\ &\sim \sqrt{\frac{2I+1}{\Omega}} \sum_{m>0} P_I(m/j)\, a^+(m)\, a^+(\tilde{m}) |0\rangle \qquad (I = 0, 2) . \end{aligned} \tag{36}$$

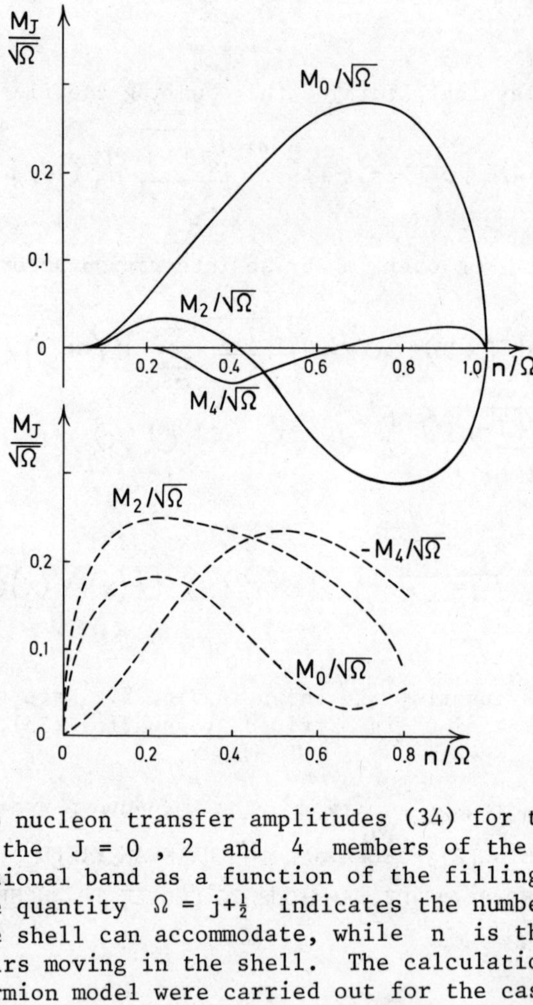

Fig. 5. Two nucleon transfer amplitudes (34) for the excitation of the $J = 0$, 2 and 4 members of the ground state rotational band as a function of the filling of the shell. The quantity $\Omega = j+\frac{1}{2}$ indicates the number of pairs that the shell can accommodate, while n is the number of pairs moving in the shell. The calculations of the aligned fermion model were carried out for the case of a $j = 41/2$ shell and for a constant deformation $\beta = -0.4$. The results are shown as continuous curves. The results corresponding to the alignment of pairs of fermions are shown as dotted curves. For more details cf. caption to Fig. 4.

In writing the last expression use of the semiclassical limit (33) of the vector coupling coefficient has been made.

The intrinsic state for an n-pair system is, in this basis, given by

$$|\text{intrinsic state}\rangle_B \sim \left(\sum_{m>0} c(m) a^{+}(m)\, a^{+}(\tilde{m})\right)^n |0\rangle, \tag{37}$$

where

$$c(m) = \frac{1}{\sqrt{\Omega}} \left(\alpha + \sqrt{5}\, P_2(m/j)\, \beta\right). \tag{38}$$

From the structure of the projected wavefunction (20) and of the normalization condition

$$U^2(m) + V^2(m) = 1, \tag{39}$$

we obtain

$$V(m) = \frac{\beta' c(m)}{\sqrt{1 + \beta'^2 c^2(m)}}, \tag{40}$$

and

$$U(m) = \frac{1}{\sqrt{1 + \beta'^2 c^2(m)}}. \tag{41}$$

The quantity β' is a normalization constant determined by the BCS number equation

$$N = 2 \sum_{m>0} V^2(m). \tag{42}$$

For a small number of particles $(N/2 \ll \Omega)$

$$N \sim 2\Omega \beta'^2 \int_0^1 dx\; c^2(x) = 2\beta'^2 (\alpha^2 + \beta^2), \tag{43}$$

$$\beta'^2 \sim \frac{N}{2} = n = \text{number of pairs}. \tag{44}$$

In this case $(\beta'^2 \ll \Omega)$ the static quadrupole moment of the system can be written as (cf. Eq. (29))

$$(Q(2))_B = 2\, Q_{sp}(2)\, \beta'^2 \sum_{m>0} \left(\frac{\beta'^2 c^2(m)}{1+\beta'^2 c^2(m)} \right) P_2(m/j)$$
$$\sim 2\, Q_{sp}(2)\, \beta'^2 \sum_{m>0} c^2(m)\; P_2(m/j)$$
$$\sim 2\, Q_{sp}(2)\, \beta'^2 \Omega \int_0^1 dx \left[\frac{1}{\sqrt{\Omega}} (\alpha + \sqrt{5}\, P_2(x)\beta) \right]^2 P_2(x) \qquad (45)$$
$$\sim 4\, Q_{sp}(2)\, \beta'^2 \left(\frac{\alpha\beta}{\sqrt{5}} + \frac{\beta^2}{7} \right).$$

The constants α and β can be obtained by determining the extremes of $(Q(2))_B$, that is,

$$\delta \left\{ \frac{(Q(2))_B}{4\, Q_{sp}(2)\, \beta'^2} - \lambda(\alpha^2 + \beta^2) \right\} = 0. \qquad (46)$$

which leads to

$$\lambda = 0.31 \quad \begin{cases} \alpha = 0.59 \\ \beta = 0.81 \end{cases} \quad (Q(2))_B \sim 1.23\, \beta'^2 Q_{sp}(2)\ \text{(oblate)}, \qquad (47)$$

$$\lambda = -0.16 \quad \begin{cases} \alpha = -0.81 \\ \beta = 0.59 \end{cases} \quad (Q(2))_B \sim -0.66\, \beta'^2 Q_{sp}(2)\ \text{(prolate)}. \qquad (48)$$

As expected, the model predicts that at the beginning of a j-shell the system deforms with a negative quadrupole moment. However, the absolute value is only $\sim 60\%$ of that predicted by aligning the fermions. Note also that the values of the coefficients α and β are very close to the SU(3) prediction $\alpha = \sqrt{\frac{1}{3}} = 0.58$ and $\beta = \sqrt{\frac{2}{3}} = 0.82$. The phenomenon of deformation is however more general, the SU(3) prediction for α and β being a single point in the entire deformation plane.

Making use of Eqs. (40), (41) and (47) we can now calculate the different properties of the system under study. The occupation parameters are exemplified in Figs. 2 and 3 while in Figs. 4 and 5 we show the quadrupole moments and the two-nucleon transfer amplitudes.

Comparison Between the Two Models

1) The essential difference between the two models manifests itself in the occupation amplitudes V(m) (cf. Figs. 2 and 3). The region over which $V^2(m)$ goes from 90% to 10% of the value achieved for $m/j \sim 1$, is about twice as large for the case in which $\lambda = 0$ and $\lambda = 2$ pairs are aligned than for the case of the fermion

alignment. That is, $(2\Delta)_F < \frac{1}{2}(2\Delta)_B$ $(\frac{n}{\Omega} \sim 0.3)$. This is because Δ_B receives contributions from both the monopole and the quadrupole pairing components. The effective quadrupole pairing contribution to the gap parameter (cf. e.g. Ref. 13) is very large in this model. Note also that the amplitude $(V(m))_B$ changes sign.

Note also that the wavefunction (20) with occupation parameters (40) and (41) does not display symmetry between interchange of particles and holes. In fact, while $V(m)/U(m)$ contains only a monopole and a quadrupole component, $U(m)/V(m)$ contains all multipolarities. This difficulty can also be seen from a comparison between the levels associated with n pairs coupled to 0 and to 2 and those associated with the complementary space of $(\Omega-n)$ hole pairs. In particular the former contains states with I-values up to $2n$ while the latter yields states with spin ranging from $I = 0$ to $I = 2(\Omega-n)$.

2) The alignment of pairs lead to values of the static quadrupole moment which are systematically smaller than those obtained by aligning the fermions (cf. Fig. 4).

3) The alignment of pairs model predicts cross sections for two-particle transfer reactions to the 2^+ member of the ground state band which are systematically larger than the cross section associate with the transition leading to the ground state. This prediction is opposite to that of the Nilsson plus BCS model. For $n/\Omega \sim 0.2$ one obtains

$$\frac{\sigma(gs \rightarrow gs)}{\sigma(gs \rightarrow 2^+)} \sim \begin{cases} 0.5 & \text{pair alignment,} \\ 4.0 & \text{fermion alignment .} \end{cases} \tag{49}$$

Experimental evidence (cf. e.g. Ref. 5) indicates that*

$$\frac{\sigma(gs \rightarrow gs)}{\sigma(gs \rightarrow 2^+)} \sim 3\text{-}5 \; . \tag{50}$$

Note also the very different behaviour of $M_J/\sqrt{\Omega}$ predicted

* The experience accumulated during the last ten years indicates that the contribution of two-step processes to the excitation of the lowest 2^+ state can contribute to this ratio by at most a factor of two (cf. e.g. Refs. 14 and 15 and references therein).

by the two models. For $J = 0$ this difference can be understood as follows:

In the case of fermion alignment and for $1 \ll n \ll \Omega$, the gap parameter Δ is $\gtrsim$ than $\delta\varepsilon$. For example for $n/\Omega \sim 0.3$, $\Delta \sim 1.2$ MeV while the splitting between the single-particle levels close to the Fermi surface is ~ 0.8 MeV. Under these conditions one can use the continuous model of pairing correlation[1] to calculate the different properties of the system. In this model

$$\frac{M_0}{\sqrt{\Omega}} \sim \frac{\Delta}{G\Omega} \sim \exp\left\{-\frac{\delta\varepsilon}{G}\right\} \tag{51}$$

The quantity $\delta\varepsilon$ decreases almost linearly with n/Ω. Thus, for small values of n/Ω one expects Δ to depend exponentially on n/Ω. As n/Ω increases, the Pauli principle becomes more important and Δ saturates.

For the case of pair alignment, the effective pairing gap is very large already for small values of n/Ω (cf. Figs. 2 and 3; e.g. for $n/\Omega \sim 0.1$, $\Delta_{eff} \sim \frac{1}{2}(\Delta\varepsilon) \sim 5$ MeV). One can thus in this case utilize the degenerate model[1)], which leads to

$$\frac{M_0}{\sqrt{\Omega}} \sim \frac{\Delta}{G\Omega} \sim \sqrt{\frac{n}{\Omega}\left(1-\frac{n}{\Omega}\right)} \; . \tag{52}$$

For $1 \ll n \ll \Omega$,

$$\frac{\Delta}{G\Omega} \sim \sqrt{\frac{n}{\Omega}} \, , \tag{53}$$

i.e. a square root behaviour is expected.

Note that for $n \sim \Omega$, the density of levels in the Nilsson scheme associated with the fermion alignment model is very large. In fact in this case $\Delta/\delta\varepsilon \gg 1$. We then expect for $n \sim \Omega$ $M_0/\sqrt{\Omega}$ behaves like $\sqrt{n/\Omega}$. This is borne from the calculations.

4) The rotational bands built on particle configurations of finite dimensions terminate at an I_{max}, which for a $(j)^n$ configuration is

$I_{max} = 2n(\Omega-n)$ particle alignment

$I_{max} = 2n$ pair alignment

The ground state rotational band of ^{238}U has been studied up to $I = 32\,\hbar$ by multipole Coulomb excitation[16)]. The quadrupole electromagnetic transition probabilities show no deviation from the rigid rotational pattern, within the experimental errors. In the pair aligned mode, ^{238}U is viewed as 15 boson moving around the ^{208}Pb core. Strong deviation from the rotational pattern are expected in this model already at spin $I \sim 20\,\hbar$.

Possible Expansion of the Boson Subspace

An immediate question the previous comparison rises is concerning the relative importance of the higher multipolarities to which pairs are allowed to couple. In Table 1 the importance of allowing pairs of fermions to couple to angular momentum 4 is studied for the case of the static quadrupole and hexadecapole moment of a system of two particles. Including a g-boson within the pair aligned scheme brings the different predictions closer to the fermion alignment predictions. The improvement is however small, pointing to a slow convergence.

CONCLUSIONS

The predictions of the model based on the alignment of fermions and of pairs of fermions coupled to angular momentum 0 and 2 are, for strongly deformed nuclei, very different. In particular the occupation probabilities of the single-particle levels of odd-nuclei predicted by the pair aligned model are in disagreement with the predictions of the Nilsson model, which in turn are in excellent agreement with experiment.

Two-nucleon transfer reactions to the different members of the ground state rotational band probe the static deformations of the levels around the Fermi surface (multipole pairing distortions). The rotational model provides an accurate description of the observed relative cross sections. The predictions of the pair aligned model are at variance with the rotational model predictions. The static quadrupole moment gives a measure of the deformation of rotational nuclei and determines, among other quantities, the position of the Nilsson levels. The pair aligned model predicts a value which is only $\sim 60\%$ of the rotational model prediction.

In conclusions, pairs of fermions coupled to angular momentum 0 and 2 seem not to have the degrees of freedom needed to describe strongly deformed nuclei.

Table 1 The predictions of the pair aligned model for a two-particle system and for the quadrupole and hexadecapole moments are shown. The calculations were carried out allowing also the particles to couple to angular momentum $\lambda = 4$. The results of the particle aligned model are also shown.

	B[a)]		F[b)]
	s,d	s,d,g	
Prolate	$-0.81\|(j)^2_0\rangle+0.59\ (j)^2_2\rangle$	$-0.68\|(j)^2_0\rangle+0.63\|(j)^2_2\rangle-0.36\|(j)^2_4\rangle$	
$Q_2/Q_{sp}(2)$	0.66	0.82	1.0
$Q_4/Q_{sp}(4)$	0.20	0.64	0.75
Oblate	$0.59\|(j)^2_0\rangle+0.81\|(j)^2_2\rangle$	$0.41\|(j)^2_0\rangle+0.74\|(j)^2_2\rangle+0.52\|(j)^2_4\rangle$	
$Q_2/Q_{sp}(2)$	-1.23	-1.59	-2
$Q_4/Q_{sp}(4)$	0.37	0.84	2

a) Pair aligned model.
b) Particle aligned model.

ACKNOWLEDGEMENTS

Discussions with M. Baldo, A. Bohr, P. F. Bortignon, E. Maglione, B. R. Mottelson, A. Vitturi and A. Winther are gratefully acknowledged.

REFERENCES

1) A. Bohr and B. R. Mottelson, Nuclear Structure, Vol. II, Addison Wesley, Reading, Massachusetts (1975)
2) S. G. Nilsson, Mat. Fys. Medd. Dan. Vid. Selsk. 29, No. 16 (1955)
3) B. Elbek and P. O. Tjøm, Adv. in Nucl. Phys., Vol. 3 (1969) 259
4) R. A. Broglia, C. Riedel and T. Udagawa, Nucl. Phys. A135 (1969) 561

5) R. A. Broglia, O. Hansen and C. Riedel, Adv. in Nucl. Phys., Eds. M. Baranger and E. Vogt, Plenum Press, Vol. 6, p. 287, N.Y. (1973), and references therein
6) D. Janssen, R. V. Jolos and F. Dönau, Nucl. Phys. A224 (1974) 93; Yaderna Viz. 22 (1975) 965; Sov. J. of Part. Nucl. 8 (1977) 138;
Interacting Bosons in Nuclear Physics, Edited by F. Iachello, Plenum Press, N.Y. 1979
7) A. Bohr and B. R. Mottelson, preprint NORDITA 80/19
8) J. P. Elliott, Collective Motion in Nuclei, Series of lectures delivered at the Department of Physics, University of Rochester, Rochester, New York, AT(30-1)-875, 1958
9) A. Bohr and B. R. Mottelson, Nuclear Structure, Vol. I, Benjamin, N.Y. (1969)
10) D. R. Bes and R. A. Sorensen, Adv. in Nuclear Physics, Vol. 3, p. 129, Plenum Press, N.Y. (1969)
11) B. F. Bayman, Lectures on Seniority, Quasi-particles, and collective vibrations, November 1960, Princeton (unpublished)
12) S. T. Beliaev, Proceedings of the École d'été de Physique Théorique on the Many-Body Problem, Dunod, Paris (1959) 373
13) J. Ragnarsson and R. A. Broglia, Nuclear Physics A236 (1975)
14) J. S. Vaagen, Thesis, University of Bergen, Norway (1976), unpublished
15) R. J. Ascuitto and J. S. Vaagen, Proceedings of the International Conference on Reactions between Complex Nuclei; Eds. R. L. Robinson, F. K. McGowan, J. B. Ball and J. Hamilton, North-Holland, Amsterdam, Vol. 2, p. 303 (1974)
16) O. Schwalm et al., to be published

a) Research sponsored by the Division of Basic Energy Sciences, U.S. Department of Energy, under contract W-7405-eng-26 with the Union Carbide Corp.

INTRUDER STATES IN THE INTERACTING BOSON MODEL

Pieter Van Isacker

Instituut voor Nukleaire Wetenschappen

B-9000 Gent, Belgium

INTRODUCTION

Although the description of nuclear properties of low-lying collective states in terms of s and d bosons has been extremely succesfull and fruitful, one of its limitations has become apparent in recent years. Indeed, in many nuclei of various mass regions, low-lying $J^{\pi}= 0^{+}$ states or $K^{\pi} = 0^{+}$ bands are observed, which are not reproduced by the IBM nor by any other purely collective model. The subject of this talk will concentrate on a possibility to include such intruder states in the IBM. First, I will indicate a possible shell-model description of the excitation mechanisms which can give rise to such intruder states. Next, I will present some experimental evidence for the intruder states. Finally, I will discuss a possible description of intruder states in the framework of the IBM in the particular case of the nucleus $^{156}Gd_{92}$.

POSSIBLE ORIGIN OF THE INTRUDER STATES

If we ask ourselves the question of the possible origin of the intruder states, we simultaneously pose the problem of some of the limitations of the IBM. We know already that the IBM description of collective states breaks down for nuclei near shell closure, where it is expected that proton or neutron pair excitations become important. Moreover, we know that the detailed single particle level structure in a shell cannot be described completely and adequately by the IBM and that, as a result, at a certain excitation energy discrepancies will occur between theory and experiment. In addition, we have to keep in mind that the IBM only includes monopole and quadrupole degrees of freedom and that, at some excitation energy, other degrees of freedom (for instance hexadecupole) can become important.

Within this study, we will try to incorporate the effects discussed above into the current boson formalism in order to extend the scope and the range of applicability of the IBM.

It thus seems reasonable to explain the origin of low-lying $J^{\pi} = 0^{+}$ ($K^{\pi} = 0^{+}$) states in either of the following two ways.
(a) Two-particle two-hole excitations through a closed shell. In fig. 1 an illustration of this excitation mechanism is given, in the particular case of the Sn isotopes (breaking of the major shell at Z = 50). We show, as a function of the quadrupole deformation ε_2, the total potential energy of both the ground state of ^{116}Sn and the Nilsson intrinsic state $\{1/2^{+}[431]^{2}\ 9/2^{+}[404]^{2}\}_{\Omega=0}$. These energies are calculated using the Strutinsky[1] procedure, starting from a macroscopic liquid drop term and incorporating shell and pairing corrections coming from local density fluctuations of single particle states around the Fermi level. More details of this type of calculations can be found in refs. 2-3. From fig. 1 it is clear that at a certain deformation ($\varepsilon_2 \simeq .11$), the total energy of the intrinsic state $\{1/2^{+}[431]^{2}\ 9/2^{+}[404]^{2}\}_{\Omega=0}$ reaches a sharp minimum and that the two-particle two-hole excitation energy becomes much lower compared with the corresponding spherical configuration.

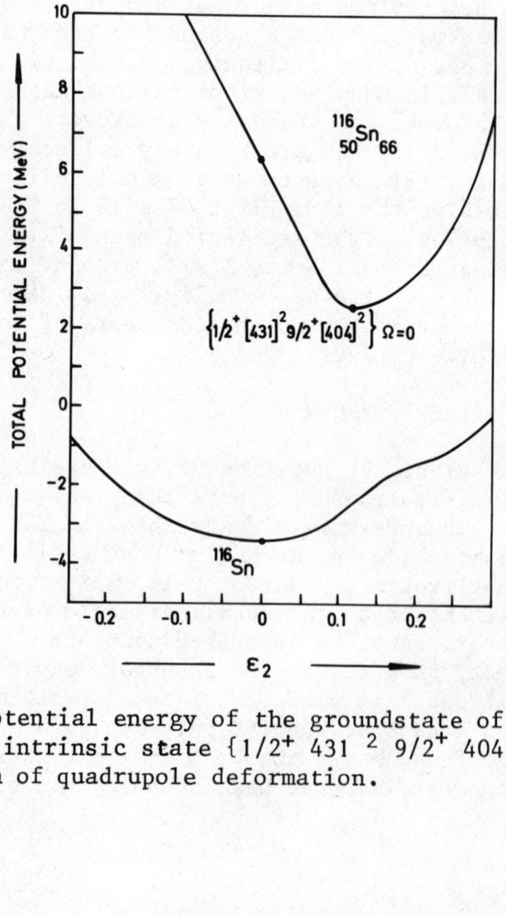

Fig.1 Total potential energy of the groundstate of ^{116}Sn and the Nilsson intrinsic state $\{1/2^{+}\ 431\ ^{2}\ 9/2^{+}\ 404\ ^{2}\}_{\Omega=0}$ as a function of quadrupole deformation.

How can we include such two-particle two-hole excitations in the formalism of s and d bosons? We have to consider, besides states with n s and d bosons (denoted by $|(s\ d)^n\rangle$), also states of the type $|(s'd')(s\ d)^{n+1}\rangle$. Here, s' or d' describes a pair of holes in the Z = 28-50 shell, whereas also the additional pair of particles in the Z = 50-82 shell is considered (n+1 s or d bosons). It thus follows that, when considering excitations through a major shell closure, the number of bosons is not the same in all basic states and consequently, the Hamiltonian will include number non-conserving terms.

(b) Subshell effects. In the case of real nuclei (many non-degenerated j-shells), discrepancies can occur between realistic calculations on the one hand and calculations with s and d bosons only on the other hand. For instance, low-lying $J^\pi = 0^+$ states could come from a subshell closure in the valence shell, as that occurring at proton number 64 in the 50-82 shell.

In the IBM formalism, the latter excitations could be taken into account by considering also s' and/or d' bosons, which have a higher energy than the s and d bosons. However, in this case one constructs, besides the normal states $|(s\ d)^n\rangle$, also states of the type $|(s'd')(s\ d)^{n-1}\rangle$. We remark that here all basic states have the same number of bosons and hence the Hamiltonian will be boson number conserving.

EXPERIMENTAL EVIDENCE

In this section, I will discuss a few examples of intruder states. First, I will show some results for the $_{44}$Ru and $_{46}$Pd isotopes. In fig.2 I compare the calculated energies of the members of the groundstate band with experiment. The agreement is excellent, except for the higher spin states ($J^\pi \geq 6^+$) in the Pd isotopes with few neutrons ($N \leq 56$). The latter discrepancy can be explained by the particle nature of those states, whereas the IBM only predicts collective states. Also other nuclear properties (B(E2) values and ratios, the quadrupole moments of the first excited 2^+ state, δ-ratios, magnetic moments, binding energies, nuclear radii, (p,t) and (t,p) transfer amplitudes) have been calculated for the Ru and Pd isotopes[4], and the overall agreement is good. However, the systematic occurrence of low-lying $J^\pi = 0^+$ states which are not reproduced by IBM, remains a problem. This is illustrated in fig.3, from where it can be seen that for the Ru and Pd isotopes often two $J^\pi = 0^+$ states are observed experimentally. The nature of one of these states is probably of type (a) discussed in the previous section, although mixing of the two $J^\pi = 0^+$ states can certainly not be excluded.

A second example is provided by the Gd isotopes, for instance nucleus $^{156}Gd_{92}$, which will be discussed extensively in this workshop by dr. Bäcklin. Although a large number of positive and negative parity bands of this nucleus can be explained by IBM[5], it seems that the $K^\pi = 0^+$ band with bandhead energy E_x = 1.168 MeV lies outside the

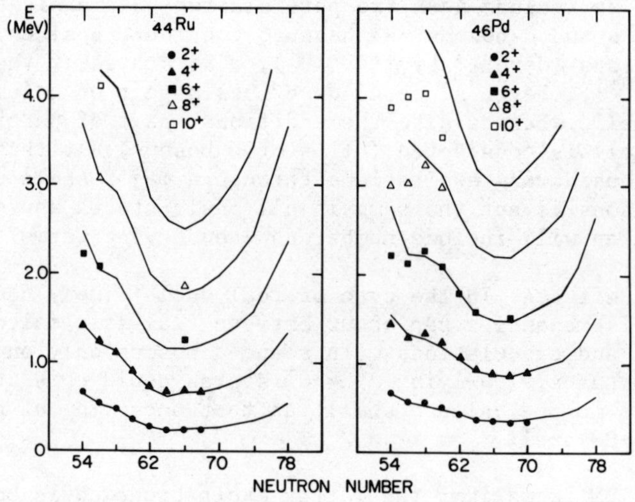

Fig.2 Comparison between experimental and calculated energies of the ground states band members in the Ru and Pd isotopes. Experimental values (points) are taken from ref.7.

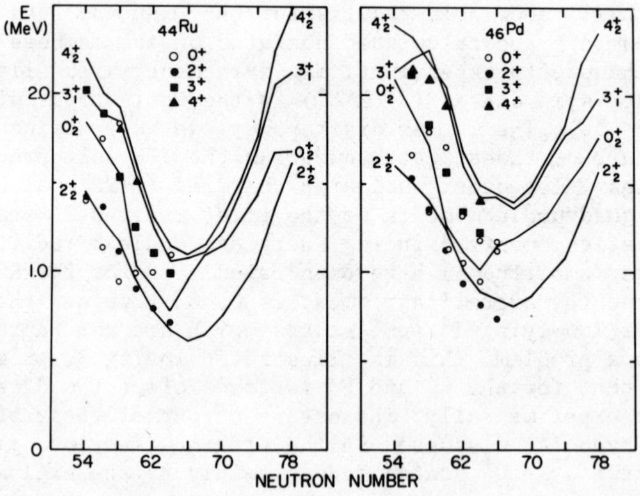

Fig.3 Comparison between experimental and calculated energies for some states of the quasi-beta and quasi-gamma bands in the Ru and Pd isotopes. Experimental values (points) are taken from ref.7.

model space. Also, the $K^\pi = 4^+$ band at $E_x = 1.511$ MeV is not reproduced by the IBM and can possibly be explained through the coupling of an $L^\pi = 4^+$ excitation to $(n-1)$ s and d bosons.

As further examples of intruder states, I can also mention the $K^\pi = 0^+$ rotational bands in the Sn isotopes with bandhead energy around $E_x = 2$ MeV. Finally, a last example is provided by the nuclei in the Zr region in which very lowlying $J^\pi = 0^+$ states are commonly observed.

AN EXAMPLE : THE NUCLEUS ^{156}Gd

I have tested the idea of including s', d' and g bosons on the nucleus ^{156}Gd. Since in this nucleus a $K^\pi = 0^+$ intruder band as well as a $K^\pi = 4^+$ band are observed, I attempted to describe this nucleus by coupling an s' or d' or g boson to $(n-1)$ s and d bosons, using the Hamiltonian

$$H = H_{sd} + \varepsilon_{s'} s'^\dagger s' + \varepsilon_{d'} d'^\dagger \cdot \tilde{d}' + \varepsilon_g g^\dagger \cdot \tilde{g} - \kappa' Q^{(2)} \cdot Q^{(2)'}$$

$$- \kappa_g [Q^{(2)} (g^\dagger \tilde{g})^{(2)}]^{(0)} - \Lambda :[(g^\dagger \tilde{d})^{(4)} (d^\dagger \tilde{g})^{(4)}]^{(0)}: \quad . \qquad (1)$$

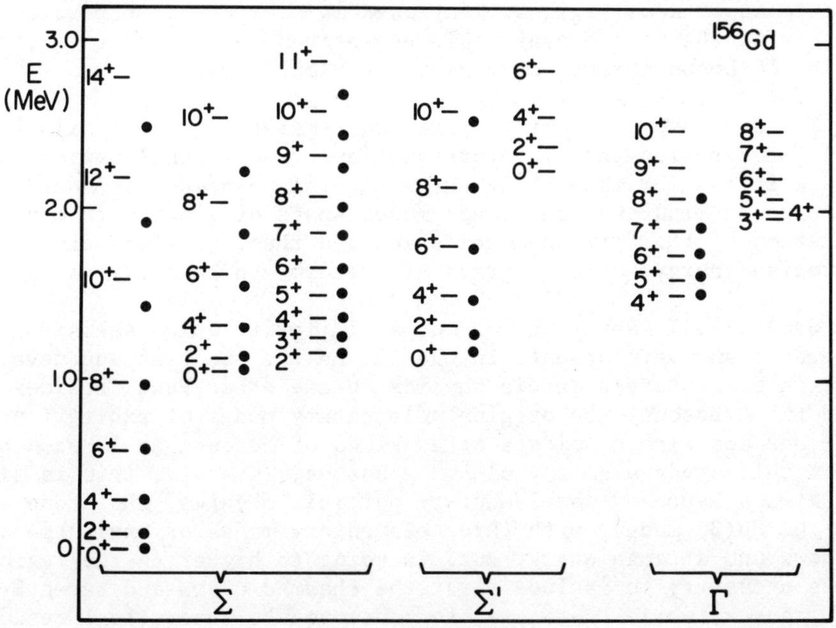

Fig.4 Comparison between experimental (dots) and calculated (lines) energies in ^{156}Gd. The Σ-states arise from s and d bosons, the Σ'-states from the coupling of an s' or d' boson to n-1 s and d bosons and the Γ-states from the coupling of a g boson.

Here, H_{sd} represents the IBM Hamiltonian in terms of s and d bosons. Furthermore, $Q^{(2)}$ is the quadrupole operator in terms of s and d bosons i.e.

$$Q^{(2)} = (s^\dagger \tilde{d} + d^\dagger s)^{(2)} - \frac{\sqrt{7}}{2} (d^\dagger \tilde{d})^{(2)} \quad ,$$

and $Q^{(2)\prime}$ is the quadrupole operator with respect to the s' and d' bosons. The last term is the exchange term, which can be included in analogy with the boson-fermion system[6]. However, in all practical calculations until now we have put $\Lambda = 0$. I remark that (1) is not the most general Hamiltonian. In particular, I have made no distinction between protons and neutrons and even then I have constructed a simplified, schematic expression in order to reduce the number of parameters. The result of this calculation for the energy spectrum of ^{156}Gd, is compared with experiment in fig.4.

CONCLUDING REMARKS

A necessary and more severe test of this model will be the calculation of electromagnetic properties, for instance for the Gd isopes. In order to give a full description of transition properties, it will be necessary to consider mixing between the different band systems. This work is in progress. Also, one can think of applying this model to other mass regions, mentioned in section 3. In particular, the case of the Sn isotopes would be interesting in order to study the problem of boson number non-conserving Hamiltonians.

In conclusion, I want to make two remarks. First of all, it should be stressed that the introduction of additional degrees of freedom in the IBM is nothing but a natural phenomenon. Indeed, one must bare in mind that the boson model space originates from a severe truncation of the fermion model space and that, at some point in excitation energy, other degrees of freedom can become important.

Secondly, I want to point out a similarity which seems to exist between recent developments in the IBM on the one hand and developments in elementary particle physics on the other hand. Besides the s and the d bosons - the original elementary modes of excitation of the IBM - one now also considers other kinds of bosons: an f boson and within this study also s', d' and g bosons. In a way, this is similar to what has happened in elementary particle physics, where one started with the SU(3) model, with three elementary modes of excitation: the up, down and strange quark. But, in going to higher energy regions, it was necessary to include first the charmed quark and recently the bottom quark, while there are, for the moment, theoretical reasons to believe in a sixth top quark.

ACKNOWLEDGEMENTS

I am very much indebted to Prof. F.Iachello for many ideas and useful discussions. I also want to thank the people of the KVI, Groningen and of the Nuclear Structure Lab at Yale for their warm hospitality. The NFWO is acknowledged for financial support.

REFERENCES

1. M.Brack,J.Damgaard,A.S.Jensen,H.C.Pauli,V.M.Strutinsky and C.Y. Wong, Rev. Mod. Phys. 44 (1972) 320
2. K.Heyde,M.Waroquier,H.Vincx and P.Van Isacker, Phys. Lett. 64B (1976) 135
3. K.Heyde,M.Waroquier,P.Van Isacker,H.Vincx and G.Wenes, to be publ.
4. P.Van Isacker and G.Puddu, accepted for publ. in Nucl. Phys.
5. J.Konijn,F.W.N. de Boer,H.Verheul and O.Scolten, proc. of the int. conf. on "Band Structure and Nuclear Dynamics", New Orleans
6. contribution of I.Talmi in this workshop; contribution of O.Scholten in this workshop.
7. M.Sakai and Y.Gono, "Quasi-ground,quasi-beta and quasi-gamma bands", preprint INS-J-160 (1979)

ON THE RENORMALIZATION OF THE IBA BY THE *g* BOSON*

Bruce R. Barrett and Keith A. Sage

Department of Physics
University of Arizona and
Tucson, AZ 85721

Theoretical Physics Division
University of Arizona, PAS 81
Tucson, AZ 85721

INTRODUCTION AND PERTURBATION THEORY FOR THE *g* BOSON

The IBA-2 proton-neutron Hamiltonian is given by[1-2]

$$H_{IBA2} = \varepsilon(n_{d_\pi} + n_{d_\nu}) + \kappa Q_\pi \cdot Q_\nu + M_{\pi\nu} + V_{\pi\pi} + V_{\nu\nu} \quad (1)$$

where $\kappa < 0$

$$Q_{\pi(\nu)} = (d^\dagger s + s^\dagger \tilde{d})^{(2)}_{\pi(\nu)} + \chi_{\pi(\nu)} (d^\dagger \tilde{d})^{(2)}_{\pi(\nu)}$$

$$V_{\pi\pi(\nu\nu)} = \sum_{L=0,2,4} \frac{1}{2} (2L+1)^{\frac{1}{2}} C_L^{\pi(\nu)} \left[(d^\dagger d^\dagger)^{(L)}_{\pi(\nu)} \times (\tilde{d}\tilde{d})^{(L)}_{\pi(\nu)} \right]^0_0 \quad (2)$$

and $M_{\pi\nu}$ is the Majorana operator which separates the totally symmetric proton-neutron states from states of mixed symmetry. This Hamiltonian has been used by several authors[1-4] to do phenomenological analyses of nuclei in the mass region Z = 50-82 and N = 50 to N $\geqslant$ 126, in which the parameters ε, κ, χ_π, χ_ν (and perhaps the C_L's) are determined by fitting the low-lying energy spectra for an individual nucleus. One would like to obtain values for these parameters which vary smoothly with neutron (or proton) number and then to understand the physical origins of these parameters. Another interesting question is how these parameters are affected by terms left out of the model Hamiltonian, such as the *g* boson, i.e. proton or neutron pairs coupled to J = 4. One approach to studying this problem is to expand the symmetry group of the IBA from SU(6) to SU(15) in order to include the degrees of freedom associated with a *g* boson. One

*Research supported in part by NSF Grant No. PHY-7902654.

can avoid this difficult task by including the effects of a *g* boson on the *s*-*d* boson model space through second order perturbation theory.

In examining the effect of a *g* boson on the IBA Hamiltonian, we work in a model space constructed from correlated pairs of fermions because the boson model space is spanned by *s* and *d* boson states only, and calculate the second-order perturbative correction to Eq. (1) for an interaction between correlated fermion pairs of the form $\kappa Q_\pi \cdot Q_\nu$, $\kappa < 0$. Here we assume that the quadrupole operator $Q_{\pi(\nu)}$ is a two-body operator for fermions, as given in Ref. 5. The completely general algebraic expression for v_2 in second quantized notation, of which Fig. 1(a) is one term, is given by

$$v_2 = (1/4)^4 \kappa^2 \sum \langle(\alpha\beta)_\pi(\gamma\delta)_\nu | Q_\pi \cdot Q_\nu | (\rho\sigma)_\pi(\tau\xi)_\nu\rangle_A \frac{P}{\Delta E} \langle(\alpha'\beta')_\pi (\gamma'\delta')_\nu | Q_\pi \cdot Q_\nu | (\rho'\sigma')_\pi(\tau'\xi')_\nu\rangle_A \tag{3}$$

$$\{a^\dagger_\alpha a^\dagger_\beta a^\dagger_\gamma a^\dagger_\delta a_\xi a_\tau a_\sigma a_\rho\}\{a^\dagger_{\alpha'} a^\dagger_{\beta'} a^\dagger_{\gamma'} a^\dagger_{\delta'} a_{\xi'} a_{\tau'} a_{\sigma'} a_{\rho'}\}$$

where the first two state labels in the bras and kets refer to protons and the last two refer to neutrons. The subscript "A" on the kets means that the kets are independently antisymmetrized for protons and neutrons. The sum is taken over all state labels of the fermion creation and annihilation operators, except that the intermediate states $|(\rho\sigma)_\pi(\tau\xi)_\nu\rangle$ and $\langle(\alpha'\beta')_\pi(\gamma'\delta')_\nu|$ are restricted by the projection operator P to a specific set of states lying outside the subspace of *S* and *D* paired fermion states (corresponding to *s* and *d* boson states, respectively). The grouping of single fermion

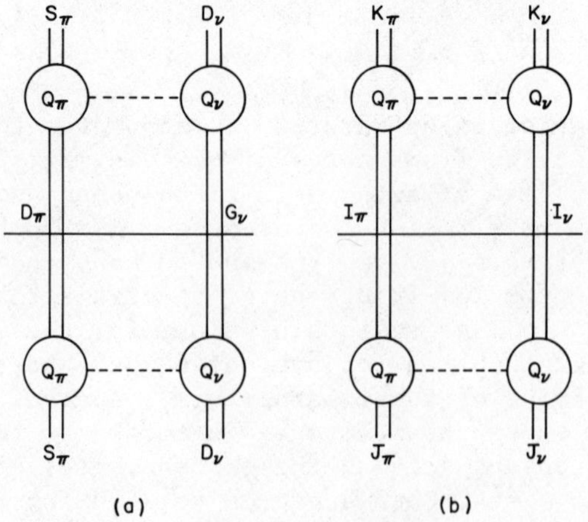

Figure 1: (a) Typical diagram of v_2 with pair of fermion labels.
(b) Diagram of v_2 showing notation used in text.

labels as in $(\alpha\beta)_\pi(\gamma\delta)_\nu$ indicates the correlation between fermions making up a pair. The states in Eq. (3) are then the unperturbed, paired-fermion states in the complete model space. Because we use Rayleigh-Schrödinger perturbation theory, the energy denominator ΔE is the difference between the unperturbed energy of the ground state and the unperturbed energy of the excited intermediate state.

After contracting the operators corresponding to the intermediate state and performing considerable Racah algebra, we obtain for the perturbative correction, assuming single j shells for protons and for neutrons,

$$v_2 = \kappa^2 \sum_{J} \sum_{K_\pi K_\nu} \sum_{I_\pi I_\nu} \sum_{J_\pi J_\nu} \langle (K_\pi K_\nu)J \| Q_\pi \cdot Q_\nu \| (I_\pi I_\nu)J \rangle_A \frac{P}{\Delta E}$$

$$\langle (I_\pi I_\nu)J \| Q_\pi \cdot Q_\nu \| (J_\pi J_\nu)J \rangle_A (-1)^J (2J+1) \sum_L \begin{Bmatrix} K_\pi & K_\nu & J \\ J_\nu & J_\pi & L \end{Bmatrix} T_L(K_\pi, J_\pi) \cdot \tag{4}$$

$$T_L(K_\nu, J_\nu)$$

where $T_{LM}(K,J) = \left[A^\dagger_K \tilde{A}_J\right]^L_M = \sum_{M'M''} (KJM'M''|LM)\, A^\dagger_{KM'}\, \tilde{A}_{JM''}$ (5)

and $A^\dagger_{JM} = A^\dagger_{JM}(j_\alpha j_\beta) = [1+\delta_{\alpha\beta}]^{-\frac{1}{2}} \left[a^\dagger_{j_\alpha}\, a^\dagger_{j_\beta}\right]^J_M$ (6)

$$\tilde{A}_{JM} = (-)^{J-M+1}\, A_{J-M}$$

The angular momentum labels in Eq. (4) are illustrated by the diagram in Fig. 1(b). The action of the projection operator P in this multipole-multipole form of a proton-neutron interaction is to require that there be one and only one I = 4 fermion pair in the intermediate states. With the initial and final states made up only of *S* and *D* paired fermions, L can be 0,1,2,3 or 4. We examine only the L = 0 part of Eq. (4) in detail. Defining the quantities

$$\hat{N}_{J_\pi} = \sum_{M_\pi} A^\dagger_{J_\pi M_\pi} A_{J_\pi M_\pi}, \quad \hat{N}_{J_\nu} = \sum_{M_\nu} A^\dagger_{J_\nu M_\nu} A_{J_\nu M_\nu}, \quad g^J_{J_\pi J_\nu} = \frac{(2J+1)}{(2J_\pi+1)(2J_\nu+1)} \tag{7}$$

we have the simple form of the monopole-monopole part of the perturbation,

$$v_2(L=0) = \kappa^2 \sum \left| \langle (J_\pi J_\nu)J \| Q_\pi \cdot Q_\nu \| (I_\pi I_\nu)J \rangle_A \right|^2 \frac{P}{\Delta E}\, g^J_{J_\pi J_\nu}\, \hat{N}_{J_\pi} \hat{N}_{J_\nu} \tag{8}$$

If we now compare the $\hat{N}_J$ operator defined above to the fermion number operator, we see that the two operators are identical in form, but the creation and annihilation operators in Eq. (8) relate to pairs of fermions; hence, while the fermion number operator is a one-body operator, $\hat{N}_J$ is a one-pair operator. Another difference is that the fermion operators obey anti-commutation rules, but the pair operators have more complex commutation rules.[5] However, if we make the approximation that the paired-fermion operators do commute, then the operator $\hat{N}_J$ can be interpreted as a number operator counting the number of fermion pairs coupled to angular momentum J. We also apply the "simple correspondence" argument of Ginocchio and Talmi[6] to map the resulting correction v_2 into a boson space spanned by s and d bosons,

$$v_{2,B}(L=0) = \kappa^2 \sum \left| \langle (J_\pi J_\nu) J \| Q_\pi \cdot Q_\nu \| (I_\pi I_\nu) J \rangle_A \right|^2 \frac{P}{\Delta E} g^J_{J_\pi J_\nu} \hat{n}_{J_\pi} \hat{n}_{J_\nu} \quad (9)$$

That is, we simply replace the paired-fermion operators with analogous boson operators $\hat{n}_J$ and consider v_2 to be a correction to the boson Hamiltonian. The energy denominator in Eq. (9) is determined in terms of the unperturbed boson states. The matrix elements of $Q_\pi \cdot Q_\nu$ have been given elsewhere.[5]

CALCULATIONS FOR THE BARIUM ISOTOPES

We have evaluated Eq. (9) for the Ba isotopes, using experimental 2^+ and 4^+ energies for the d and g boson energies (the s boson energy is set to zero) in ΔE. We obtain for the correction to H_{IBA2} (in MeV)

$$v_{2,B}(L=0) = C_1(\hat{n}_\pi \hat{n}_{d_\nu} + \hat{n}_{d_\pi} \hat{n}_\nu) + C_2 \hat{n}_{d_\pi} \hat{n}_{d_\nu}$$
$$C_1 \cong -\kappa^2 A^{4/3}\ 6\mathrm{x}10^{-6}\ (\mathrm{barns})^4/\mathrm{MeV} \quad C_2 \cong -\kappa^2 A^{4/3}\ 2\mathrm{x}10^{-5}\ (\mathrm{barns})^4/\mathrm{MeV} \quad (10)$$

where A is the mass number and the $\hat{n}$'s are number operators for the total number of proton or neutron bosons or the number of proton or neutron d bosons. Combining Eq. (10) with the unperturbed boson Hamiltonian in Eq. (1), we may clearly see the renormalization of the single boson energy ε,

$$H_{IBA2} + v_{2,B}(L=0) = (\varepsilon + C_1 \hat{n}_\nu)\hat{n}_{d_\pi} + (\varepsilon + C_1 \hat{n}_\pi)\hat{n}_{d_\nu} + C_2 \hat{n}_{d_\pi} \hat{n}_{d_\nu}$$
$$+\kappa Q_\pi \cdot Q_\nu + M_{\pi\nu} + V_{\pi\pi} + V_{\nu\nu} \quad (11)$$

Using the parameterization of Kisslinger and Sorensen[7] we find that $\kappa \cong -116\ A^{-2/3}$ MeV/barns^2 for A = 130-140 in a single j shell with $j = 31/2$.

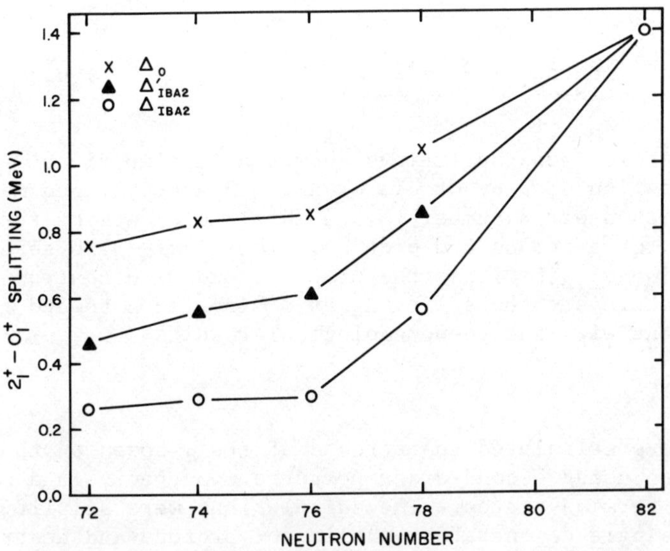

Figure 2: Calculated splittings of the 2_1^+ state from the ground state for several Ba isotopes.

Otsuka, et al.[1] have reported that it is necessary to renormalize the single boson energy ε as a function of the number of valence bosons (or in this case, boson holes), when fitting data from the Ba isotopes, as reflected in the empirical curve in Fig. 2. Because our objective is to provide a microscopic explanation of this renormalization, we have calculated the splitting of the first excited 2^+ state from the ground state for the isotopes of barium with A = 128, 130, 132, 134, and 138, which we compare to the phenomenological IBA splittings rather than experimental data. In this comparison we ignore the effect of all but the first two terms in H_{IBA2} in determining the empirical curve (for which we have used the parametrization of Otsuka, et al.[1]). The perturbed Hamiltonian obtained from Eq. (11) is then

$$H'_{IBA2} = (\varepsilon_0 + C_1\hat{n}_\nu)\hat{n}_{d_\pi} + (\varepsilon_0 + C_1\hat{n}_\pi)\hat{n}_{d_\nu} + C_2\hat{n}_{d_\pi}\hat{n}_{d_\nu} \quad (12)$$

where ε_0 = 1.4 MeV, the value at the closed shell. The corresponding splittings are given by

$$\Delta_{IBA2} = \langle 2_1^+|\varepsilon(\hat{n}_{d_\pi} + \hat{n}_{d_\nu})|2_1^+\rangle - \langle 0_1^+|\varepsilon(\hat{n}_{d_\pi} + \hat{n}_{d_\nu})|0_1^+\rangle \quad (13)$$

$$\Delta'_{IBA2} = \langle 2_1^+|H'_{IBA2}|2_1^+\rangle - \langle 0_1^+|H'_{IBA2}|0_1^+\rangle \quad (14)$$

We have also plotted in the figure the unrenormalized, unperturbed splitting

$$\Delta_0 = \langle 2_1^+ | \varepsilon_0 (\hat{n}_{d_\pi} + \hat{n}_{d_\nu}) | 2_1^+ \rangle - \langle 0_1^+ | \varepsilon_0 (\hat{n}_{d_\pi} + \hat{n}_{d_\nu}) | 0_1^+ \rangle \tag{15}$$

These unrenormalized splittings show that a significant part of the renormalization seen by Otsuka et al.[1] is accomplished merely by changes with neutron number in the matrix elements of the number operators for neutrons and protons. The figure also shows that the perturbation $v_{2,B}(L=0)$ has the proper trend with neutron number, i.e., it decreases the splitting as n_ν increases toward mid-shell, in agreement with the phenomenological results.

CONCLUSIONS

We have calculated the effects of the *g* boson on the IBA Hamiltonian using second-order perturbation theory in a model space of paired-fermion states. The calculations were simplified by assuming single degenerate *j* shells for protons and neutrons and by approximating the operators of Eq. (7) as number operators. In a specific application of the resulting correction term, the agreement between the predicted $2_1^+ - 0_1^+$ splittings and the phenomenological IBA values for several Ba isotopes suggests that the perturbative approach to including the *g* boson in the IBA Hamiltonian accounts for part of the observed renormalization of the phenomenological single boson energies in the IBA.

REFERENCES

1. T. Otsuka, A. Arima, F. Iachello and I. Talmi, Phys. Lett. 76B, 139 (1978).
2. A. Arima and F. Iachello, Ann. Phys. (N.Y.) 99, 253 (1976); 111, 201 (1978).
3. F. Iachello in Interacting Bosons in Nuclear Physics, ed. F. Iachello (Plenum Press, New York, 1979), p. 1.
4. O. Scholten, F. Iachello and A. Arima, Ann. Phys. (N.Y.) 115, 325 (1978).
5. K. Sage and B. R. Barrett, to be published in Physical Review C (1980).
6. J. Ginocchio and I. Talmi, Nucl. Phys. A337, 431 (1980).
7. L. Kisslinger and R. A. Sorensen, Rev. Mod. Phys. 35, 853 (1963).

BOSON CUTOFF AND BAND CROSSING*

Adrian Gelberg[+] and Amos Zemel[++]

[+] Institut für Kernphysik der Universität zu Köln, D-5000 Köln 41, W.-Germany
[++] Weizmann Institute of Science, Rehovot, Israel

The fact that the total number of s- and d-bosons $N = n_s + n_d$ is finite represents one of the most important features of the Interacting Boson Model (IBM). This boson cutoff has a strong influence on the behaviour of electric quadrupole transitions.

Let us consider an E2-transition between two levels belonging to the ground state band (gsb) of an SU(5)-type nucleus. The initial state has n_d d-bosons and $N-n_d$ s-bosons; the respective numbers for the final state are n_d-1 and $N-n_d+1$. The quadrupole operator is [1]

$$T(E2) = \alpha_2 \,(d^+s + s^+\tilde{d})^{(2)} + \beta_2 (d^+\tilde{d})^{(2)} \qquad (1)$$

During the transition a d-boson is annihilated and an s-boson is created; therefore the transition matrix element will be proportional to the product of two oscillator matrix elements

$$\langle f \| T(E2) \| i \rangle \sim \left[n_d (N - n_d + 1) \right]^{1/2}$$

The exact expression of the reduced transition probability is [1]

$$B(E2;\ I \to I-2)/B(E2;\ 2 \to 0) = \frac{1}{4}\, I(2N - I+2)/N \qquad (2)$$

where I is the spin of the initial state. Similar formulas have been calculated for the SU(3) and SO(6) cases[1]. It is convenient to use B(E2)'s normalized to the leading order rotational values[2]

$$R = \frac{B(E2;\ I \to I-2)/B(E2;\ 2 \to 0)}{\left[B(E2;\ I \to I-2)/B(E2;\ 2 \to 0)\right]_{rot}} \qquad (3)$$

A plot of R vs. I is given in fig. 1 for the three limiting cases; curves obtained from the diagonalization of the full Hamiltonian interpolate between the three limits.

What do the experimental data look like? At present the situation is ambiguous. Lifetimes in ^{78}Kr have been measured by H. P. Hellmeister et al. [3,4]. The B(E2)'s display the cutoff effect, at least below spin 10, where a band crossing probably occurs[4]. Experimental results in ^{194}Pt[5] also showed a behaviour similar to that predicted by IBM; a recent reevaluation[6] of the data lead to results showing no influence of the cutoff. Data from the Ba-Ce-region will be discussed later on. It is clear that more experiments on transition nuclei are needed.

If we now examine B(E2) values in strongly deformed nuclei, we cannot observe the decrease at high spin predicted by the model. Rather reliable data exist e.g. for ^{162}Dy, ^{174,176}Yb, and ^{232}Th[7]. The normalized B(E2)'s in the gsb of ^{162}Dy are shown in fig. 2. The full line corresponds to the SU(3) limit and it deviates from experiment, which is rather consistent with R = 1 (rotor). If we now introduce g-bosons with angular momentum 4 besides s- and d-bosons, we obtain the dashed line which is in better agreement with experiment. The introduction of g-bosons in the SU(3) limit represents a natural extension of the model in its present form[8].

It is interesting to draw a comparison with some light nuclei, although it is by no means clear that IBM could be applied to them. The band termination effect predicted by the Elliot model[9] has been observed in ^{20}Ne (fig.2). If we consider only valence nucleons, N = 2 and the band should terminate at I = 4, contrary to experiment [10]. If we again introduce g-bosons, the agreement with experiment will be excellent. If we now examine ^{24}Mg, which has an O(6) like energy spectrum, we see that the normalized B(E2)'s calculated in this limit with only

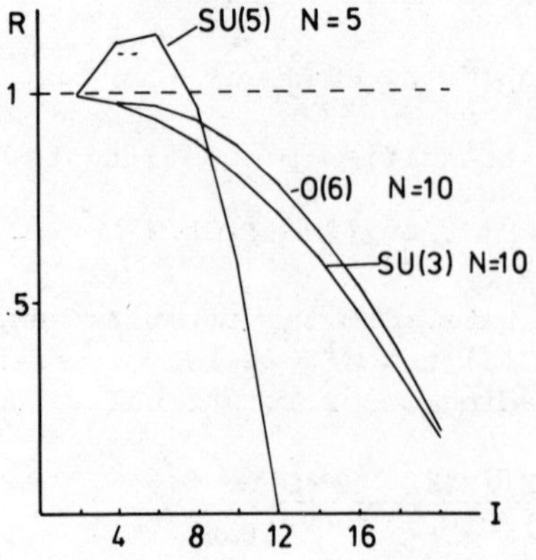

Fig. 1:
Effect of boson cutoff on E2 transition probabilities

s- and d-bosons are in fairly good agreement with experiment[11]. This comparison suggests that the introduction of the g-boson could be necessary in the SU(3) limit if we want to describe properties of high spin states.

Moreover, one should carefully examine the possibility of defining N for a strongly deformed nucleus. The spreading of the Nilsson levels with increasing deformation leads to a decrease of the energy gap between shells; it is therefore difficult to define the number of valence nucleons. One possible way-out would be the introduction of an effective value of N.

Although the E2-transition probabilities are directly and in a rather dramatic way affected by the boson cutoff, there are other ways through which this effect manifests itself. R. Casten has shown[12] that the properties of the excited 0^+ levels in the O(6) limit are strongly dependent on N. Another possibility is provided by odd nuclei. One of the terms of the boson-fermion coupling hamiltonian[13] is proportional to the boson quadrupole operator, which will be attenuated at large values of n_d in the same way as in E2-transitions.

Although the best way of studying the boson cutoff is to measure E2 transition strengths, it is not easy to find good cases in which the collective band stays pure up to high spin without being

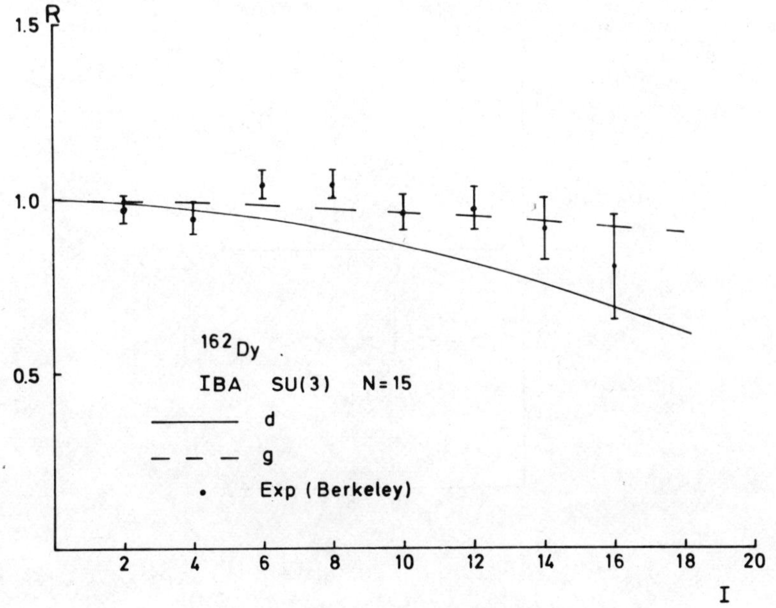

Fig. 2:
Normalized B(E2)'s in ^{162}Dy

interrupted by a band crossing (backbending) and at the same time N is not too large. It is hence interesting to know to what extent the effect of the boson cutoff can also be observed in the presence of band crossing.

F. Stephens et al. have shown that backbending can be interpreted in the framework of the rotational model as the crossing of the gsb and a rotation aligned two-particle band[14]. The properties of bands based on a two-particle state have been examined in IBM[1] in a simple approximation without band mixing. These calculations have recently been extended[15,16] to the case of band mixing.

The Hamiltonian of the system consisting of a boson core and two particles[15] has been written in the SU(3) approximation, whose great advantage is simplicity. As far as the B(E2)'s are concerned, the results will be similar to those of the O(6) approximation. The calculation made by A. Arima[16] is not subject to this limitation.

The two particles are described by a boson like operator

$$b^{+}_{\ell m} = \sum_{m_1 m_2} \langle j m_1 j m_2 | \ell m \rangle \, a^{+}_{j m_1} a^{+}_{j m_2} \qquad (4)$$

if we assume that the two particles belong to the same orbital j and are coupled to angular momentum ℓ.

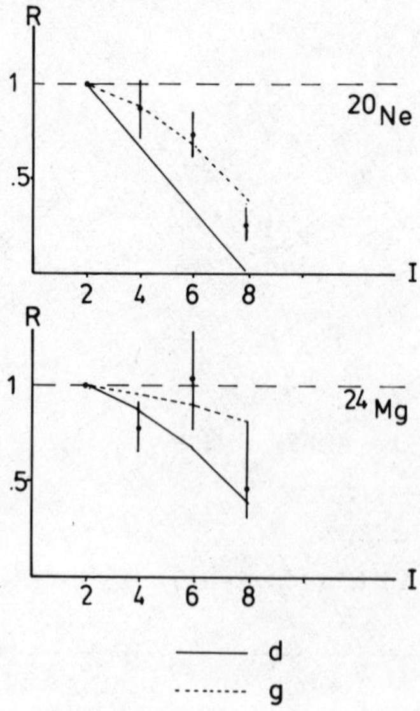

Fig. 3:
Normalized B(E2)'s in ^{20}Ne and ^{24}Mg

The Hamiltonian of the system can be written

$$H = \alpha L(L+1) + \varepsilon_\ell + y' \left[(b^+_\ell \tilde{b}_\ell)^{(1)} L^{(1)} \right]^{(0)}$$

$$+ z' \left[(b^+_\ell \tilde{b}_\ell)^{(2)} Q^{(2)} \right]^{(0)} + H_{mix}$$

where $$\tilde{b}_{\ell m} = (-1)^{\ell - m}\, b_{\ell - m}$$

L is the core angular momentum, ε_ℓ is the energy of the two-particle state, and α, y', z' are free parameters; only the gsb of the core has been considered. The operators $L^{(1)}$ and $Q^{(2)}$ are the generators of SU(3) and are given in ref. 1. H_{mix} has matrix element connecting different two-particle states. For the sake of simplicity we will consider only $\ell = 0$ and $\ell = J$.

The reduced matrix elements of $L^{(1)}$ and $Q^{(2)}$ can easily be calculated; the reduced matrix elements of $(b^+\tilde{b}\ )^{(k)}$ will be absorbed into the parameters, e.g. $z = z' \langle J || (b^+_\ell \tilde{b}_\ell)^{(2)} || J \rangle$. We will assume that the matrix elements of H_{mix} are constant:

$$\langle L'\ell' I | H | L \ell I \rangle = \beta\alpha \qquad \ell \neq \ell'$$

The state I = J will thus be a mixture of $\ell = 0$ and $\ell = J$, the mixing amplitude will primarily determine the E2 transition strengths. The influence of H_{mix} on states with $I \geq J+2$ is weaker if we assume that $\beta\alpha \lesssim 100$ keV. On the other hand the mixing of several L values in the two-particle band can be quite strong.

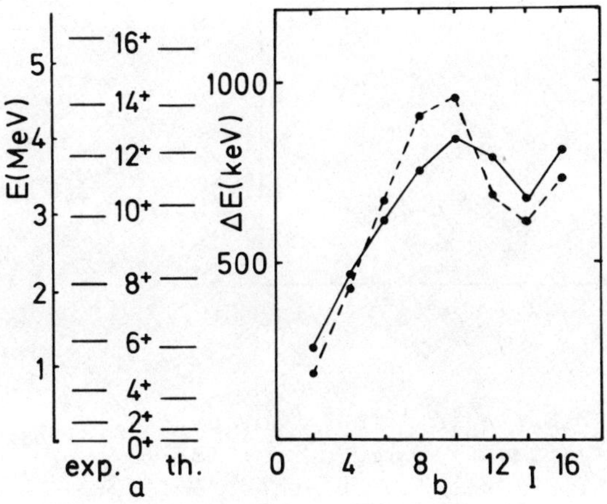

Fig. 4:
a) Experimental and theoretical energy levels in ^{126}Ba.
b) Energy intervals vs. spin in ^{126}Ba. Solid curve, experiment; dashed curve, theory. Parameters: α = 30 keV, β = 4, z = -2, ε_J = 3480 keV, N = 8, y = 0.

The wave function of a state with spin I

$$|I\rangle = \sum_{L\ell} C_{L\ell}(I)\,|(L,\ell)I\rangle \qquad (6)$$

is obtained by diagonalizing the Hamiltonian (5). The amplitudes $C_{L\ell}$ have been used to calculate B(E2)'s in the yrast cascade; the particle contribution has been neglected.

Energies and B(E2)'s have been calculated for several Ce and Ba isotopes. In this case the two-particle configuration is $(h_{11/2})^2_{10+}$ [17].

The calculation of energy levels succeeds in reproducing the main trends. The comparison between calculated and experimental energies[18] in ^{126}Ba is given in fig. 4. The main source of discrepancy comes from the constant inertial parameter α.

Experimental and theoretical normalized B(E2)'s in the yrast cascade of ^{126}Ba are given in fig. 5. The experiment[18] shows that R already decreases at spin 4 or 6 and reaches a minimum in the band crossing region. Nilsson model calculations[19] in which four bands are mixed, predict a purely rotational behaviour except at the band crossing (fig. 5). The fact that a drop in B(E2) happens over a wide range of spins has also been observed in experiments on $^{130-134}$Ce[20].

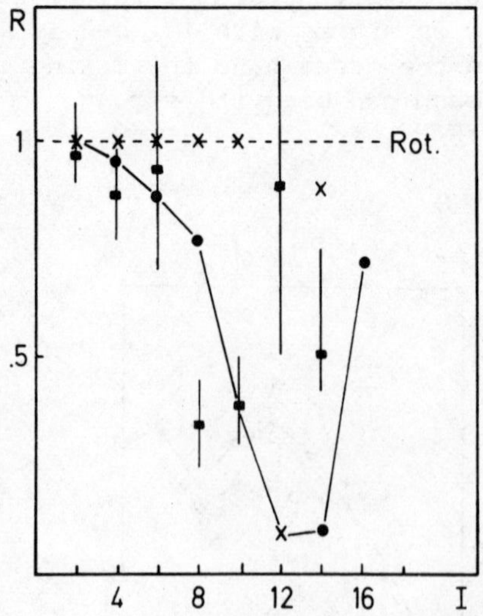

Fig. 5:
Normalized B(E2)'s in ^{126}Ba. Closed squares: experiment; solid curve: IBM calculation; crosses: calculation ref. 19.

A microscopic calculation of ^{126}Ba has been carried out by R. D. Ratna Raju et al.[21]; a pseudo-SU(3) method has been used. This calculation reproduces fairly well energies and B(E2) values.

Although our simple calculation cannot pretend to accuracy, it describes correctly the general behaviour of B(E2). This behaviour is the result of the interplay of a collective phenomenon, i.e. the boson cutoff and the crossing of two bands.

The authors would like to thank A. Arima, P. von Brentano, R. Casten, K. Hecht, F. Iachello, and I. Talmi for interesting and stimulating discussions.

REFERENCES

1. A. Arima and F. Iachello, Ann. Phys. (NY) 99, 253 (1976); 111, 201 (1978) and 123, 468 (1979)
2. A. Bohr and B. Mottelson, "Nuclear Structure", Vol. II, Benjamin, Reading, Mass. (1975)
3. H. P. Hellmeister, U. Kaup, J. Keinonen, K. P. Lieb, R. Rascher, R. Ballini, J. Delaunay, and H. Dumont, Phys. Lett. 85B, 34 (1979)
4. K. P. Lieb, in these proceedings
5. K. Stelzer, F. Rauch, Th. W. Elze, Ch. E. Gould, J. Idzko, G. E. Mitchell, H. P. Nottrodt, R. Zoller, H. Wollersheim, and H. Emling, Phys. Lett. 70B, 297 (1977)
6. J. Idzko, K. Stelzer, Th. W. Elze, H. Ower, H. J. Wollersheim, H. Emling, P. Fuchs, E. Grosse, R. Piercey, and D. Schwalm, Contribution to the Int. Conf. on Nuclear Behaviour at High Angular Momentum, Strasbourg 1980
7. M. W. Guidry, I. Y. Lee, N. R. Johnson, P. A. Butler, D. Cline, P. Colombani, R. M. Diamond, and F. S. Stephens, Phys. Rev. C20, 1814 (1979)
8. F. Iachello, in: Proc. of the Conf. on "Structure of Medium-Heavy Nuclei", Rhodos 1979, The Institute of Physics, Bristol 1980, p. 161
9. J. P. Elliott, Proc. Roy. Soc. A245, 128 (1958), and A245, 562 (1958)
10. F. Ajzenberg-Selove, Nucl. Phys. A300, 1 (1978)
11. P. M. Endt and C. Van der Leun, Nucl. Phys. A310, 1 (1978)
12. R. Casten, in these proceedings
13. O. Scholten, in these proceedings
14. F. Stephens, Rev. Mod. Phys. 47, 43 (1975), and references therein
15. A. Gelberg and A. Zemel (Phys. Rev. C 22, in print)
16. A. Arima, in these proceedings
17. C. F. Flaum, D. Cline, A. W. Sunyar, and Q. C. Kistner, Phys. Rev. Lett. 33, 973 (1974)

18. G. Seiler-Clark, D. Husar, R. Novotny, H. Gräf, and D. Pelte, Phys. Lett. 80B, 345 (197) and references therein
19. M. Reinecke, H. Ruder, Z. f. Phys. A 292, 267 (1977)
20. D. Husar, S. J. Mills, H. Gräf, U. Neumann, D. Pelte, and G. Seiler-Clark, Nucl. Phys. A292, 267 (1977)
21. R. D. Ratna Raju, K. T. Hecht, B. D. Chang, and J. P. Drayer, Phys. Rev. C20, 2397 (1979)

* Supported by BMFT

TWO-NEUTRON TRANSFER IN Pt AND Os NUCLEI AND THE IBA*

J. A. Cizewski†

Los Alamos Scientific Laboratory
Los Alamos, New Mexico 87545

ABSTRACT

Two-neutron transfer studies in Pt and Os were investigated. The extracted relative enhancement factors in Pt were found to be in excellent agreement with the O(6) limit predictions. The differences in transfer strength between Pt and Os (t,p) could not be reproduced within the present framework of the Interacting Boson Approximation Model.

EXPERIMENTAL ANALYSIS

To test the predictions of the Interacting Boson Approximation[1] (IBA) model of Arima and Iachello, we have recently investigated[2] the Pt, Os(t,p) reactions at E_t=17 MeV. Given the positive ground-state Q-values for these reactions, the reaction protons are being emitted well above the Coulomb barrier, making it possible to extract relatively reliable nuclear spectroscopic strengths.

The enhancement factors ϵ given by

$$\sigma_{exp} = \epsilon N \sigma_{DW} \tag{1}$$

were obtained[2] from distorted wave calculations using standard optical model parameters. In extracting spectroscopic information from our (t,p) measurements, we have assumed a simple one-step DWBA mechanism for the (t,p) reaction, neglecting the effects due to inelastic scattering combined with two-neutron transfer or sequential transfer. Neglecting coupled-channel effects in this mass region should be reasonable when one restricts the investigation to the dominant ground-state to ground-state transitions. Because coupled-channel effects will be more important for excited states, we have not investigated L=2 transitions and will only briefly discuss excited L=0 transitions.

Another simplification in our analysis is the use of the same $(2p_{3/2})^2$ form factor in our DWBA calculations. Because the IBA deals with collective excitations, it was felt a more microscopic form factor, with several components which varied from nucleus to nucleus as the single particle structure near the Fermi surface changed, would be inappropriate for an initial comparison of IBA predicted and empirical two-neutron transfer strengths. For this reason it may be more accurate to call our ϵ-values experimental cross sections from which Q-value and A dependences have been removed.

TWO-NEUTRON TRANSFER STRENGTHS

In order to calculate two-neutron transfer strengths within the IBA model, Arima and Iachello have chosen to define the L=0 two-neutron transfer operators[3] as essentially neutron s-boson creation and annihilation operators that properly account for the finite dimensionality of the boson space. In the limiting symmetries of the IBA-1 model (which does not distinguish between neutron and proton degrees of freedom) they obtained the following analytical expressions for the ground-state to ground-state transition strengths[1,3]

$$I^{SU(5)}(N_\nu \rightarrow N_\nu + 1) = \alpha_\nu^2 (N_\nu + 1)(\Omega_\nu - N_\nu) \tag{2}$$

$$I^{SU(3)}(N_\nu \rightarrow N_\nu + 1) = \alpha_\nu^2 (N_\nu + 1)\left(\frac{2N+3}{3(2N+1)}\right)\left(\Omega_\nu - N_\nu - \frac{4(N-1)N_\nu}{3(2N-1)}\right) \tag{3}$$

$$I^{O(6)}(N_\nu \rightarrow N_\nu + 1) = \alpha_\nu^2 \frac{(N+4)(N_\nu+1)}{2(N+2)}\left(\Omega_\nu - N_\nu - \frac{(N-1)N_\nu}{2(N+1)}\right) \tag{4}$$

Here N_ν is the total number of neutron bosons, N is the total number of bosons, and Ω_ν is the boson degeneracy of the shell (for the 82-126 interval $\Omega_\nu = 22$). The strength of the operator is incorporated in α_ν^2. In addition to the analytical expressions, there are selection rules which help determine the allowed transitions to excited 0^+ states. These are illustrated by the arrows in Fig. 1. One sees that transitions to excited states are forbidden in the SU(5) limit, while one excited 0^+ state will be populated in (t,p) reactions (at beginning of shell) in the SU(3) and O(6)

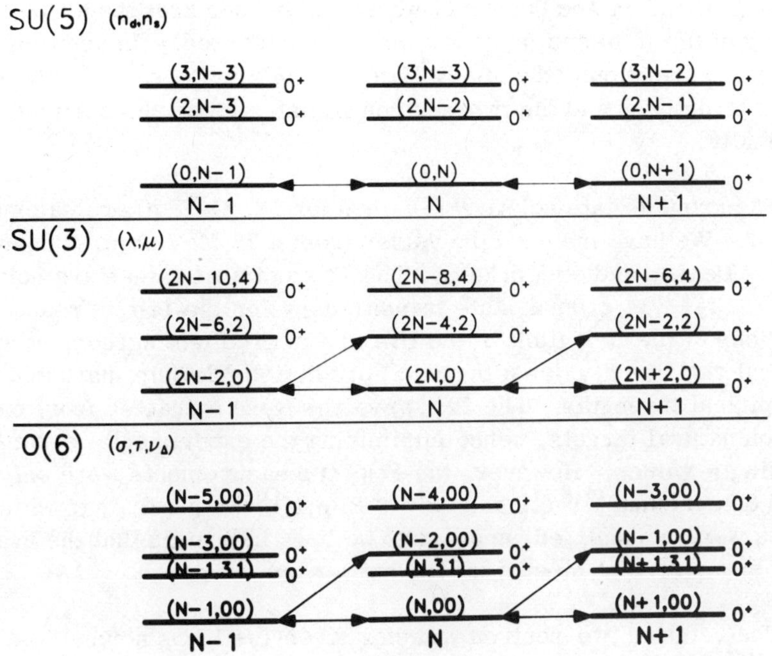

Fig. 1. Selection rules for two-particle transfer in IBA limits.

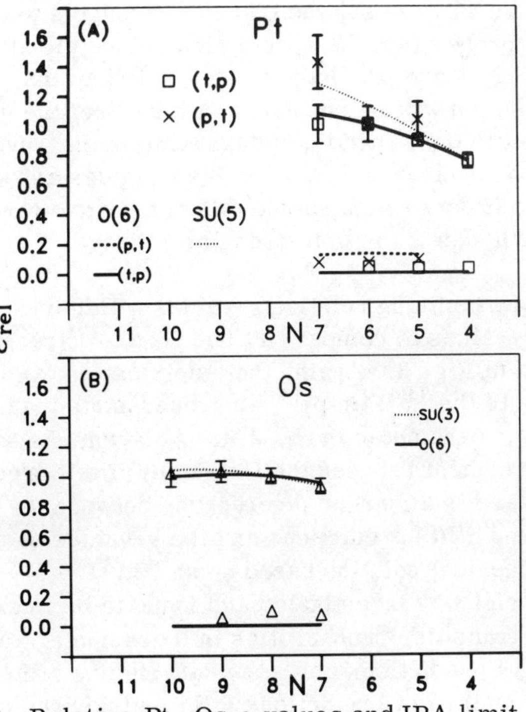

Fig. 2. Relative Pt, Os ϵ-values and IBA limit predictions.

limits. (For the Pt and Os nuclei where the bosons are "hole" bosons the roles of the (t, p) and (p, t) reactions are reversed.) In addition to the analytical expressions, the IBA-1 Hamiltonian may be solved[4] for intermediate structures and the two-neutron transfer strengths obtained numerically.

The ground-state relative ϵ-values for Pt, Os (t, p) are summarized in Fig. 2. We have included the values from a 35 MeV Pt(p, t) measurement[5] by Deason and co-workers. The Pt empirical values are normalized to the ^{194}Pt-^{196}Pt ground-state transitions. For the Pt(t, p) reaction the predictions of the O(6) limit of the IBA are in excellent agreement with the empirical relative ϵ-values; the SU(5) predictions deviate markedly from the empirical strengths. The Pt(t, p) values were extracted from measurements on natural targets, hence minimizing the experimental errors in extracting ϵ-values. However, the Pt(p, t) measurements were only performed on enriched targets and the ^{196}Pt(p, t)^{194}Pt ϵ-value, to which theory and data were normalized, appears to be low, indicating that the available Pt(p, t) measurement may be inadequate for our purposes.

The relative two-neutron transfer strengths in Os normalized to the ^{190}Os-^{192}Os g. s. value are fairly insensitive to the structure of these nuclei, with the SU(3), O(6) and any intermediate structure giving essentially identical results. Also, the transfer strengths to excited 0^+ states are not a sensitive probe of the IBA predictions since the allowed transitions illustrated in Fig. 1 are predicted to be ~8-10% of the ground-state strength while most 0^+ states will be populated with 2-3% of the ground-state strength due to two quasi-particle admixtures in the wave functions. This is in contrast to absolute and relative B(E2) values associated with excited 0^+ states where orders of magnitude differences are observed between allowed and forbidden transition strengths.

Instead of examining relative ϵ-values within one element, we have also obtained ϵ-values to compare Pt and Os (t, p) strengths to each other, as summarized in Fig. 3. Again, the empirical and O(6) predicted strengths are normalized to the ^{194}Pt(t, p)^{196}Pt ground-state transition, effectively obtaining the α_ν^2 parameter in eq. 2-4. Although, as seen in Fig. 2, there is excellent agreement between the IBA-1 O(6) limit predictions and the Pt ϵ-values, there is a marked discrepancy between the values expected from the O(6) and SU(3) predictions and the ϵ-values obtained for Os (t, p). Recently, a numerical solution based on an IBA-1 O(6) $\rightarrow$ SU(3) transition[6] in the Pt-Os nuclei was investigated and found to be in excellent agreement with the γ-ray transition probabilities in these nuclei. While the two-neutron transfer strength predictions using the parameters of Ref. 6 are essentially identical to the O(6) limit predictions in Pt and do predict two-neutron transfer strengths intermediate between the O(6) and SU(3) limit values for Os, these calculations are also in marked disagreement with the Os empirical ϵ-values. Recent IBA-2 calculations[7] in which the neutron and proton degrees of freedom are explicitly included, and, hence, should provide a more accurate test of two-neutron transfer strengths than the one-boson IBA-1 model, have been performed. Despite the excellent agreement Bijker and co-workers obtained for the other properties of the Pt-Os nuclei, they fail to reproduce the differences between Pt and Os ground-state transfer strengths.

The inadequacy of the IBA to reproduce two-neutron "absolute" strengths may be due to several effects. Possibly the method used to extract the ϵ-values in Fig. 3 does not correspond to an "absolute" strength and is simply a measure of Q-reduced cross sections and, therefore, the comparison between Pt and Os in Fig. 3 is meaningless. If one assumes the method is reasonable, then one is left with an inadequacy of the IBA model as investigated so far to understand two-particle transfer. It is well known that certain single-particle orbitals such as $p_{3/2}$ give large (t,p) and (p,t) cross sections. If the relative location of the Fermi surfaces in Pt and Os are such that Pt is closer to the "hot orbitals" while Os is further

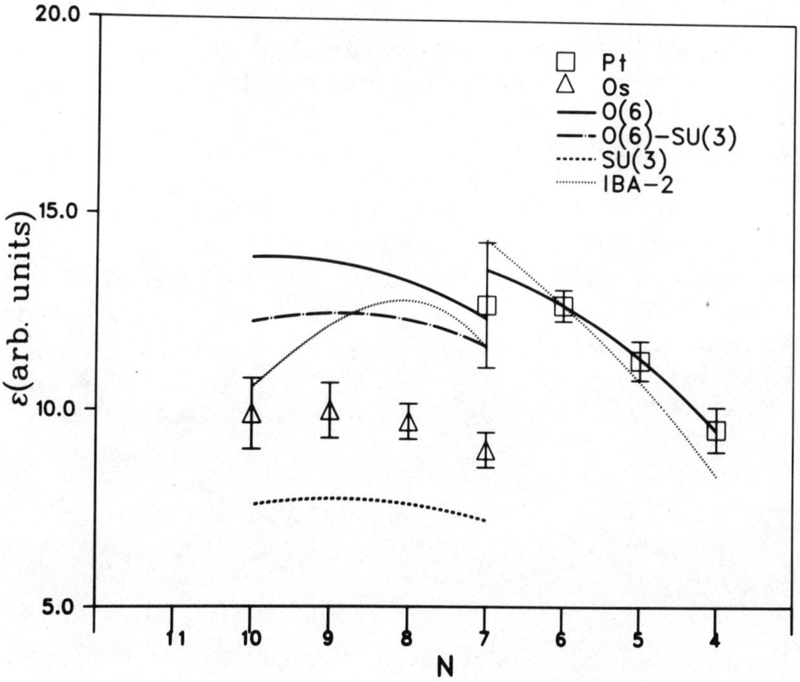

Fig. 3. Comparison between Pt, Os ϵ-values and IBA.

away, one would expect the Pt ϵ-values to be larger than those in Os. Since the IBA-1 and IBA-2 models only deal with the collectivity of the g. s. wave functions and the single-particle components of the bosons are not available, maybe it is unrealistic to expect the IBA to reproduce all trends in two-neutron transfer strengths. However, the success the IBA has in reproducing relative ϵ-values, especially in Pt, may indicate that the importance of particular orbitals is less critical for a given isotope chain and, therefore, it is valid to compare IBA predictions to relative ϵ-values for each element.

ACKNOWLEDGMENTS

I am greatly indebted to my colleagues at LASL, E.R. Flynn, R.E. Brown, and J.W. Sunier for their essential roles in the data acquisition and analysis reported here. I would also like to thank A. Arima, F. Iachello, R.F. Casten, R. Broglia and J. Ginocchio for highly stimulating discussions, and A.E.L. Dieperink for communicating results prior to publication. Finally, I would like to thank the staff at WNSL for their assistance in the preparation of this manuscript.

REFERENCES

*Work supported by U. S. Department of Energy.

† Present address, Wright Nuclear Structure Laboratory, New Haven, Connecticut, 06511

1. A. Arima and F. Iachello, Ann. Phys. (N.Y.) 99, 253 (1976); 111, 201 (1978); 123, 468 (1979).
2. J.A. Cizewski, E.R. Flynn, R.E. Brown, and J.W. Sunier, Phys. Lett. 88B, 207 (1979), and to be published.
3. A. Arima and F. Iachello, Phys. Rev. C16, 2085 (1977).
4. IBA codes PHINT and FTPT written by O. Scholten, unpublished.
5. P.T. Deason, et al., Phys. Rev. C20, 927 (1979).
6. R.F. Casten and J.A. Cizewski, Nucl. Phys. A309, 477 (1978).
7. R. Bijker, et al., KVI preprint 224; A.E.L. Dieperink, private communication.

A TEST OF THE INTERACTING BOSON MODEL USING PION-NUCLEUS INELASTIC SCATTERING

T.-S. H. Lee

Argonne National Laboratory, Argonne, Ill. 60439

ABSTRACT

It is shown that the comparison between π^+ and π^- cross sections can be used to test the IBA of Arima and Iachello.

Recent works[1,2] have demonstrated the value of the pion-nucleus inelastic scattering (π,π') as a probe of nuclear structure. In this paper , we point out that the (π,π') study can play an important role in the study of the Interacting Boson Approximation (IBA) model of Arima and Iachello.[3]

Extensive studies have shown that the pion-nucleus inelastic scattering can be satisfactorily described by the distorted-wave impulse approximation[1,2] (DWIA). In the DWIA model, the inelastic scattering transition amplitude is written as

$$T_{fi}(\vec{k}_0',\vec{k}_0,E) = \int \chi^{(-)*}_{\vec{k}_0'}(k')U_{fi}(k',k,E)\chi^{(+)}_{\vec{k}_0}(\vec{k})d\vec{k}'d\vec{k} \qquad (1)$$

The distorted waves $\chi^{(\pm)}$ are generated from an appropriate optical potential fitted to the pion-nucleus elastic scattering cross section. The transition interaction U_{fi} is related to the nuclear structure by

$$U_{fi}(\vec{k}',\vec{k},E) = \sum_{LM} \begin{pmatrix} J_f & J_i & L \\ M_f & M_i & M \end{pmatrix} [t_{\pi n}(\vec{k}',\vec{k},w) F^n_{LM}(\vec{q}) + t_{\pi p}(\vec{k}',\vec{k},w)\, F^p_{LM}(\vec{q})] \ , \qquad (2)$$

where $t_{\pi n(p)}$ is the pion-neutron (proton) scattering t matrix, and $F^{n(p)}_{LM}$ is the nuclear transition form factor for the neutron (proton) excitation. With suitable approximations, $t_{\pi n}$ and $t_{\pi p}$ can be generated from the known pion-nucleon (πN) scattering phase-shifts. In this way, the (π,π') cross section can be used as a direct measure of the nuclear transition form factors.

The most important feature of (π,π') scattering is that at energy near the (3,3) resonance, the $\pi^+p(\pi^-n)$ interaction strength ($|t_{\pi N}|^2$) is 9 times stronger than the $\pi^+n(\pi^-p)$ interaction strength. Consequently, we can see from eq. (2) that the relative strength between the neutron and proton form factors can be sensitively studied by comparing the π^+ and π^- inelastic scattering cross sections. This interesting property has been utilized[2] to explore many important nuclear structure information, such as the core polarization in O^{18} and the pure neutron excitation states in C^{13}. We now show its usefulness in the study of the IBA model.

To be specific, we consider the application of the IBA-2 model[4] to the Xe, Ba and Ce isotopes recently done by Puddu, Scholten and Otsuka (PSO)[5]. The model Hamiltonian is assumed to be

$$H = H_n + H_p + V_{np} \ ,$$

where H_n and H_p are respectively the IBA Hamiltonian for the neutron and the proton, V_{np} is interaction between the neutron boson and the proton boson. The parameters of the PSO model Hamiltonian have been determined to give an excellent description of the low-lying spectra of the isotope chains of Xe^{54}, Ba^{56} and Ce^{58}. From their model wave functions, we can construct the transition form factors F^n_{LM} and F^p_{LM} for the (π,π') calculations. The IBA transition form factors are the Fourier-transform of the following transition densities

$$\rho^N_{LM}(\vec{r}) = [a^N_L \alpha^N_L(r) + b^N_L \beta^N_L(r)] Y_{LM}(\hat{r}) \ , \quad N{=}n \text{ or } p. \qquad (3)$$

The coefficients a^N_L and b^N_L are calculated from the IBA model wavefunctions

$$a_L^N = \left\langle L || [d_N^\dagger \times s_N + s_N^\dagger \times d_N]^L || 0^+ \right\rangle ,$$
$$b_L^N = \left\langle L || [d_N^\dagger \times d_N]^L || 0^+ \right\rangle , \qquad N = n \text{ or } p. \qquad (4)$$

where d_n (d_p) and s_n (s_p) are respectively the L=2 and the L=0 neutron (proton) bosons. It is important to note here that the relative strengths between the neutron coefficients (a_L^n and b_L^n) and the proton coefficients (a_L^p and b_L^p) are sensitive to the choice of the interaction V_{np}. A sensitive test of the IBA model is to see whether the calculated coefficients a_L^n, b_L^n, a_L^p, and b_L^p are consistent with the (π,π') data; in particular the difference between the π^+ and π^- cross sections.

To proceed, we must define the radial dependence $\alpha_L^N(r)$ and $\beta_L^N(r)$. As suggested by an (e,e') study[6] using the IBA model, we assume for simplicity that

$$\alpha_L^N(r) \propto \frac{\partial}{\partial r}\rho$$
$$\beta_L^N(r) \propto \frac{\partial^2}{\partial r^2}\rho \qquad (5)$$

where $\rho(r)$ is the standard Woods-Saxon density. To fix the normalizations, the integrals

$$\int_0^\infty \alpha_L^N(r) r^{2+L} dr = \alpha_{L0}^N ,$$
$$\int_0^\infty \beta_L^N(r) r^{2+L} dr = \beta_{L0}^N . \qquad (6)$$

are chosen to be the same as that of PSO model. For a given set of the normalization constants defined above, the (π,π') scattering from the entire isotope chains are then completely determined by the IBA model.

Although the radial dependences defined in eq. (5) are reasonable, they are somewhat arbitrary. Fortunately, we have learned from previous studies[7] that the magnitude of the first maximum of the (π,π') differential cross section is not sensitive to the detailed shape of the transition density. It is fixed by the proton and neutron normalization constants defined in eq. (6). Note that the proton normalization constants also fix the electromagnetic B(EL) value. Therefore, we can test the IBA model by examining whether the first maxima of ($\pi^\pm,\pi^{\pm\prime}$) cross sections and the B(EL) values in the entire isotope chain of a given nucleus can be described by a set of the four normalization constants.

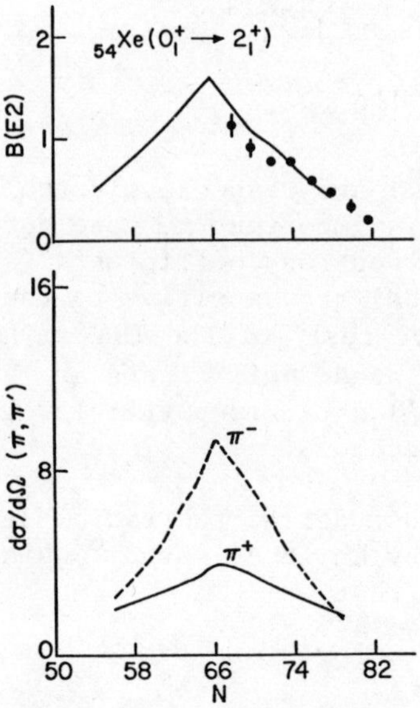

Fig. 1. The B(E2) values and the ($\pi^{\pm},\pi^{\pm'}$) cross sections at the first maximum $\theta = 19^{o}$ for the Xe^{54} isotope chain. The pion energy is 164 MeV and the unit is arbitrary.

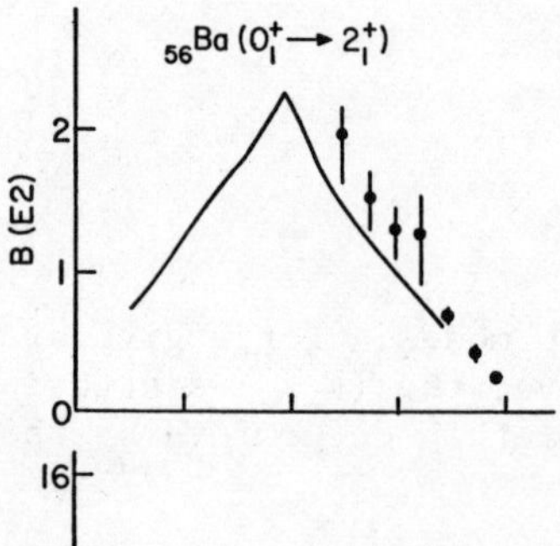

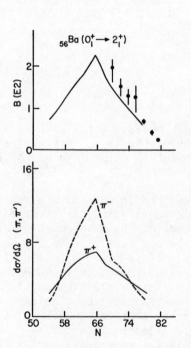

Fig. 2. Same as fig. 1, but for the Ba^{56} isotope chain.

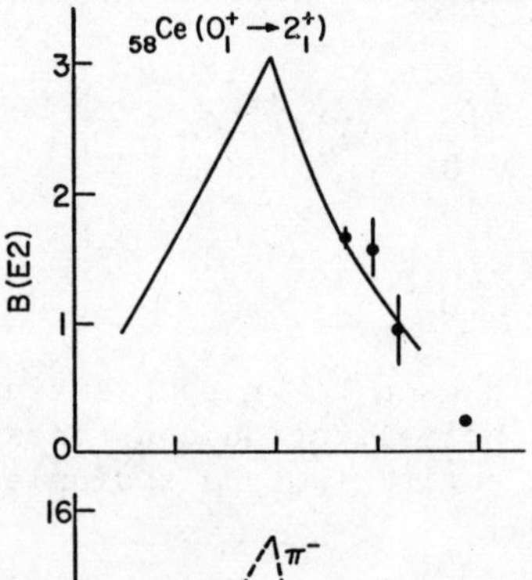

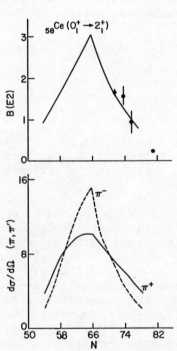

Fig. 3. Same as fig. 1, but for the Ce^{58} isotope chain.

In figures 1-3, we show our calculation results of the quadrupole L=2 excitation cross sections at the first maxima $\theta=19°$, and compared them with the calculated B(E2) values. Here we use the normalization constants defined in ref. 4. It is seen that the variations of the π^+ and π^- with respect to the change of the neutron number are very different. Measurements to check our predictions would provide helpful tests of the IBA model. In particular, the measurements of the ratios between the π^+ and π^- cross sections are most interesting.

To conclude, we emphasize that the IBA model can describe the inelastic scattering from the entire isotope chain with very few parameters. Experimental verification of our predictions of the $(\pi^\pm,\pi^{\pm\prime})$ from Xe^{54}, Ba^{56}, and Ce^{58} is an important step toward this direction. It is also an important test of the IBA model.

This work supported in part by the U. S. Dept. of Energy.

REFERENCES

1. T.-S. H. Lee and D. Kurath, Phys. Rev. C21, 293 (1980).
2. T.-S. H. Lee and R. D. Lawson, Phys. Rev. C21, 679 (1980).
3. A. Arima and F. Iachello, Ann. of Phys. 99 253 (1976); 111, 201 (1978); 123, 468 (1979).
4. A. Arima, T. Otsuka, F. Iachello and I. Talmi, Phys. Lett. 66B, 205 (1977); T. Otsuka, A. Arima, F. Iachello and I. Talmi, Phys. Lett. 76B, 139 (1978).
5. G. Puddu, O. Scholten and T. Otsuka, Nucl. Phys., in press
6. A. E. L. Dieperink, F. Iachello, A. Rinat and C. Creswell, Phys. Lett. 76B, 135 (1978).
7. T.-S. H. Lee and F. Tabakin, Nucl. Phys. A226, 253 (1974).

GEOMETRIC AND DYNAMICAL PROPERTIES OF THE INTERACTING BOSON MODEL†

Robert Gilmore

Institute for Defense Analyses
400 Army Navy Drive, Arlington, VA 22202

Da Hsuan Feng

Department of Physics and Atmospheric Science
Drexel University, Philadelphia, PA 19104

A continuous coordinate system is introduced to complement the discrete coordinate system used for the description of the Interacting Boson Model. The description of ground state energy phase transitions and dynamical processes are facilitated in this representation. The duality between the discrete and continuous representations can be used to compare nuclear models presented in either an algebraic or a geometric framework.

1. SCOPE

The Interacting Boson Model[1] (IBM) has had widespread successes in organizing and interpreting the systematics of energy level spectra for a large number of medium and heavy even-even nuclei.[2-5] These successes suggest that the study of this model be continued along lines traditional for other successful quantum mechanical models. Such studies fall into two broad categories, as follows:

Static Properties

1. Determination of ground state energy phase transitions.
2. Determination of thermodynamic phase transitions.

†Work partly supported by the National Science Foundation under grant # PHY-7908402

3. Canonical requantization of the classical limit of the original model.

Dynamic Properties

1. Determination of the approximate time-dependent wave-functions (TDHF approximation).
2. Determination of the transition operator and its limit, the S-matrix.
3. Bohr-Sommerfeld-Maslov requantization from the classical orbital structure.

A close parallel can be made between the study of the static and dynamic properties of quantum mechanical systems. Roughly speaking, the dynamic properties can be obtained by analytic continuation from the static properties. The six studies indicated above have been initiated for the IBM, but the present work will be confined primarily to a discussion of its dynamical properties.

2. MODELS

The IBM possesses two features that are intrinsic to a number of other important quantum mechanical models:

1. A mean field assumption is made.
2. A simple algebraic structure is present.

These assumptions generally manifest themselves as follows. A quantum system is assumed to consist of N identical subsystems, each of which possesses r internal states. Suppose $e_{ij}^{(\alpha)}$ $(1 \leq \alpha \leq N,\ 1 \leq i,j \leq r)$ is the operator which produces a transition from state j to state i in the α-th subsystem. Then the collective operator E_{ij} is defined by

$$E_{ij} = \sum_{\alpha=1}^{N} e_{ij}^{(\alpha)} \tag{2.1}$$

For such systems, these two assumptions have the following content:

1. Mean field. The Hamiltonian $\hat{H}$ is a function only of the intensive collective operators E_{ij}/N:

$$\hat{H}/N = h(E_{ij}/N) \tag{2.2}$$

2. Algebraic structure. The single particle operators $e_{ij}^{(\alpha)}$ and the collective operators E_{ij} obey commutation relations of the Lie algebra u(r):[6]

$$[e_{ij}^{(\alpha)}, e_{k\ell}^{(\beta)}] = \delta^{\alpha,\beta} (e_{i\ell}^{(\alpha)} \delta_{jk} - e_{kj}^{(\alpha)} \delta_{\ell i}) \tag{2.3a}$$

$$[E_{ij}, E_{k\ell}] = E_{i\ell}\,\delta_{jk} - E_{kj}\,\delta_{\ell i} \tag{2.3b}$$

Models which satisfy these two assumptions belong to a broader class of models ("bialgebraic") about which a great deal is known.[7] A number of powerful analytic tools have been developed for the study of such models. They have been applied to the Dicke model of atomic physics (r=2)[8], and more recently to the Lipkin-Meshkov-Glick pseudo-spin model (r=2)[9]. These tools are used below to study the static and dynamic properties of the IBM.

3. COHERENT STATES

The principal tools for this study are Lie Group Theory and Catastrophe Theory. Catastrophe Theory[10] enters primarily through its big brother, Bifurication Theory. Lie Group Theory enters primarily through the introduction of coherent states as a working basis in the appropriate Hilbert space.

Coherent states are defined by means of the following mathematical structures:[11-14]

1. A dynamical group G, with Lie algebra $\mathfrak{g}$;
2. An invariant subspace V^Λ carrying an irreducible unitary representation Γ^Λ;
3. An extremal state $|\Lambda, \text{ext}\rangle$;
4. A stability subgroup $H \subset G$, with Lie algebra $\mathfrak{h} \subset \mathfrak{g}$.

Then for an arbitrary dynamical group element $g \in G$, it is always possible to find a unique (coset) decomposition[6] of the form $g=\Omega h$ with $h \in H$, $\Omega \in G/H$. Then

$$g\left|\begin{matrix}\Lambda\\ \text{ext}\end{matrix}\right\rangle = \Omega h\left|\begin{matrix}\Lambda\\ \text{ext}\end{matrix}\right\rangle = \Omega\left|\begin{matrix}\Lambda\\ \text{ext}\end{matrix}\right\rangle e^{i\phi(h)} = \left|\begin{matrix}\Lambda\\ \Omega\end{matrix}\right\rangle e^{i\phi(h)} \tag{3.1}$$

Here $|\Lambda,\Omega\rangle$ is a coherent state, and $\phi(h)$ is a phase factor. The complex coordinates represented by Ω will be described in the examples.

Example 1: For the angular momentum group SU(2) these four mathematical structures are[11]

1. $G = SU(2)$, $\mathfrak{g} = su(2)$
2. $V^\Lambda = V^{2j+1}$ $\quad \Gamma^\Lambda = \mathcal{D}^{2j+1}$ (Wigner representation)
3. $|\Lambda, \text{ext}\rangle = |j, j\rangle$
4. $H = U(1)$

For $g(\phi,\theta,\psi) = e^{i\phi J_z}\, e^{i\theta J_y}\, e^{i\psi J_z}$

$$g(\phi,\theta,\psi)\left|\begin{matrix}j\\ j\end{matrix}\right\rangle = \left|\begin{matrix}j\\ \theta\,\phi\end{matrix}\right\rangle e^{ij\psi} \tag{3.2}$$

and $|j;\theta\phi\rangle$ are the SU(2) coherent states.

Remark 1: For SU(2) the coherent states are defined on the surface of a sphere. This suggests that the spherical harmonics and the coherent state matrix elements are in some way related. This relationship is best seen through the Wigner $\mathcal{D}$ representation matrix elements[15] for $g(\phi,\theta,\psi)$:

$$\left\langle {\ell \atop m} \middle| g(\phi,\theta,\psi) \middle| {\ell \atop 0} \right\rangle = \mathcal{D}^{\ell}_{m0}(\phi,\theta) \simeq Y^{\ell}_{m}(\theta,\phi) \qquad (3.3a)$$

$$\left\langle {j \atop m} \middle| g(\phi,\theta,\psi) \middle| {j \atop j} \right\rangle = \mathcal{D}^{j}_{mj}(\phi,\theta,\psi) = \left\langle {j \atop m} \middle| {j \atop \Omega} \right\rangle e^{ij\psi} \qquad (3.3b)$$

Coherent state matrix elements may be treated in the same way that spherical harmonics may be treated, although they are not spherical harmonics.

Example 2: The coherent states of use for the IBM are defined by[12,14]

1. $G = SU(6)$
2. Γ^{Λ} = fully symmetric representation $\{N\}$ of dimension $(N+6-1)!/N!(6-1)!$
3. $|\Lambda,\text{ext}\rangle$ = state which minimizes the Hamiltonian
$$H = \sum_{i=1}^{6} \varepsilon_i E_{ii}, \quad \varepsilon_1 < \varepsilon_2 < \ldots < \varepsilon_6$$
4. $H = U(5)$

The 5 complex parameters Ω which characterize the SU(6) coherent states may be taken as z_2, z_3, z_4, z_5, z_6, where

$$\sum_{i=2}^{6} z_i^* z_i \leq 1 \qquad (3.4)$$

It is useful to define $z_1 = +[1 - \sum_2^6 z_i^* z_i]^{\frac{1}{2}}$ Then $(z_1, z_2, z_3, z_4, z_5, z_6)$ may be interpreted as the wavefunction of a single particle with 6 internal degrees of freedom. The space characterizing SU(6) coherent states is S^{10}, the 10-dimensional sphere.

Remark 2: In the event that a distinction is made between protons and neutrons,[16,17] Γ^{Λ} is no longer generally the fully symmetric representation, $|\Lambda,\text{ext}\rangle$ no longer generally minimizes the diagonal Hamiltonian, $H = U(4) \otimes U(1)$, and 9 complex parameters are required to describe the relevant coherent states.[14]

Remark 3: Two other coordinate systems for coherent states have been widely used in the past: the algebraic and the projective.[18] Projective coordinates for SU(6) have been used by Dieperink, Scholten and Iachello[19] (preceding paper) and by Ginocchio and Kirson[20] (following paper).

Coherent states $|\Lambda,\Omega\rangle$ form an overcomplete set of vectors in the space V^{Λ}. Any state $|\psi\rangle$ can be expressed in terms of the coherent states

$$\left|\psi\right\rangle = \int \left|\begin{matrix}\Lambda\\ \Omega\end{matrix}\right\rangle \left\langle\begin{matrix}\Lambda\\ \Omega\end{matrix}\middle|\psi\right\rangle d\Omega \tag{3.5}$$

Here $d\Omega$ is the volume element[6] on G/H. The weight function $\langle\Lambda,\Omega|\psi\rangle$ is not unique. This nonuniqueness may be used to find a relatively simple form for $\langle\Lambda,\Omega|\psi\rangle$.

An arbitrary state $|\psi\rangle$ can be expressed in terms of an orthonormal system of basis vectors in V^{Λ}. A Gel'fand-Tsetlein[21] basis $|\Lambda, "M"\rangle$ may be used, or any other basis which diagonalizes a complete set of commuting operators. Three such bases have been constructed for the three limits of the IBM. The representation of a state in terms of an orthonormal basis (e.g. Gel'fand-Tsetlein basis) is

$$\left|\psi\right\rangle = \sum_{"M"} \left|\begin{matrix}\Lambda\\ "M"\end{matrix}\right\rangle \left\langle\begin{matrix}\Lambda\\ "M"\end{matrix}\middle|\psi\right\rangle \tag{3.6}$$

The weight function for this state is

$$\left\langle\begin{matrix}\Lambda\\ \Omega\end{matrix}\middle|\psi\right\rangle = \sum_{"M"} \left\langle\begin{matrix}\Lambda\\ \Omega\end{matrix}\middle|\begin{matrix}\Lambda\\ "M"\end{matrix}\right\rangle \left\langle\begin{matrix}\Lambda\\ "M"\end{matrix}\middle|\psi\right\rangle \tag{3.7}$$

The overlap integrals $\langle\Lambda, \Omega|\Lambda"M"\rangle$ have been constructed explicitly for all symmetric representations for SU(r).[22]

Expectation values of collective operators E_{ij} as well as functions f(E) of these operators can be taken in the coherent state basis relatively easily. The function f(E) is expanded in terms of irreducible tensor operators. The expectation value of an irreducible tensor operator is the corresponding irreducible tensor function (i.e., spherical harmonic), multiplied by a numerical constant of known value.[23]

In many instances it is convenient to carry out calculations in the classical limit $(N \to \infty)$. Results of these calculations differ only by order 1/N from results of calculations for finite N. Two useful simplifications arise in the classical limit.[24]

1. Expectation values are obtained simply from the substitution

   $E_{ij}/N \to z_i^* z_j$.

2. Weight functions $\langle \Lambda, \Omega | \psi \rangle$ are often very well approximated by gaussians. The state $|\psi\rangle$ can then be characterized simply by a centroid and a covariance matrix.

Example: For SU(2) the classical limit takes the form

$$\lim_{N = 2J \to \infty} \left\langle \begin{matrix} N \\ \theta \ \phi \end{matrix} \middle| J_{\pm}/N \middle| \begin{matrix} N \\ \theta \ \phi \end{matrix} \right\rangle = \tfrac{1}{2} \sin\theta \, e^{\pm i\phi} \tag{3.8}$$

where $z_{\hat{2}} = e^{i\phi} \sin\theta/2$.

4. GROUND STATE DEFORMATION

Before carrying out a study of the dynamical properties of the IBM, it is necessary to determine the ground state. This is so because the "topography" of the expectation value $h(z_i^* z_j) = \langle \hat{H}/N \rangle$ in the coherent state basis severely constrains the types and properties of TDHF orbits (these are constant energy orbits).

Consider first the simple MGL model[25].

$$H = \varepsilon J_3 + \frac{V}{2N} (J_+^2 + J_-^2) \tag{4.1}$$

The factor 1/N has been included for thermodynamic reasons.[26] The classical limit $(N = 2J \to \infty)$ of this model is

$$h(\theta,\phi) = \left\langle \begin{matrix} N \\ \theta \ \phi \end{matrix} \middle| H/N \middle| \begin{matrix} N \\ \theta \ \phi \end{matrix} \right\rangle = \varepsilon \cos\theta$$

$$+ \ 1/2 \ V \sin^2\theta \ (e^{2i\phi} + e^{-2i\phi}) \tag{4.2}$$

The ground state energy occurs at $\theta = \pi$ if $|V| < \varepsilon$, and at $\theta \neq \pi$ if $|V| > \varepsilon$. In this latter case, the function $h(\theta,\phi)$ has the shape of a deformed sphere as shown in Fig.1.[26] The minimum value of (4.2) is rigorously the ground state energy per nucleon in the classical limit. The MGL model exhibits a second-order ground state energy phase transition as the normalized interaction strength $|V|/\varepsilon$ increases through 1.

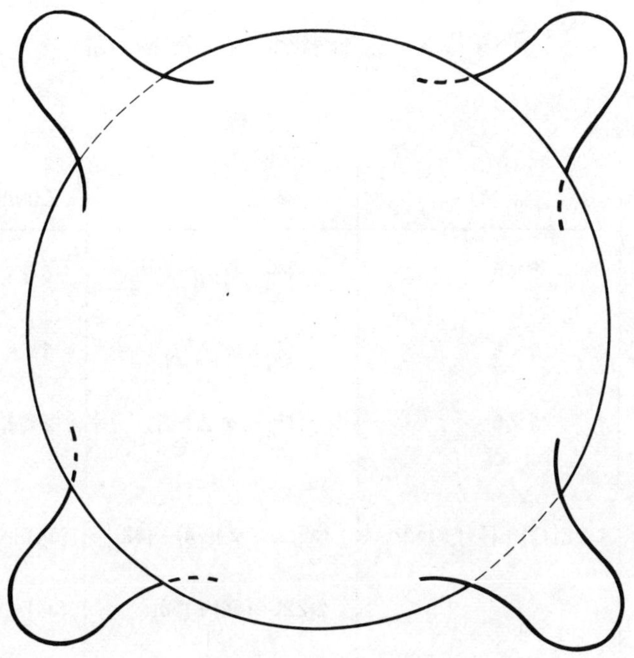

Fig.1 In the deformed regime, the classical limit of the MGL Hamiltonian has the functional form of a deformed sphere.

The ground state properties of the IBM are determined similarly. The model Hamiltonian may be taken as

$$H/N \;=\; \varepsilon\, C^1[U(5)]/N + \kappa C^2\,[SU(3)]/N^2 + \kappa' C^2\,[SO(6)]/N^2$$

$$+\,\mu C^2\,[SO(5)]/N^2 + \mu' C^2\,[SO(3)]/N^2 \qquad (4.3)$$

Here $C^i[G]$ is the ith (first or second) Casimir invariant of the Lie group G. In the classical limit the expectation value of each intensive Casimir invariant $C^j[G]/N^j$ is given by an expression of the form Tr $M^\dagger M$. The matrices M for each group G are given in Table 1.

The IBM with $\mu=0$, $\mu'=0$ exhibits a second order ground state energy phase transition from vibrational to γ-unstable rotor as a function of increasing κ'/ε (for $\kappa=0$). A first order phase transition from vibrational to rigid axial symmetric rotor occurs as a function of increasing κ/ε (for $\kappa'=0$). Fig.2 shows the minimum energy as a function of the parameters ε,κ' ($\kappa=\mu=\mu'=0$). The presence

$$\left\langle {}^{N}_{Z} \right| \mathcal{H}/N \left| {}^{N}_{Z} \right\rangle = \Sigma \text{ coefficient} \times \mathrm{Tr}\, M^{\dagger} M$$

GROUP	Dim M	M	CONDITIONS
U(5)	5×5	$M_{ij} = Z_i \delta_{ij}$	$2 \leqslant i,\ j \leqslant 6$
$SU(4) \simeq SO(6)$	6×6	$M_{ij} = \mathrm{Im}\ Z_i^* Z_j$	$1 \leqslant i,\ j \leqslant 6$
SO(5)	5×5	$M_{ij} = \mathrm{Im}\ Z_i^* Z_j$	$2 \leqslant i,\ j \leqslant 6$
SU(3)	$2(11)+(44)+(55)$	$(56)+\sqrt{2}\,[(14)+(42)]$ $2(22)+(44)+(66)$	$(64)+\sqrt{2}\,[(35)+(51)]$ $(54)+\sqrt{2}\,[(36)+(62)]$ $2(33)+(55)+(66)$
	$(ij) = Z_i^* Z_j$		
SO(3)	3×3	$M(SO(3)) = \mathrm{Im}\ \ M(SU(3))$	

Table 1. In the classical limit, the expectation value of the Casimir invariants in the IBM are given by simple matrix expressions.

of a second order phase transition is clearly indicated by the continuity in slope and discontinuity in curvature along the diagonal behind the vertical axis.

5. DYNAMICS

The dynamical equations of motion are derived from the variational principal[27]

$$\delta \int_{t_1}^{t_2} L(q,\dot{q},t)\ dt = 0 \tag{5.1a}$$

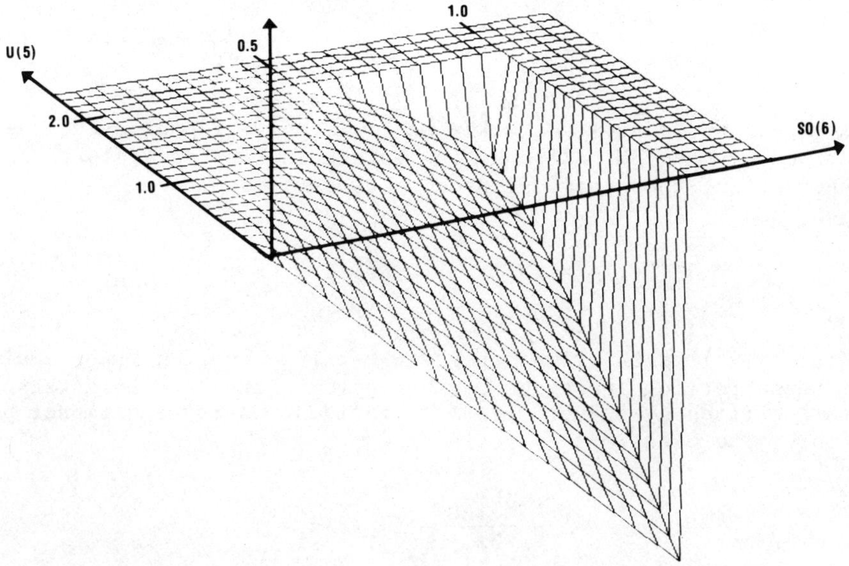

Fig.2. The ground state energy per nucleon is shown as a function of the two competing dynamical processes indicated.

where $$L\,(q,\dot{q},t) = \left\langle \psi(q) \middle| i\frac{\partial}{\partial t} - H \middle| \psi(q) \right\rangle \tag{5.1b}$$

Configuration space is the set of coordinates q used to parameterize the set of trial states used in the variational calculation.

When coherent states $|N,z\rangle$ are used as trial states, the Lagrangian is

$$L(z,\dot{z},t) = N\,\frac{i}{2}\sum_{j=2} z_j^*\dot{z}_j - z_j\dot{z}_j^* - Nh(z^*,z) \tag{5.2}$$

Here $Nh(z^*,z) = \langle N,z|H|N,z\rangle$. The Euler-Lagrange equations of motion are[28]

$$i\dot{z}_j = \frac{\partial h}{\partial z_j^*} \qquad i\dot{z}_j^* = -\frac{\partial h}{\partial z_j} \tag{5.3a}$$

These equations are transformed into canonical hamiltonian form by the substitution $z_j = (q_j + ip_j)/\sqrt{2}$

$$\frac{dq_j}{dt} = \frac{\partial h}{\partial p_j} \quad , \quad \frac{dp_j}{dt} = -\frac{\partial h}{\partial q_j} \tag{5.3b}$$

Example. We consider here the Coulomb excitation of a nucleus described by the IBM. If only the valence nuclei participate in this interaction, the dynamical process is described by

$$H(t) = H_{IBM} + f_{ji}(t)\, E_{ij} \tag{5.4}$$

Here $f_{ji}(t)$ is the time dependent forcing function. The amplitude of these forcing functions may be approximated by a Lorentzian characterized by the projectile's incident energy and impact parameter. The equations of motion are

$$i\dot{z}_j^* = -\frac{\partial h_{IBM}}{\partial z_j} - f_{ji}(t) z_i^* \tag{5.5}$$

This system of coupled ordinary nonlinear differential equations on the sphere S^{10} can be integrated numerically. If this integral extends from $t=-\infty$ to $t=+\infty$, and the system state at $t=-\infty$ is a coherent state approximating the IBM ground state, then the final state is a coherent state describing the superposition of states into which the nucleus has been excited (assuming no decay).

Remark. If gaussians are used as trial states for the variational principal (5.1), then the centroid evolves according to the TDHF equations (5.3). The evolution of the covariance matrix is governed by a system of transport equations. Use of pure coherent states as trial states is equivalent to using gaussians with a zero covariance matrix (and vanishing transport equation).

6. CLASSICAL CONNECTION

The TDHF equations (5.3b) suggest a close connection with classical mechanics. The equations of motion in the classical and quantum cases are derived from Lagrangians of the form

Classical: $$L = \frac{1}{2} M_{ij}\dot{q}_i\dot{q}_j - V(q) \tag{6.1 CM}$$

Quantum: $$L = \frac{1}{2} A_{ij}\dot{q}_i q_j - \langle \psi(q) | H | \psi(q) \rangle \tag{6.1 QM}$$

In both cases the q_i are real coordinates for the configuration space. In the classical case the mass tensor M is symmetric: $M^t = +M$. In the quantum case the real tensor A is antisymmetric (i.e. symplectic): $A^t = -A$.

The momenta conjugate to the independent real coordinates are

Classical: $$p_j = \frac{\partial L}{\partial \dot{q}_j} = M_{ji}\dot{q}_i \qquad (6.2\ \mathrm{CM})$$

Quantum: $$p_j = \frac{\partial L}{\partial \dot{q}_j} = A_{ji}q_i \qquad (6.2\ \mathrm{QM})$$

In the classical case the momenta are superpositions of the velocities while in the quantum case they are superpositions of the coordinates. This creates the following problem. In the classical case phase space has twice the dimension as configuration space while in the quantum case, phase space may be identified with configuration space.

This dilemma is only apparent. In the quantum case phase space has twice the dimension as configuration, but there are exactly as many conservation laws as coordinates ($\partial h/\partial \dot{I}_j = 0$, where $z_j = \sqrt{I_j}\, e^{i\phi_j}$), thus effectively halving the dimension of phase space.

The differences between the classical and quantum phase spaces occur for a natural reason. Phase space contains exactly enough information to specify the initial conditions for integrating the equations of motion. In the classical case, Newton's equation is second order, while in the quantum case, Schrodinger's equation is first order.

7. QUALITATIVE DYNAMICS

To each IBM there is a corresponding classical hamiltonian. Thus, the enormous body of knowledge associated with classical mechanics can be brought to bear on the study of the IBM and its dynamical properties. For example, the TDHF orbits and trajectories can be determined simply by studying those of the classical counterpart.

Example. The TDHF orbits of the MGL model are simply the intersections of the plane E = constant with the surface shown in Fig.1. For energies slightly above the ground state energy, there are two disconnected and approximately circular orbits, one surrounding each minimum. As the energy increases towards that of the saddle along south polar axis, the two disjoint orbits become distended, approach each other, and coalesce in a figure 8 as the saddle point is reached.

Slightly above the saddle there is a single orbit which is extremely elongated. This process of elongation as a critical point of $\langle H \rangle$ is traversed is called critical elongation. The MGL model, the IBM, and the other models of bialgebraic type exhibit a large number of other critical phenomena. These phenomena are predicted by Catastrophe Theory.[29]

8. TRANSITION OPERATOR

The time evolution of a wave function is given by

$$\left|\psi(t_b)\right\rangle = e^{-\frac{i}{\hbar}H(t_b-t_a)}\left|\psi(t_a)\right\rangle \tag{8.1}$$

The time evolution of the weight function $\langle\Omega|\psi\rangle$ is

$$\left\langle\Omega_b\middle|\psi(t_b)\right\rangle = \left\langle\Omega_b\right|e^{-\frac{i}{\hbar}H(t_b-t_a)}\left|\Omega_a\right\rangle\left\langle\Omega_a\middle|\psi(t_a)\right\rangle$$

$$= T\,(\Omega_b t_b, \Omega_a t_a)\left\langle\Omega_a\middle|\psi(t_a)\right\rangle \tag{8.2}$$

The transition operator T can be written as the product of infinitesimal operators[30-32]

$$T = \Pi \text{ (infinitesimal operators)}$$

Each infinitesimal operator has the form

$$\left\langle\Omega_{i+1}\right|e^{-\frac{i}{\hbar}H\Delta t}\left|\Omega_i\right\rangle = \left\langle\Omega_{i+1}\middle|\Omega_i\right\rangle\left\{1 - \frac{\left\langle\Omega_{i+1}\right|H\left|\Omega_i\right\rangle}{\left\langle\Omega_{i+1}\middle|\Omega_i\right\rangle}\Delta t\right\} \tag{8.3}$$

This is simplified using the continuity of the coherent states and neglecting terms above first order

$$= 1 - \Delta t\left\{\frac{i}{\hbar}\left\langle\Omega\right|H\left|\Omega\right\rangle + \frac{d\Omega}{dt}\right\} = e^{i\Delta S/\hbar} \tag{8.4}$$

$$\Delta S = \int_{t}^{t+\Delta t} \left\langle \Omega \middle| i\hbar \frac{\partial}{\partial t} - H \middle| \Omega \right\rangle dt \tag{8.5}$$

The transition operator is therefore expressed as a path integral

$$T(\Omega_b t_b, \Omega_a t_a) = \int e^{iS/\hbar} \, d\mu(\text{paths}) \tag{8.6a}$$

$$S(\Omega_b t_b, \Omega_a t_a) = \int_{t_a}^{t_b} L(\Omega,\dot{\Omega},t)\,dt \tag{8.6b}$$

$$L(\Omega,\dot{\Omega},t) = \left\langle \Omega \middle| i\hbar \frac{\partial}{\partial t} - H \middle| \Omega \right\rangle \tag{8.6c}$$

The integral in (8.6b) is along a path $\Omega(t)$ joining $\Omega_a = \Omega(t_a)$ with $\Omega_b = \Omega(t_b)$.

If the action is expanded

$$S = S_o + \delta S + \delta^2 S + \ldots \tag{8.7}$$

then the path integral can be estimated. The principal contribution comes from paths for which

$$\delta S = \delta \int_{t_a}^{t_b} \left\langle \Omega \middle| i\hbar \frac{\partial}{\partial t} - H \middle| \Omega \right\rangle dt = 0 \tag{8.8}$$

Along these paths the second variation $\delta^2 S$ contributes a phase factor, so that the estimate for the transition operator is [33,34]

$$T(\Omega_b t_b, \Omega_a t_a) \simeq \sum K_j \exp i(S_j/\hbar + \#_j \pi/4) \tag{8.9}$$

Here the sum extends over each allowed trajectory ($\delta S = 0$), and the amplitude K, action S, and Morse index # are computed for each allowed trajectory.

Remark. A single valuedness requirement on T for closed trajectories leads to the Bohr-Sommerfeld-Maslov quantization condition.[33,35]

9. CONSTRUCTION OF EIGENFUNCTIONS

A. Algebraic Approach

An IBM hamiltonian may be diagonalized in any convenient orthonormal basis

$$H|\psi\rangle = E|\psi\rangle \tag{9.1a}$$

$$\sum_{M'} \langle M|H|M'\rangle \langle M'|\psi\rangle = E\langle M|\psi\rangle \tag{9.1b}$$

Eigenstates $|\psi_E\rangle$ can be expressed in the form (3.6), and the associated weight functions $\langle\Omega|\psi_E\rangle$ in the form (3.7)

$$\langle\Omega|\psi_E\rangle = \sum_{"M"} \langle\Omega|"M"\rangle \langle "M"|\psi_E\rangle \tag{9.2}$$

The coefficients $\langle "M"|\psi_E\rangle$ are outputs of a matrix diagonalization routine and the functions $\langle\Omega|"M"\rangle$ are known. Therefore these weight functions can be computed for any IBM hamiltonian.

B. Geometric Approach 1. Differential

The Schrodinger equation may be solved in a continuous as well as a discrete representation

$$H|\psi\rangle = E|\psi\rangle \tag{9.3a}$$

$$\int \langle\Omega|H|\Omega'\rangle \, d\Omega' \langle\Omega'|\psi\rangle = E\langle\Omega|\psi\rangle \tag{9.3b}$$

As in the nonrelativistic quantum mechanics, the matrix elements of H involve differential operators

$$\langle\Omega|H|\Omega'\rangle = \mathcal{D}(H;\Omega)\, \delta(\Omega - \Omega') \tag{9.4}$$

The differential realizations of operators associated with u(r) have previously been constructed.[18] For the IBM, $\mathcal{D}(H,\Omega)$ is a second order differential operator since H involves nothing more complicated than second order Casimir operators.

In the coherent state representation the Schrodinger equation (9.3b) reduces to

$$\mathcal{D}(H;\Omega)\langle\Omega|\psi\rangle = E\langle\Omega|\psi\rangle \tag{9.5}$$

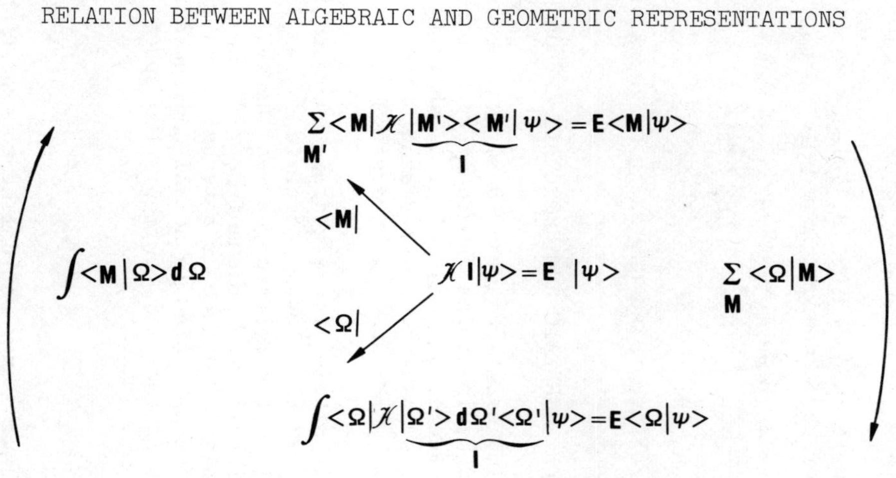

Fig.3 The SU(6) coherent states are the matrix elements in the unitary transformation between the algebraic(discrete) and geometric(continuous) representations.

The second order differential operator $\mathcal{D}(H;\Omega)$ and the wave function $\langle\Omega|\psi\rangle$ are defined on the sphere S^{10}.

Remark. The coherent state representation makes it possible to compare the IBM with the Bohr-Mottelson hamiltonian.[36] A direct comparison is possible once the 5 complex parameters characterizing U(6) coherent states have been reduced to 5 real parameters, and then the three Euler angles are removed.

C. Geometric Approach 2. Integral

Use of the transition operator (8.9) leads to the following integral equation for the eigenfunctions

$$\langle\Omega|\psi(t+\Delta t)\rangle = \int T(\Omega,\ t+\Delta t;\ \Omega' t)\langle\Omega'|\psi(t)\rangle\, d\Omega' \qquad (9.6)$$

The solutions of (9.6) determine the energy eigenvalue E through $\Delta t = h/E$.

10. RELATION BETWEEN ALGEBRAIC AND GEOMETRIC APPROACHES

A simple relation exists between these two approaches. The relation is obtained by inserting the identity operator $I = \sum|M\rangle\langle M| = \int|\Omega\rangle\langle\Omega|\,d\Omega$ in the Schrodinger equation, as shown in Fig.3. The transformation between these two representations is effected by the transformation matrix $\langle\Omega|M\rangle$ and its inverse $\langle M|\Omega\rangle$.

The relationship between the algebraic approach, the differential geometric approach, and the global geometric to the description of the IBM is identical to the relation between Heisenberg's matrix mechanics,[37] Schrodinger's wave mechanics,[38,39] and Feynman's path integrals.[40] The only difference is a technical detail. The continuous space for wave mechanics is normal configuration space, while for the IBM it is the U(6) configuration space (the coherent state space S^{10}).

ACKNOWLEDGEMENTS

One of us (Da Hsuan Feng) would like to express his gratitude to Professor Aage Bohr for his hospitality. Very useful and informative discussions on this work were conducted with Franco Iachello on various occasions. Finally, we thank Mr. Antony Joseph for his expert preparation of the manuscript.

REFEFENCES

1. A. Arima and F. Iachello, Phys. Rev. Lett. 35, 1069 (1975).
2. A. Arima and F. Iachello, Ann. Phys. (N.Y.) 99, 253 (1976).

3. A. Arima and F. Iachello, Ann. Phys. (N.Y.) 111:201(1978).
4. A. Arima and F. Iachello, Phys. Rev. Lett. 40:385(1978).
5. O. Scholten, F. Iachello, and A. Arima, Ann. Phys. (N.Y.) 115:325(1978).
6. R. Gilmore, "Lie Groups, Lie Algebras and Some of Their Applications", Wiley, N.Y., 1974.
7. R. Gilmore, in: "Journees Relativistes", M. Cahan, R. Debever, and J. Geheniau, eds., Université Libre, Bruxelles, 1976, p.71.
8. R.H. Dicke, Phys. Rev. 93:99(1954).
9. H.J. Lipkin, N. Meshkov, and A.J. Glick, Nucl. Phys. 62:188 (1965), 199(1965), 211(1965).
10. T. Poston and I.N. Stewart, "Catastrophe Theory and Its Applications", Pitman, London, 1978.
11. F.T. Arecchi, E. Courtens, R. Gilmore, and H. Thomas, Phys. Rev. A6:2211(1972).
12. R. Gilmore, Ann. Phys. (N.Y.) 74:391(1972).
13. A.M. Perelomov, Commun. Math. Phys. 26:22(1972).
14. R. Gilmore, Rev. Mex. de Fisica 23:143(1974).
15. E.P. Wigner, "Group Theory and Its Application to the Quantum Mechanics of Atomic Spectra", Academic, N.Y., 1959.
16. T. Otsuka, A. Arima, F. Iachello, and I. Talmi, Phys. Lett. 76B:139(1978).
17. T. Otsuka, A.Arima, and F. Iachello, Nucl. Phys. A309:1(1978).
18. R. Gilmore, C.M. Bowden, and L.M. Narducci, Phys. Rev. A12: 1019(1975).
19. A.E.L. Dieperink, O.Scholten, and F. Iachello, Phys. Rev. Lett. 44:1747(1980).
20. J.N. Ginocchio and M.W. Kirson, Phys. Rev. Lett. 44:1744(1980).
21. I.M. Gel'fand and M.L. Tsetlein, Dokl. Akad. Naak. SSSR 71: 840(1950), 71:1017(1950).
22. F.T. Arecchi, R. Gilmore, and D.M. Kim, Lett. Nuovo Cimento 6:219(1973).
23. R. Gilmore, J. Phys. A9:L65(1976).
24. R. Gilmore, J. Math. Phys. 20:891(1979).
25. R. Gilmore and D.H. Feng, Phys. Lett. 76B:26(1978).
26. R. Gilmore and D.H. Feng, Nucl. Phys. A301:189(1978); D.H. Feng, R. Gilmore and L.M. Narducci, Phys. Rev. C19:1119(1979).
27. P.A.M. Dirac, Proc. Camb. Phil. Soc. 26:367(1930).
28. D.H. Feng and R. Gilmore, Phys. Lett. 90B, 327(1980).
29. R. Gilmore, "Catastrophe Theory for Scientists and Engineers", Wiley, N.Y., 1981, Chapter 15.
30. H. Kuratsuji and T. Suzuki, J. Math. Phys. 21:472(1980).
31. H. Kuratsuji and T. Suzuki, Kyoto University Preprint 514 (unpublished).
32. H. Kuratsuji and Y. Mizobuchi, Kyoto University Preprint 519 (unpublished).
33. V.P. Maslov, "Theorie des perturbations et methods asymptotiques", Dunod, Paris, 1972.
34. V. Guillemin and S. Sternberg, "Geometric Asymptotics", American Mathematical Society, Providence, 1977.

35. V.I. Arnol'd, Fun. Aral. Appl. 1:1(1967).
36. A. Bohr and B.R. Mottelson, Mat. Fys. Medd. Dan. Vid. Selsk. 27:(no.16) (1953).
37. W. Heisenberg, Zeits. fur Phys. 33: 879(1925).
38. E. Schrodinger, Ann. der Phys. 79: 361(1926).
39. E. Schrodinger, Ann. der Phys. 79: 734(1926).
40. R.P. Feynman, Rev. Mod. Phys. 20: 367(1948).

SHAPES AND SHAPE PHASE TRANSITIONS IN THE INTERACTING BOSON MODEL

A.E.L. Dieperink and O. Scholten

Kernfysisch Versneller Instituut, University of Groningen, The Netherlands

1. INTRODUCTION

The interacting boson approximation (IBA) has been shown to provide a rather successful description of collective nuclear properties. Since the model is formulated in an elegant but rather abstract algebraic way its precise relation to geometrical models like the Bohr-Hamiltonian describing a liquid drop with quadrupole surface oscillations is by no means obvious. Up to now the connection with the geometrical approach has only been made indirectly: it has been inferred on the basis of the similarity of the expressions for energy levels and E2 transition rates that the three dynamical symmetries[1,2,3] that occur in IBA, namely SU(5), SU(3) and O(6), correspond (for $N \to \infty$) to the anharmonic vibrator, the axially symmetric rigid rotor and the gamma unstable rotor, respectively. To establish a more direct connection between the two pictures one needs to express the IBA hamiltonian in terms of shape variables, thus allowing one to associate a geometry with the IBA formulation.

We stress that this is not a purely academic problem but is also of practical interest. It often happens that as an input in the analysis of experimental data information about the shape of the nucleus is required, as for example in nuclear energy level density formulas, or in the analysis of muonic X-ray data.

In Fig. 1 a schematic representation of the IBA is shown. In the first part of this contribution the classical limit will be constructed for the three extreme cases (represented by the

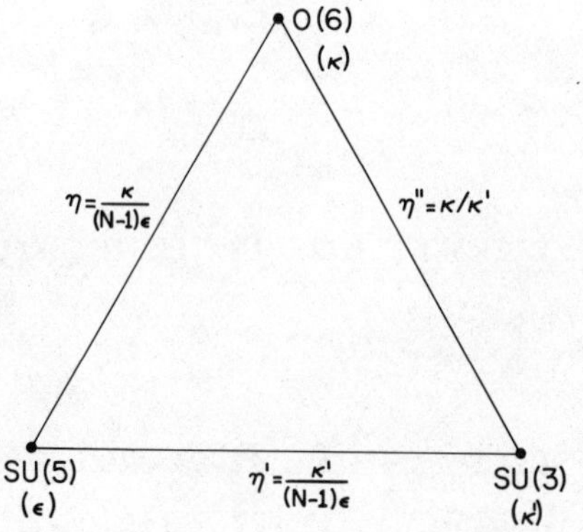

Fig. 1. Schematic representation of the structure of the IBA SU(6) hamiltonian. The three corners of the triangle represent the dynamical symmetries and the three legs the interplay between them as a function of the strength parameters ε, κ and κ', respectively.

corners of the triangle) using a method recently developed by Gilmore and Feng[4,5] that can be applied to interacting many-body systems that are invariant under a dynamical symmetry group. In the second part this method will be extended to the three transitional region of IBA (represented by the legs of the triangle), where an interplay between two shape phases occurs. Here the interesting question arises whether it is possible to describe the properties in these regions by simple many-body approximation methods (like e.g. mean field theories). In this respect the SU(6) IBA hamiltonian provides a welcome generalization of the SU(2) Lipkin model[6]. The latter has been used extensively[7,8,9] as an exactly solvable schematic model to test the accuracy of many-body theories. Whereas the Lipkin model is purely schematic the IBA model has the virtue of being realistic in addition to having a much richer group structure.

2. SHAPE PHASES

In general the correspondence between classical and quantum variables is ambiguous. However, recently it was shown by Gilmore[4] that the classical limit of any operator $\hat{A}$ that belongs to a compact Lie algebra G can be constructed unambiguously. This method, which is also applicable to the case of the IBA which has

SU(6) as a dynamical symmetry group, consists of the following steps: (i) construction of coherent states for the totally symmetric representation [N] of SU(6) (where N is the number of bosons) as a working basis, (ii) calculation of the expectation value of the hamiltonian in the coherent state representation, (iii) the minimum of this energy surface converges to the true ground state energy for $N \to \infty$.

2.1 Coherent States for SU(n)

A coherent state can be expressed[4] as $\{U(g)|\tilde{0}\rangle,\ g \in G\}$, where U(g) is a unitary representation of the Lie group G acting on the Hilbert space, $|\tilde{0}\rangle$ is a given vector in this space (extremal state) and H the corresponding stability subgroup. In the case of the group G=SU(n) one has H=U(n-1), and U(g) can be parametrized as

$$U(g(\underset{\sim}{\eta})) = \exp\left(\sum_{j=2}^{n} (\eta_j E_{1j} - \eta_j^* E_{j1})\right), \tag{2.1}$$

where E_{ij} are the shift operators of SU(n).
Specifying to the totally symmetric representations of SU(6) and taking for the extremal state the vector containing only s bosons, $|\tilde{0}\rangle = |s^N\rangle$, it can be shown that the normalized coherent state can be expressed as

$$|N\,\alpha_\mu\rangle = (1+|\alpha|^2)^{-N/2} \exp\left(\sum_\mu \alpha_\mu d_\mu^\dagger s\right)|s^N\rangle. \tag{2.2}$$

The parameters α_μ are sometimes referred to as projective coordinates[10] since they are obtained by stereographic projection from a (10 dimensional) sphere defined by the realization $\eta_\mu = \frac{1}{2}\theta_\mu\, e^{-i\phi_\mu}$, namely

$$\alpha_\mu = \eta_\mu/|\eta_\mu| \tan|\eta_\mu| = \exp(-i\phi_\mu) \tan(\tfrac{1}{2}\theta_\mu). \tag{2.3}$$

By rewriting Eq. (2.2) in the form

$$|N\alpha_\mu\rangle = (N!(1+|\alpha|^2)^N)^{-1/2}(s^\dagger + \sum_\mu \alpha_\mu d_\mu^\dagger)^N|0\rangle \tag{2.4}$$

it is clear that $|N\alpha_\mu\rangle$ can also be interpreted as a trial wave function describing the N-bosons as moving in a mean field.

We note that in the present case only five (complex) variables enter in Eq. (2.4); the sixth variable is eliminated since we insisted on staying within the totally symmetric representation [N] of SU(6).

An important property of coherent states is their over-completeness. This property has been used by Ginocchio and Kirson[11] to convert the secular equation of the IBA model into a differential equation which resembles that of the Bohr hamiltonian.

2.2 Classical Limit of the Hamiltonian

We consider the expectation value of the hamiltonian H in the coherent state representation, which is conventionally referred to as Q-representative of H

$$Q_N(H,\alpha_\mu) \equiv \langle N\alpha_\mu | H | N\alpha_\mu \rangle . \tag{2.5}$$

It is convenient to scale the hamiltonian in such a way that the binding energy per particle is proportional to N. This amounts to a scaling of the strength parameter κ for the two-body interactions in H(like $\kappa\ s^\dagger s^\dagger ss$) by a factor 1/N. The resulting hamiltonian will be denoted by $\overline{H}$. It can be shown that $Q_N(\overline{H},\alpha_\mu)$ has the simple property that for $N \to \infty$ the minimum of $Q_N(\overline{H},\alpha_\mu)$ - which obviously represents an upper bound - converges to the exact binding energy per boson, E_g:

$$E_g = \min_{\{\alpha_\mu\}} Q_N(\overline{H},\alpha_\mu)/N + O(1/N). \tag{2.6}$$

An elegant proof of this property has been given by Gilmore and Feng[5] and is based upon the existence of a lower bound ("P-representative") defined by the integral equation

$$\overline{H} = \frac{N+1}{4\pi} \int d\Omega_{\alpha_\mu} \, |N\alpha_\mu\rangle \, P_N(\overline{H},\alpha_\mu) \, \langle N\alpha_\mu| , \tag{2.7}$$

where $\int d\Omega_{\alpha_\mu}$ is the appropriate measure, such that

$$\min_{\{\alpha_\mu\}} \, [P_N(\overline{H},\alpha_\mu) - Q_N(\overline{H},\alpha_\mu)]/N \sim O(1/N). \tag{2.8}$$

Eq. (2.6) then follows from the observation that the exact binding energy is enclosed by P_N and Q_N.

2.3 Application to SU(6)

First we consider the three limiting cases of the IBA hamiltonian in which the eigenstates can be classified according to one of the following group chains[1,2,3]

A) $SU(6) \supset SU(5) \supset O(5) \supset O(3)$, (2.9a)
B) $SU(6) \supset O(6) \supset O(5) \supset O(3)$, (2.9b)
C) $SU(6) \supset SU(3) \supset O(3)$. (2.9c)

For the corresponding hamiltonian we take for the limit A:

$$H_A = \varepsilon\, d^\dagger . \tilde{d} \equiv \varepsilon n_d ; \tag{2.10a}$$

for B the pairing operator of O(6):

$$H_B = \kappa (d^\dagger . d^\dagger - s^\dagger s^\dagger)(\tilde{d}.\tilde{d} - ss) \equiv \kappa\, P_6 ; \tag{2.10b}$$

and for C the quadratic Casimir operator of SU(3) (up to a constant):

$$H_C = -\kappa' (2Q^{\pm}.Q^{\pm} + \frac{3}{4} L.L \;\; - 4N^2 - 6N) \equiv -\kappa'\, \overline{C}_3 , \tag{2.10c}$$

where $Q^{\pm}_\mu = d^\dagger_\mu s + s^\dagger \tilde{d}_\mu \pm \frac{\sqrt{7}}{2} (d^\dagger \tilde{d})^{(2)}_\mu$, and $L_\mu = \sqrt{10} (d^\dagger \tilde{d})^{(1)}_\mu$.

It is convenient to make use of the rotational invariance of the hamiltonian by seeking static solutions (α_μ real) of the form

$$|N\alpha_\mu\rangle = R(\Omega)\; \psi_N(a_\mu), \tag{2.11}$$

by rotating to the intrinsic frame: $\alpha_\mu = \sum_\nu D^2_{\mu\nu}(\Omega) a_\nu$, where $\Omega = \{\theta, \phi, \psi\}$ represent the three Euler angles. For the two independent intrinsic variables we take the conventional parametrization:

$$a_0 = \beta \cos \gamma,\; a_{-2} = a_2 = \beta/\sqrt{2} \sin \gamma,\; a_{-1} = a_1 = 0. \tag{2.12}$$

The actual calculation of the matrix elements (2.5) can be carried out in several ways; a straightforward method consists of using relations like

$$\langle N\alpha_\mu | d^\dagger_\mu s | N\alpha_\mu \rangle = \langle N\alpha_\mu | (1+|\alpha|^2)^{-N/2} \frac{\partial}{\partial \alpha_\mu} \; e^{\sum_{\mu'} \alpha_{\mu'} d^\dagger_{\mu'} s} | s^N \rangle = \frac{N\alpha_\mu}{1+\alpha^2} . \tag{2.13}$$

The resulting Q-representations are given by

$$Q_N(H_A,\beta,\gamma) = N\varepsilon \frac{\beta^2}{1+\beta^2} , \tag{2.14a}$$

$$Q_N(H_B,\beta,\gamma) = \kappa N(N-1)\left(\frac{1-\beta^2}{1-\beta^2}\right)^2 , \tag{2.14b}$$

$$Q_N(H_C,\beta,\gamma) = 4\kappa' N(N-1) \frac{1+\sqrt{2}\beta^3\cos 3\gamma+3/4\beta^4}{(1+\beta^2)^2} . \tag{2.14c}$$

The equilibrium shapes (β_0,γ_0) are obtained by minimizing $Q_N(H,\beta,\gamma)$ with respect to β and γ. One then finds that the SU(5) limit is characterized by $\beta_0 = 0$ and the O(6) limit by $\beta_0 = 1$, both independent of γ; the SU(3) limit has $\beta_0 = \sqrt{2}$ and $\gamma = 0^\circ$ or 60° (depending upon the choice of the sign of the one d-boson changing term in the quadrupole operator).

We can thus associate the following geometry[12] with the ground states of the three dynamical symmetries of the IBA: SU(5): spherical, O(6): γ-unstable rotor with $\beta_0 = 1$, SU(3): axially symmetric rigid rotor (prolate or oblate) with $\beta_0 = \sqrt{2}$, $\gamma_0 = 0^\circ, 60^\circ$. The corresponding wave functions $|N\beta_0\gamma_0\rangle$, which are eigenstates of the various hamiltonians for all N, can be interpreted as intrinsic states from which states with good angular momentum can be projected out.

The present results can be generalized in several ways:

(i) Instead of considering only the Casimir operators of the largest groups in the chains (2.9) one could examine the effect of adding additional terms such as the Casimir operators of the groups O(5) and O(3). However, it can easily be shown that the latter, which are expressed in terms of odd multipole operators $(d^\dagger\tilde{d})^{(\lambda)}.(d^\dagger\tilde{d})^{(\lambda)}$, $(\lambda=1,3)$ give a contribution to Q_N proportional to N, in contrast with the even operators (λ=2) which go like N^2, and therefore do not contribute in leading order.

(ii) While in the above analysis no stable triaxial shapes ($\gamma_0 \neq 0^\circ, 60^\circ$) were found one could inquire whether there exists a generalized hamiltonian written in terms of the SU(6) generators which has triaxial minima. An analysis similar to that carried out above shows that cubic terms of the type $\kappa_\lambda(d^\dagger d^\dagger d^\dagger)^{(\lambda)}.(\tilde{d}\tilde{d}\tilde{d})^{(\lambda)}$ give rise to a Q-representation containing terms of the type $\cos^2 3\gamma$, which are sufficient to produce a triaxial minimum. Since during this conference several pieces of experimental evidence for triaxial features (especially in the Pt isotopes) were presented[13] it seems interesting to investigate whether the addition of such cubic terms to the O(6) hamiltonian would improve the agreement with experiment.

(iii) Excited states and excitation energies can directly be constructed for the three limiting cases by acting on the ground state with the appropriate ladder operators. For example in the case of O(6) we find

$$\psi_N(\sigma) \sim (d^\dagger.d^\dagger - s^\dagger s^\dagger)^{(N-\sigma)/2}(s^\dagger + \cos\gamma\, d_0^\dagger + \frac{1}{\sqrt{2}}\sin\gamma(d_2^\dagger + d_{-2}^\dagger))^\sigma|0\rangle \tag{2.15}$$

and corresponding energies levels:

$$E_N(\sigma) = \kappa(N-\sigma)(N+\sigma+4) \ ,$$

where $\sigma = N, N-2, \ldots$.

3. PROPERTIES OF TRANSITIONAL REGIONS

We turn now to the description of the properties of transitional regions where there exists a competition between the three shape phases defined in section 2. To this end we consider the hamiltonian[12]

$$\overline{H} = H_A + H_B/(N-1) + H_C/(N-1) = \varepsilon n_d + \kappa/(N-1)P_6 - \kappa'/(N-1)\overline{C}_3 \tag{3.1}$$

where H_A, H_B and H_C are defined in Eq. (2.10).
For simplicity we consider only the competition between two extreme cases (represented by the legs of the triangle in Fig. 1) by taking in turn one of the parameters ε, κ, κ' in Eq. (3.1) equal to zero and varying the ratio of the remaining two (but with the restriction to physical values $\varepsilon, \kappa, \kappa' \geq 0$).

The Q-representative connected with Eq. (3.1) can easily be constructed. It is convenient to define the strength parameters $\eta = 4\kappa/\varepsilon$, $\eta' = 2\kappa'/\varepsilon$, and $\eta'' = \kappa/\kappa'$. In Fig. 2 the energy surfaces $Q_N(\overline{H},\beta,\gamma)$ are drawn schematically for the case $\gamma=0$ for several values of η, η' and η'' in each of the three transitional regions. They allow one to draw direct conclusions about the nature of the shape phase transition. For example in the case of SU(5) - O(6) (described by the variation of the strength parameter η) for values of $\eta \leq 1$ the minimum occurs at $\beta_0 = 0$ with corresponding ground state energy $E_N/N\varepsilon = \eta/2$; for $\eta > 1$ we find $\beta_0 = \sqrt{(1-\eta)/(1+\eta)}$ and $E_N/N\varepsilon=(\eta-1)/2\eta$. At the critical value $\eta = \eta_c = 1$ a phase transition occurs: the second derivative of E_N changes discontinuously. (Of course this picture describes reality only for $N \to \infty$; for finite N the behaviour of the true binding energy as a function of η is expected to be smoother.)

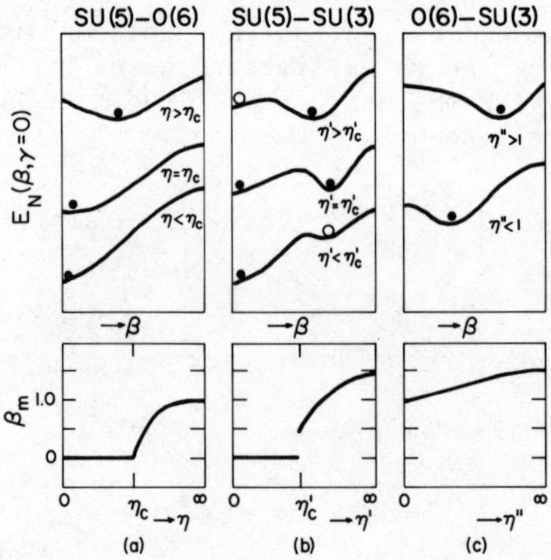

Fig. 2. Schematic representation of the classical energy functional $E_N(\beta,\gamma=0)$ (above) and the static value of β as a function of the strength parameters in the three transitional regions (below). Fig. 2a represents a second order shape phase transition (discontinuity in $\partial^2E_N/\partial\eta^2$), fig. 2b a first order transition (discontinuity in $\partial E_N/\partial\eta$), and fig. 2c a continuous transition. The full circles indicate an absolute minimum, and the open circles a local minimum.

In the case of the SU(5) → SU(3) transition (described by η') a different situation occurs: for values of $\eta' \approx \eta_c' = 8/9$ two competing minima occur, one spherical at $\beta_0 = 0$ and a deformed one for $\beta_0 \sim 1/9$. At $\eta' = \eta_c'$ the first derivative of the ground state energy changes discontinuously, giving rise (for $N \to \infty$) to a first order phase transition. Finally in the case of the O(6) → SU(3) transition the minimum in Q moves completely smoothly from $\beta_0 = 1$ to $\beta_0 = \sqrt{2}$ (and from γ-independent to $\gamma_0 = 0^o$ or 60°). In summary we find that the three transitional regions of SU(6) are described by three shape phase transitions of different nature.

It is tempting to compare these predictions (which for finite N should be considered as qualitative) with experimental information. Two regions have indeed been suggested as examples of SU(6) shape phase transitional regions, namely the Sm isotopes (as an example of an SU(5) - SU(3) transition) and the Pt/Os isotopes (as an example of an O(6) - SU(3) transition). Fig. 3 shows that the behaviour of the two-neutron separation energies, S_{2n}, is drastically different for these two regions. It is seen that the behaviour of S_{2n}, that can be estimated from the above results by using

$$S_{2n} = E_N - E_{N-1} = \frac{\partial E_N}{\partial \eta}\;\frac{\partial \eta}{\partial N}\,, \qquad (3.2)$$

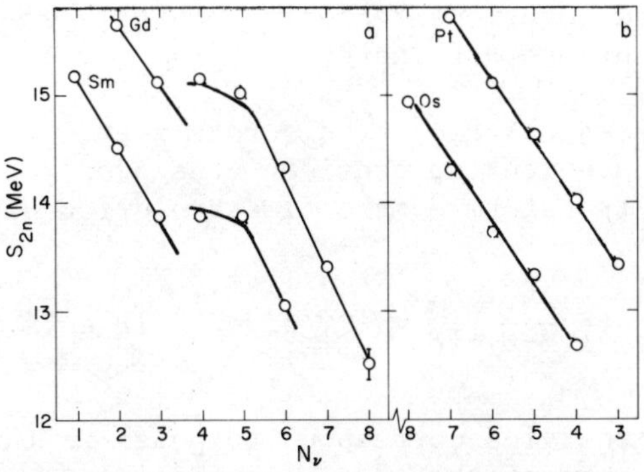

Fig. 3. The two-neutron separation energies, S_{2n}, in the SU(5)-SU(3) (left) and SU(3)-O(6) (right) transitional regions. The circles represent the experimental data and the full lines a fit based upon eq. (3.2).

and assuming $\frac{\partial\eta}{\partial N}$ to be constant, qualitatively agrees with the observed trend.

4. EXCITED STATES IN THE TRANSITIONAL REGIONS

The calculation of excitation energies in the transitional regions requires a generalization of the method presented above. For nuclear many-body systems the most general method for the description of excited states is the time dependent Hartree-Fock or mean field approach. In the present case in which the hamiltonian exhibits a dynamical symmetry it has been shown[14] that the mean field equations can be formulated rather conveniently in terms of the coherent state formalism. The starting point is the solution of the variational equation

$$\delta \int_{t_1}^{t_2} dt < \psi | i \frac{\partial}{\partial t} - H | \psi > = 0 \tag{4.1}$$

in the coherent state approximation for the trial wave functions

$$\psi = |N\alpha_\mu > \sim (s^\dagger + \Sigma\alpha_\mu d_\mu^\dagger)^N |0> . \tag{4.2}$$

The time dependent parameters α_μ are complex, i.e. $\alpha_\mu = |\alpha_\mu(t)| e^{i\phi_\mu(t)}$.

Substitution of eq. (4.2) in (4.1) leads to coupled equations in the variables α_μ, $\dot{\alpha}_\mu$, ϕ_μ and $\dot{\phi}_\mu$. Since the equations are conservative the solutions represent orbitals on the surface $\langle N\alpha_\mu|H|N\alpha_\mu\rangle$ with constant energy E.

For the calculation of discrete energies of excited states it is convenient to consider[7] the propagator G(E) which is the Fourier transform of the time evolution operator e^{iHT}:

$$G(E) = \mathrm{Tr}\,\frac{1}{E-H} = \sum_m \frac{1}{E-E_m+i\eta} = i\int_0^\infty dT\, e^{iET}\,\mathrm{Tr}\, e^{iHT} \ . \tag{4.3}$$

The excitation energies correspond to poles of the propagator G(E). The matrix elements of the time evolution operator can be calculated in the path integral formulation:

$$\langle\alpha''|e^{iHT}|\alpha'\rangle = \int_{\alpha'=\alpha(0)}^{\alpha''=\alpha(T)} D[\alpha]\ \exp\left[i\int_0^T dt\ \langle N,\alpha(t)|i\frac{\partial}{\partial t} - H|N,\alpha(t)\rangle\right] \tag{4.4}$$

By using the coherent state representation for α(t) we have

$$\mathrm{Tr}\,\hat{A} = \int d\Omega_\alpha\ \langle N,\alpha(t)|\hat{A}|N,\alpha(t)\rangle \ . \tag{4.5}$$

Successive application of the stationary phase approximation (SPA) to the three integrals which appear in eqs. (4.3-5) has the following result

(i) $\int D[\alpha] \rightarrow \delta\int dt\ \langle N,\alpha(t)|i\frac{\partial}{\partial t} - H|N,\alpha(t)\rangle = 0$, (4.6)

i.e. the mean field equations are recovered which can be interpreted as classical equations of motion;

(ii) $\int d\Omega_\alpha \rightarrow$ orbitals are periodic, i.e. $\alpha(0) = \alpha(T)$ and

$\dot{\alpha}(0) = \dot{\alpha}(T)$, (4.7)

(iii) $\int dT \rightarrow$ requantization rule for the energy spectrum:

$$\int dt\ \langle N,\alpha(t)|i\frac{\partial}{\partial t}|N,\alpha(t)\rangle = 2\pi n \ . \tag{4.8}$$

Actual calculations using this formalism have been restricted[7,15] up to now to the case of the SU(2) Lipkin hamiltonian[6]. As an illustration we consider here a slightly different case, namely the hamiltonian, that is obtained from the IBA SU(6) hamiltonian by replacing the d-boson by a second monopole boson r:

$$H = \varepsilon\, r^\dagger r + \kappa/(N-1)(r^\dagger r^\dagger - s^\dagger s^\dagger)(rr-ss) \ . \tag{4.9}$$

The solutions of the TDHF equation (4.6) must in general be obtained numerically. However, in the extreme cases $\kappa=0$ or $\varepsilon=0$ a closed solution can easily be found. Indeed it can be checked that in these cases the requantization rule (4.8) yields the exact energy spectra, i.e. $E_n = n\varepsilon(n=0,1,\ldots N)$, and $E_{\bar{n}} = 4\kappa\bar{n}(N-\bar{n})/(N-1)$ $(\bar{n}=0,1,\ldots\frac{1}{2}N)$, respectively.
Finally for $N\to\infty$ the intermediate situation can be solved[16] exactly. The resulting energy levels E_n are given by

$$E_n = \begin{cases} n\,\varepsilon\sqrt{1-\eta} & \text{if } \eta \leq 1, \\ \frac{1}{2} n\,\varepsilon\sqrt{\eta^2-1} & \text{if } \eta \geq 1,\ n \text{ even},\ n \ll N, \\ \frac{1}{2}(n-1)\,\varepsilon\sqrt{\eta^2-1} & \text{if } \eta \geq 1,\ n \text{ odd},\ n \ll N. \end{cases}$$

where $\eta = 4\kappa/\varepsilon$ is a dimensionless strength parameter. The energy levels are shown in fig. 4. Note that in this approximation the excitation energies at the critical point $\eta_c = 1$, where a second order phase transition occurs, are zero.

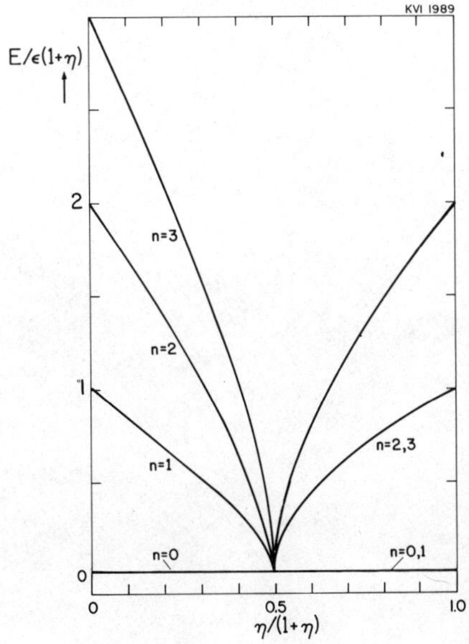

Fig. 4. Excitation energies calculated for the hamiltonian (4.9) in the limit $N\to\infty$, as a function of $\eta/(1+\eta)$, where $\eta=4\kappa/\varepsilon$.

REFERENCES

1. A. Arima and F. Iachello, Ann. Phys. 99 (1976) 253.
2. A. Arima and F. Iachello, Ann. Phys. 111 (1978) 201.
3. A. Arima and F. Iachello, Ann. Phys. 123 (1979) 468.
4. R. Gilmore, J. Math. Phys. 20 (1979) 891, and references therein.
5. R. Gilmore and D.H. Feng, Nucl. Phys. A301 (1978) 189; D.H. Feng, R. Gilmore and L.M. Narducci, Phys. Rev. C19 (1979) 1119.
6. H. Lipkin, N. Meshkov and A.J. Glick, Nucl. Phys. 62 (1965) 188.
7. S. Levit, J.W. Negele and Z. Paltiel, Phys. Rev. C21 (1980) 1603.
8. P. Hoodboy and J.W. Negele, Phys. Rev. C18 (1978) 2380.
9. K. K. Kan, P.C. Lichtner, M. Dworzecka and J.J. Griffin, Phys. Rev. C21 (1980) 1098.
10. R. Gilmore, C.M. Bowden and L.M. Narducci, Phys. Rev. A12 (1975) 1019.
11. J.N. Ginocchio and M.W. Kirson, Phys. Rev. Lett. 44 (1980) 1744.
12. A.E.L. Dieperink and O. Scholten, Nucl. Phys. A346 (1980) 125; A.E.L. Dieperink, O. Scholten and F. Iachello, Phys. Rev. Lett. 44 (1980) 1098.
13. D. Cline, contribution to this conference.
14. D.H. Feng and R. Gilmore, Phys. Lett. 90B (1980) 327.
15. H. Kuratsuji and T. Suzuki, J. Math. Phys. 21 (1980) 472.
16. O. van Roosmalen, private communication.

THE RELATIONSHIP BETWEEN THE BOHR COLLECTIVE MODEL AND THE INTERACTING BOSON MODEL

Joseph N. Ginocchio

Theoretical Division
Los Alamos Scientific Laboratory
University of California
Los Alamos, New Mexico 87545

1. INTRODUCTION

The Bohr-Mottelson collective model[1] has been very successful in describing the quadrupole collective degrees of freedom of heavy nuclei. As we have seen from this workshop, the interacting boson model (IBM) has also been successful in describing similar data.[2] At first glance these models seem very different. The former treats the nucleus as a rotating and vibrating liquid drop with quadrupole degrees of freedom. The latter describes the low-lying collective states of heavy nuclei by assuming that only the collective J = 0 (monopole) and J = 2 (quadrupole) pairs of valence fermions outside of closed shells play the dominant role in these states. The purpose of this talk is to make a connection between these two models. The important concept in making that connection is that of the intrinsic state for the IBM. Before discussing the intrinsic state, we give a short review of the Bohr-Mottelson model.

2. BRIEF REVIEW OF THE BOHR-MOTTELSON MODEL

The Bohr-Mottelson collective model assumes that the collective states of nuclei can be described by assuming that the nucleus behaves like a liquid drop with a quadrupole surface given by

$$R = R_o \left[1 + \overline{\beta} \left(\cos\gamma Y_0^{(2)}(\overline{\theta},\overline{\phi}) + \frac{\sin\gamma}{\sqrt{2}} \left(Y_2^{(2)}(\overline{\theta},\overline{\phi}) + Y_{-2}^{(2)}(\overline{\theta},\overline{\phi}) \right) \right) \right] \tag{2.1}$$

where the $Y^{(2)}_{\mu}$ are spherical harmonics[3], $\bar{\theta},\bar{\phi}$ are measured with respect to the body-fixed axes of the nucleus, $\bar{\beta}$ is the quadrupole deformation, and γ measures the asymmetry of the nucleus mass density about the body-fixed z axes. In the space-fixed system, there are three additional variables, the Euler angles $\Omega = (\phi,\theta,\psi)$, which specify the orientation of the body-fixed axes with respect to the space-fixed axes. The quadrupole variables in the space-fixed system are then given by

$$\bar{\alpha}_{\mu} = \bar{\beta}\left[\cos\gamma_{\mu}\mathcal{D}^{(2)}_{\mu 0}(\Omega) + \frac{\sin\gamma}{\sqrt{2}}\left(\mathcal{D}^{(2)}_{\mu 2}(\Omega) + \mathcal{D}^{(2)}_{\mu -2}(\Omega)\right)\right] \tag{2.2}$$

where $\mathcal{D}^{(2)}_{\mu\mu'}(\Omega)$ is the Wigner D-function.[3]

The wave function of the nucleus is obtained by solving a second-order differential equation in these quadrupole variables,

$$\mathcal{H}_{B} = \frac{-\hbar^{2}}{2\sqrt{B}}\Sigma_{\mu\nu}\frac{\partial}{\partial\bar{\alpha}_{\nu}}\sqrt{B}\,B^{-1}_{\mu\nu}(\bar{\alpha})\frac{\partial}{\partial\bar{\alpha}_{\nu}} + V(\bar{\alpha}) \tag{2.3}$$

where the first term is the generalized kinetic energy, the second term the potential energy, and $B_{\mu\nu}(\bar{\alpha})$ is the mass matrix which will depend on the quadrupole variables in general with

$$B(\bar{\alpha}) = \det B_{\mu\nu}(\bar{\alpha}). \tag{2.4}$$

In the harmonic limit, that is $\bar{\beta}$ small, the mass matrix becomes constant and diagonal

$$B_{\mu\nu}(\bar{\alpha}) \rightarrow B_{0}\delta_{\mu,\nu},\ \bar{\beta}\text{ small} \tag{2.5a}$$

and we have[4]

$$\mathcal{H}_{B} \rightarrow \frac{-\hbar^{2}}{2B_{0}}\left\{\frac{1}{\bar{\beta}^{4}}\frac{\partial}{\partial\bar{\beta}}\bar{\beta}^{4}\frac{\partial}{\partial\bar{\beta}} - \frac{1}{\bar{\beta}^{2}}C_{5}\right\} + V_{0}(\bar{\alpha}) \tag{2.5b}$$

where C_5 is the SO_5 Casimir operator[5]

$$C_{5} = \frac{-1}{\sin 3\gamma}\frac{\partial}{\partial\gamma}\sin 3\gamma\frac{\partial}{\partial\gamma} + \sum_{k=1}^{3}\frac{L_{k}^{2}}{2\sin^{2}\left(\gamma - \frac{2\pi k}{3}\right)} \tag{2.5c}$$

and L_k are the components of angular momenta which depend on the Euler angles. For a simple quadratic potential

$$V_0(\bar{\alpha}) \rightarrow \frac{1}{2}\,\omega^2 B_0 \bar{\beta}^2 \tag{2.6}$$

the harmonic Bohr Hamiltonian (2.5) will give the spectrum of a spherical quadrupole vibrator with equal-spaced levels which will have large degeneracies. These degeneracies are split by anharmonic terms in the potential.

Due to the totational invariance of the Bohr Hamiltonian, the rotational kinetic energy can always be separated,

$$\mathcal{H}_B = \mathcal{H}_I(\bar{\beta},\gamma) + \sum_k \frac{\hbar^2 L_k^2}{2\mathcal{I}_k(\bar{\beta},\gamma)} \tag{2.7}$$

where $\mathcal{H}_I(\bar{\beta},\gamma)$ is the intrinsic Hamiltonian which does not depend on the Euler angles and $\mathcal{I}_k(\bar{\beta},\gamma)$ are the moments of inertia which, in general, will depend on the shape variables $\bar{\beta},\gamma$.

If the intrinsic kinetic energy is much less than the rotational kinetic energy for some equilibrium values of $\bar{\beta} = \beta_0$ and $\gamma = 0$ and if $\mathcal{I}_1(\bar{\beta},0) = \mathcal{I}_2(\bar{\beta},0)$, then the nucleus has well-defined rotational bands. The excited bands will be built on intrinsic excitations; i.e., $\bar{\beta}$-bands and γ-bands. Hence for nuclei with β-small or with rotational energy large compared to the intrinsic energy, the Bohr-Mottelson model has simple solutions. However, most nuclei are in the middle of these extremes where there is strong mixing between rotational and intrinsic degrees of freedom. In that case, the Bohr Hamiltonian must be solved numerically to obtain the eigenenergies and eigenfunctions. This program has been carried out with remarkable success for a large range of nuclei.[6,7]

3. THE INTRINSIC STATE

In order to make a connection to the Bohr-Mottelson collective model of the nucleus, we define an intrinsic boson state in terms of intrinsic quadrupole variables[8,9] β,γ:

$$|N;\beta,\gamma\rangle = \frac{1}{\sqrt{N!}\,(1+\beta^2)^N}\,(B^\dagger)^N|0\rangle \tag{3.1a}$$

where

$$B^{\dagger} = s^{\dagger} + \beta \left[\cos\gamma d_0^{\dagger} + \frac{\sin \gamma}{\sqrt{2}} (d_2^{\dagger} + d_{-2}^{\dagger}) \right] \tag{3.1b}$$

with $s^{\dagger}$ the creation operator for a monopole boson (J = 0) and $d_{\mu}^{\dagger}$ the creation operator for a quadrupole boson (J = 2, projection quantum number μ), N is the number of pairs of valence nucleons, and $|0\rangle$ refers to the spherical core with the remaining A-2N nucleons.

We call this state an intrinsic state for two reasons. The first is that all the eigenstates of the most general IBM Hamiltonian discussed in this workshop can be projected out of this state by suitably averaging over the intrinsic quadrupole variables β and γ and the orientations of this intrinsic state specified by the Euler angles $\Omega = (\phi,\theta,\psi)$. That is, let $|N;\lambda\rangle$ be an eigenstate of the most general IBM Hamiltonian H which does not distinguish between neutrons and protons, so that

$$H|N;\lambda\rangle = E_{N\lambda}|N;\lambda\rangle, \tag{3.2}$$

then there exists a weighting function $f_{N\lambda}$, which is a function of the five quadrupole variables β, γ, ϕ, θ, ψ such that

$$|N,\lambda\rangle = \int d\Omega \int_0^{\infty} \beta^4 \, d\beta \int_0^{\frac{\pi}{3}} \sin 3\gamma d\gamma f_{N\lambda}(\beta,\gamma,\Omega)|N;\beta,\gamma,\Omega\rangle \tag{3.3}$$

where the intrinsic state (3.1) has been rotated through Euler angles Ω,

$$|N;\beta,\gamma,\Omega\rangle = e^{i\phi J_z} e^{i\theta J_y} e^{i\psi J_z}|N;\beta,\gamma\rangle . \tag{3.4}$$

I will not go into the details of the proof of this theorem in this talk.[9] The significance of this theorem is that the weighting functions $f_{N\lambda}$ give an alternative but complete description of the eigenstates of H in terms of the quadrupole variables. If for a series of states with different angular momenta, the f_{λ} are centered strongly around a particular value of $\beta = \beta_0$ and $\gamma = 0$, then the concept of a well-deformed rotational band makes sense, and the eigenstates in that band can be projected out from a single intrinsic state:

$$|N,\lambda\rangle \approx g_{N\lambda}(\beta_o) \int d\Omega \mathcal{D}_{M_{\lambda}K_{\lambda}}^{(J_{\lambda})}(\Omega)|N;\beta_o,0,\Omega\rangle . \tag{3.5}$$

The second reason for calling the state (2.1) an intrinsic state is that the fluctuations of the quadrupole field in this state are small, and in fact go to zero as N becomes large. The most general first-order quadrupole operator in IBM is given by

$$Q_\mu = s^\dagger \tilde{d}_\mu + d^\dagger s + \sqrt{\frac{7}{2}}\,\bar{\chi}(d^\dagger d)^{(2)}_\mu \tag{3.6}$$

where $\tilde{d}_\mu = (-1)^\mu d_{-\mu}$, $\bar{\chi}$ is a parameter, and $(d^\dagger \tilde{d})^{(2)}_\mu$ means that the quadrupole operators have been coupled to angular momentum two. The expectation value of the z-component of this quadrupole operator in the intrinsic state is

$$\langle N;\beta\gamma|Q_0|N;\beta\gamma\rangle = \frac{N}{1+\beta^2}\left(2\beta\cos\gamma - \bar{\chi}\beta^2\cos 2\gamma\right). \tag{3.7}$$

The mean variation about this average vanishes like $1/\sqrt{N}$ for any value of β. Furthermore, for equilibrium values of β which minimize the quadrupole-quadrupole interaction, these fluctuations fall off even faster,[9] more like $1/N\sqrt{N}$.

For Hamiltonians which distinguish between neutrons and protons, the intrinsic state will be a product of the neutron and proton intrinsic states with a set of intrinsic variables for neutrons and a set for protons.

4. THE ENERGY SURFACE

We define the energy surface as the expectation value of the boson Hamiltonian in the intrinsic state (3.1),

$$E_N(\beta,\gamma) = \langle N;\beta\gamma|H|N;\beta\gamma\rangle\ . \tag{4.1}$$

The minimum of this energy bounds the exact ground state energy from above and for N large is a close approximation to the ground state energy.[10] The deepness of the minima gives an indication as to how well-centered the nucleus is about the intrinsic state with the equilibrium values of $\beta = \beta_o$ and $\gamma = \gamma_o$ which minimize the energy. The behavior of this energy surface as a function of N can be analyzed to determine whether or not the nucleus goes through a phase transition,[11] and can also illustrate the change in shape of the nucleus.[8,9]

The most general IBM Hamiltonian which does not distinguish between neutrons and protons is given by

$$H = \varepsilon_s s^\dagger s + \varepsilon_d d^\dagger \cdot \tilde{d} + u_0 (s^\dagger)^2 s^2 + u_2 s^\dagger d^\dagger \cdot \tilde{d} s$$
$$+ v_0 \left[d^\dagger \cdot d^\dagger s^2 + (s^\dagger)^2 \tilde{d} \cdot \tilde{d} \right] + v_2 \left[(d^\dagger d^\dagger)^{(2)} \cdot \tilde{d} s + s^\dagger d^\dagger \cdot (\tilde{d}\tilde{d})^{(2)} \right]$$
$$+ \sum_{L=0,2,4} c_L (d^\dagger d^\dagger)^{(L)} \cdot (\tilde{d}\tilde{d})^{(L)} . \qquad (4.2)$$

We can easily calculate the expectation of this Hamiltonian using special properties of the intrinsic state.[9] Defining

$$\varepsilon = \varepsilon_d - \varepsilon_s \qquad (4.3)$$

the general energy surface is

$$E_N(\beta,\gamma) = N\varepsilon_s + \frac{N\beta^2}{1 + \beta^2}\varepsilon + \frac{N(N-1)}{\left(1 + \beta^2\right)^2} \left[u_0 + (u_2 + 2v_0)\beta^2 \right.$$
$$\left. - 2\sqrt{\frac{2}{7}}\, v_2 \beta^3 \cos 3\gamma + \left(\frac{1}{5} c_0 + \frac{2}{7} c_2 + \frac{18}{35} c_4 \right) \beta^4 \right] . \qquad (4.4)$$

This general energy surface has a very simple dependence on β and γ. The first two terms come from the single-boson terms in the Hamiltonian and are a ratio of polynomials at most quadratic in β and no dependence on γ. The remaining terms come from the interactions and are a ratio of polynomials at most quartic in β with only a cos 3γ dependence on γ. This last fact has the consequence that the minimum in the energy surface always occurs at $\gamma = \gamma_o = 0$ if $v_2 > 0$, and at $\gamma = \gamma_o = \pi/3$ if $v_2 < 0$. If $v_2 = 0$, the energy surface is independent of γ. Hence for IBM Hamiltonians which do not distinguish between neutrons and protons and which have at most interactions between pairs of bosons, the nuclei can be either prolate ($\gamma_o = 0$), oblate ($\gamma_o = \pi/3$), or gamma-unstable. No triaxial nuclei are possible. Triaxial minima can occur only for Hamiltonians with interactions between three or more bosons.

5. THE BOHR HAMILTONIAN

We can use the intrinsic state to find a second-order differential equation in the quadrupole variables which has eigenfunction corresponding to the IBM eigenstates with the exactly same spectrum. We define the wavefunction of the IBM eigenstates as the overlap of these eigenstates with the intrinsic state

$$\chi_{N\lambda}(\xi) = \langle N;\xi|N,\lambda\rangle \quad . \tag{5.1a}$$

where for simplicity we denote the set of coordinates β,γ,Ω by a single symbol ξ,

$$\xi = (\beta,\gamma,\Omega). \tag{5.1b}$$

If we multiply the eigenvalue equation (3.2) on the left side by the intrinsic state, we get the equation,

$$\langle N;\xi|H|N,\lambda\rangle = E_{N\lambda}\chi_{N\lambda}(\xi). \tag{5.2}$$

We can now use a special property of the intrinsic state to convert this eigenvalue equation into a differential equation for the eigenfunction $\chi_{N\lambda}$. Using the explicit form of the intrinsic state, the monopole boson creation operator on the intrinsic state will give,

$$s^{\dagger}|N-1;\xi\rangle = \left[\sqrt{\frac{N}{1+\beta^2}} - \sqrt{\frac{1+\beta^2}{N}}\,\beta\frac{\partial}{\partial\beta}\right]|N;\xi\rangle \quad . \tag{5.3}$$

Similar results hold for the quadrupole bosons.[9] Operating to the left onto the intrinsic state in (5.2), this eigenvalue equation becomes

$$\mathcal{H}(\xi)\chi_{N\lambda}(\xi) = E_{N\lambda}\chi_{N\lambda}(\xi) \tag{5.4}$$

where $\mathcal{H}(\xi)$ is a second-order partial differential operator which we refer to as a Bohr Hamiltonian.

For the most general IBM Hamiltonian given in (4.2), the corresponding Bohr Hamiltonian is found to be

$$\mathcal{H}(\xi) = T_N(\beta,\gamma) + V_N(\beta,\gamma) + \sum_k \frac{\hbar^2 L_k^2}{2I_k(\beta,\gamma)} \tag{5.5}$$

which is of the form (2.7) with the intrinsic Hamiltonian separated from the rotational energy. The intrinsic kinetic energy $T_N(\beta,\gamma)$ and potential energy $V_N(\beta,\gamma)$ are rather complicated, and we do not write them out in detail here.[9] The potential energy has a similar form as the energy surface (4.4) and is exactly equal to it in the leading order in N,

$$V_N(\beta,\gamma) \xrightarrow{N\to\infty} E_N(\beta,\gamma), \tag{5.6}$$

and hence the minimum of the potential energy is also a good approximation to the ground-state energy for N large. The potential energy like the energy surface is also finite as β becomes large,

$$V_N(\beta,\gamma) \xrightarrow{\beta \to \infty} N(N-2)\left(\frac{1}{5}C_0 + \frac{2}{7}C_2 + \frac{18}{35}C_4\right) + N(V_o + C_o). \quad (5.7)$$

This finiteness has the direct consequence that there are only a finite number of eigenfunctions of the Hamiltonian (5.5) which are analytic functions in the quadrupole variables. These finite number of eigenfunctions are just the eigenfunctions which have the same eigenspectrum as the IBM eigenstates. In Figure 1, we plot this potential energy as a function of β for $\gamma = 0$ for the SU_3 limit[12] of the interacting boson model. We see that this potential has a deep minimum for $\beta \approx \sqrt{2}$ and is finite for all values of β.

The moments of inertia depend on β and γ in general and are given by

$$\frac{\hbar^2}{2I_k} = -\frac{1}{7}(C_2 - C_4) - \left[v_0 + \left(\frac{1}{5}C_0 - \frac{2}{7}C_2 + \frac{3}{35}C_4\right)\beta^2 + \sqrt{\frac{2}{7}}\, v_2\beta\cos(\gamma - 2\pi k/3)\right] \Big/ \left[4\beta^2 \sin^2(\gamma - 2\pi k/3)\right]. \quad (5.8)$$

The Hamiltonian in (5.5) will not necessarily be self-adjoint, although it has a finite real eigenspectrum. However, for N large, it is self-adjoint and, as we shall see, is self-adjoint in interesting cases for all N.

The limiting cases in which the IBM Hamiltonian has an SO_6 symmetry[13] and an SU_3 symmetry[12] are of particular interest. For the SO_6 symmetry, the only parameter is the strength $-\kappa$ of the quadrupole-quadrupole interaction, and the only non-vanishing parameters in (4.2) are

$$\varepsilon_s = -5\kappa, \ \varepsilon_d = -\kappa, \ u_2 = -2\kappa, \ v_o = -\kappa. \quad (5.9)$$

The Hamiltonian in this case is self-adjoint and is given by

$$\mathcal{H}^{(6)} = \frac{1}{W(\beta)}\frac{\partial}{\partial\beta}B(\beta)\frac{\partial}{\partial\beta} + V(\beta) + \frac{\kappa C_5}{\beta^2} \quad (5.10a)$$

where

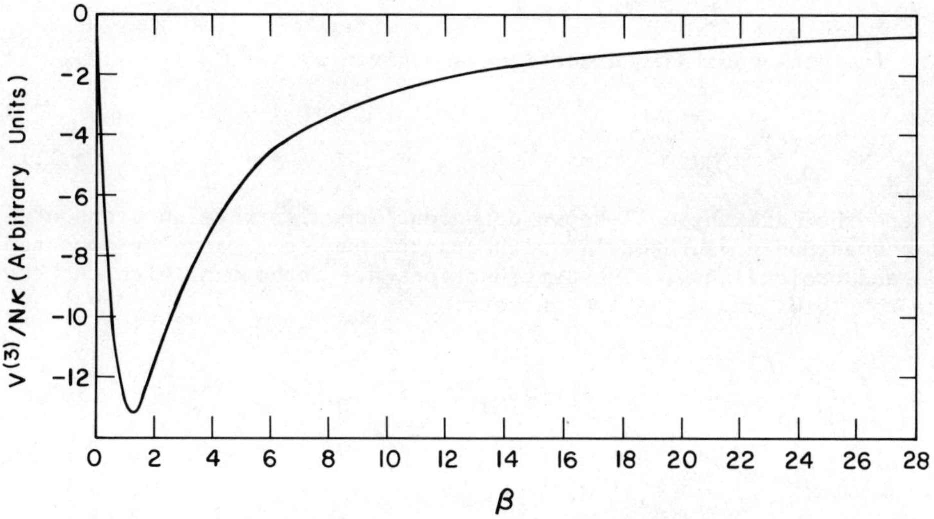

Fig. 1. The potential energy $V^{(3)}(\beta, \gamma = 0)$ in arbitrary units for the SU_3 limit with the number of pairs N = 10 versus the deformation β. The potential energy asymptotes to zero as β becomes large.

$$W(\beta) = \frac{\beta^4}{\left(1 - \beta^2\right)^3} \left[\frac{1 + \beta^2}{1 - \beta^2}\right]^N \tag{5.10b}$$

$$B(\beta) = -\kappa\left(1 - \beta^2\right)^2 W(\beta), \tag{5.10c}$$

and

$$V(\beta) = -2N\kappa \left[\frac{2N\beta^2 + 4}{\left(1 + \beta^2\right)^2} + 1\right] \cdot \tag{5.10d}$$

The orthogonal eigenfunctions are given by

$$\chi^{(6)}_{N\sigma\tau JM}(\xi) = \Phi_{N\sigma\tau}(\beta)\Psi_{\tau JM}(\gamma,\Omega) \tag{5.11}$$

where the Ψ are the well-known orthogonal angular wavefunctions of the quadrupole vibrator[5,14] with quantum number τ, angular momentum J, and projection M. The wavefunctions are orthogonal with respect to the τ, J, and M quantum numbers,

$$\int d\Omega \int_0^{\pi/3} d\gamma \ \sin\ 3\gamma \Psi^*_{\tau' J' M'}(\gamma,\Omega)\ \Psi_{\tau JM}(\gamma,\Omega)$$

$$= \delta_{\tau,\tau'}\delta_{J,J'}\delta_{M,M'} \cdot \tag{5.12}$$

The β dependence of the eigenfunctions[15] is

$$\Phi_{N\sigma\tau}(\beta) \propto \beta^\tau \frac{\left(1 - \beta^2\right)^{(N-\sigma)/2}}{\left(1 + \beta^2\right)^{N/2}} \sum_{P=0}^{\frac{1}{2}(\sigma-\tau)} A_P(\sigma\tau) \left(1 - \beta^2\right)^P \tag{5.13a}$$

where

$$A_P(\sigma\tau) = \left(\frac{-1}{4}\right)^P \frac{(\sigma + 1 - P)!}{(\sigma - \tau - 2P)!P!} \tag{5.13b}$$

and these functions are orthogonal in the σ quantum number

$$\int_0^\infty d\beta W(\beta)\Phi_{N\sigma'\tau}(\beta)\Phi_{N\sigma\tau}(\beta) \propto \delta_{\sigma,\sigma'} \cdot \tag{5.13c}$$

The allowed values of σ are $\sigma = N, N-2, \ldots$, and of τ are $\tau = \sigma, \sigma - 1, \ldots, 0$. The eigenenergies are

$$E^{(6)}_{N\sigma\tau} = -\kappa\Big(\sigma(\sigma+4) - \tau(\tau+3)\Big) . \tag{5.14}$$

For N large, the potential energy in (5.10d) has a minimum at $\beta \simeq 1$ and for the lowest band, $\sigma = N$, the eigenfunction becomes centered around $\beta = 1$,

$$\chi^{(6)}_{NN\tau JM}(\xi) \sim e^{-\frac{N}{2}(\beta-1)^2}\, \Psi_{\tau JM}(\gamma,\Omega), \text{ N large.} \tag{5.15}$$

We note also from (5.10) that the kinetic energy in the β-direction is exactly zero at $\beta = 1$, and hence this SO_6 limit corresponds to a well-developed γ unstable rotor.

In the SU_3 limit, quadrupole interactions still dominate. The potential energy in this case will depend linearly on cos 3γ, but the minimum will occur at $\gamma = 0$ and, for N large, $\beta \sim \sqrt{2}$. The ground-state band of states will be centered about these equilibrium values,

$$\chi^{(3)}_{NJM}(\xi) \sim e^{-\frac{N}{2}(\beta-\sqrt{2})^2}\, e^{-\frac{N}{2}\gamma^2}\, \mathcal{D}^{(J)}_{M0}(\Omega), \text{ N large.} \tag{5.16}$$

Also the kinetic energy in both the β and γ-direction vanishes at the equilibrium values,

$$T^{(3)}_N\ (\beta = \sqrt{2}, \gamma = 0) = 0, \tag{5.17}$$

and hence the intrinsic and rotational motion are completely decoupled and the energy is completely rotational energy

$$E^{(3)}_{NJM} = -\kappa\left(N(2N+3) - \frac{3}{8}J(J+1)\right) . \tag{5.18}$$

Also at these equilibrium values, the moments of inertia (5.8) about the x and y-directions are equal

$$I^{(3)}_1\ (\beta = \sqrt{2}, \gamma = 0) = I^{(3)}_2\,(\beta = \sqrt{2}, \gamma = 0) \tag{5.19}$$

and hence the z-axes is a symmetry axes. Furthermore, the inverse of the moment of inertia about this axis is zero

$$I_3^{(3)^{-1}} (\beta = \sqrt{2}, \gamma = 0) = 0 \tag{5.20}$$

and hence there is no rotational energy about the symmetry axes. In both the SO_6 limit and the SU_3 limit, the intrinsic state with $\gamma = 0$ and $\beta = 1$ and $\beta = \sqrt{2}$ respectively are eigenstates of the quadrupole operator and hence the fluctuations in the quadrupole field are exactly zero.

Hence we see that the SU_3 limit produces an example of the quintessential rotor!

6. SUMMARY

With the intrinsic state (3.1), we are able to make the connection with the IBM and the Bohr-Mottelson collective model. For any IBM Hamiltonian (4.2), we are able to derive a second-order differential operator in the quadrupole variables which has the same eigenspectrum as the original boson Hamiltonian. In (5.8), we have given the expression for the moments of inertia of the most general IBM Hamiltonian. We have also shown that in the SO_6 and SU_3 limits, the resulting Bohr Hamiltonian for a large number of valence nucleons produces well-developed rotors, a γ-unstable rotor for SO_6 and an axially symmetric rotor for SU_3. However, the potential energy is finite, and hence the bands do terminate which follows from the finite degrees of freedom in the IBM.

7. ACKNOWLEDGMENT

The bulk of this work was done in collaboration with Michael Kirson.

REFERENCES

*Work supported by the U. S. Department of Energy.

1. A. Bohr and B. R. Mottelson, "Nuclear Structure, Vol. II," W. A. Benjamin, Reading, Massachusetts (1975).
2. F. Iachello, ed., "Interacting Bosons in Nuclear Physics," Plenum Press, New York (1979).
3. A. R. Edmonds, "Angular Momentum in Quantum Mechanics," Princeton University Press, Princeton (1960).
4. A. Bohr, Dan. Mat. Fys. Medd. 26, no. 14, 1 (1952).
5. G. Rakavy, Nucl. Phys. 4, 4 (1957).

6. K. Kumar and M. Baranger, Nucl. Phys. A92, 608 (1967).
7. K. Kumar, J. Phys. G4, 849 (1978).
8. J. N. Ginocchio and M. W. Kirson, Phys. Rev. Lett. 44, 1744 (1980).
9. J. N. Ginocchio and M. W. Kirson, to be published in Nuclear Physics A (1980).
10. R. Gilmore and D. H. Feng, Nucl. Phys. A301, 189 (1978).
11. A. Dieperink, O. Scholten , and F. Iachello, Phys. Rev. Lett. 44, 1747 (1980).
12. A. Arima and F. Iachello, Ann. Phys. 111, 201 (1976).
13. A. Arima and F. Iachello, Phys. Rev. Lett. 40, 40 (1978).
14. E. Chacón, M. Moshinsky, and R. T. Sharp, J. Math. Phys. 17, 668 (1976).

A GENERAL FORMULA FOR THE CLASSICAL LIMIT OF THE INTERACTING BOSON HAMILTONIAN

Jin-Quan Chen¶ and Pieter Van Isacker†

Physics Department

Yale University, New Haven, CT 06511

Recently, Dieperink, Scholten and Iachello[1] studied the shape and the shape phase transitions in the Interacting Boson Model[2-4] (IBM) by using the algorithm developed by Gilmore and Feng[5]. They showed that each of the three limiting cases of the model corresponds to a certain shape phase. Each of the three transitional regions between the extreme cases are characterized by ground state shape phase transitions. However, in constructing the classical limit they used somewhat simplified versions of the three limiting Hamiltonians originally employed by Arima and Iachello[2-4]. In this note we use a simple method - the differentiation method - to calculate the expectation value of the IBM Hamiltonian in the coherent state. The classical limits of the most general IBM Hamiltonian of SU(6) as well as those of its three limiting cases used in refs. 2-4 are given.

Let $b_i^\dagger$ (b_i) be boson creation (annihilation) operators and $f(b)$ be a polynomial of $b_1^\dagger$, $b_2^\dagger$,... and b_1, b_2,... .It is well known and can be proved readily that the boson commutators can be evaluated by means of differentiation i.e.

$$[b_i, f(b)] = \frac{\partial}{\partial b_i^\dagger} f(b) \quad , \tag{1a}$$

¶ Permanent address: Physics Department, Nanjing University, Nanjing, People 's Republic of China.

† Permanent address: Laboratorium voor Kernfysika, Proeftuinstraat 86, B-9000 Gent, Belgium.

$$[f(b),b_i^\dagger] = \frac{\partial}{\partial b_i} f(b) \quad . \tag{1b}$$

Suppose the operator $\mathcal{B}$ is a linear combination of various kinds of bosons

$$\mathcal{B} = \sum_i \alpha_i b_i \quad . \tag{2}$$

Let 1, A_1, A_2 be the identity, one-body and two-body operators, respectively

$$A_1 = \sum_{ij} \xi_{ij}\, b_i^\dagger\, b_j \quad , \tag{3}$$

$$A_2 = \sum_{ijk\ell} \eta_{ijk\ell}\, b_i^\dagger\, b_j^\dagger\, b_k\, b_\ell \quad . \tag{4}$$

According to eq.(1), the matrix elements of these operators between the n-boson states $|\mathcal{B}^{\dagger n}> \equiv \mathcal{B}^{\dagger n}|0>$ (the so-called coherent states), can be written down directly:

$$(i)\ N_n \equiv <\mathcal{B}^n|1|\mathcal{B}^{\dagger n}> = <\mathcal{B}^{n-1}|\sum_i \alpha_i \frac{\partial}{\partial b_i^\dagger}|\mathcal{B}^{\dagger n}> = n\alpha^2 <\mathcal{B}^{n-1}|\mathcal{B}^{\dagger n-1}>,$$

which implies

$$N_n = n!\,\alpha^{2n} \quad , \qquad \alpha^2 = \sum_i \alpha_i^2 \quad . \tag{5}$$

$$(ii)\ <A_1> \equiv N_n^{-1} <\mathcal{B}^n|A_1|\mathcal{B}^{\dagger n}> = N_n^{-1} \sum_{ij} \xi_{ij} <\frac{\partial}{\partial b_i} \mathcal{B}^n|\frac{\partial}{\partial b_j^\dagger} \mathcal{B}^{\dagger n}>$$

$$= N_n^{-1} \sum_{ij} n^2 \alpha_i \alpha_j \xi_{ij} N_{n-1} = \frac{n}{\alpha^2} \sum_{ij} \alpha_i \alpha_j \xi_{ij} \quad . \tag{6}$$

$$(iii)\ <A_2> = \frac{n(n-1)}{\alpha^4} \sum_{ijk\ell} \alpha_i \alpha_j \alpha_k \alpha_\ell \eta_{ijk\ell} \quad . \tag{7}$$

We now apply the above results to the Interacting Boson Model. Here, eq.(2) becomes

$$\mathcal{B} = s + \sum_m \beta_m d_m \quad . \tag{8}$$

The real coefficients β_μ satisfy

$$\beta_\mu^* = (-1)^\mu \beta_{-\mu} = \beta_\mu \quad . \tag{9}$$

In the intrinsic frame we have

$$\beta_0 = \beta\cos\gamma \quad , \quad \beta_2 = \beta_{-2} = \frac{1}{\sqrt{2}}\,\beta\sin\gamma \quad , \quad \beta_{\pm1} = 0 \quad . \tag{10}$$

We first give the expectation values for some typical operators.

1. $$\langle n_s \rangle = \frac{n}{1+\beta^2} \quad . \tag{11}$$

2. $$\langle n_d \rangle = \frac{n\beta^2}{1+\beta^2} \quad . \tag{12}$$

3. $$\langle n_d^2 \rangle = n(n-1)\,\frac{\beta^4}{(1+\beta^2)^2} + \frac{n\beta^2}{1+\beta^2} \quad . \tag{13}$$

4. $$\langle (d^\dagger d^\dagger)^{(\ell)} \cdot (\tilde{d}\,\tilde{d}\,)^{(\ell)} \rangle = \frac{n(n-1)}{(1+\beta^2)^2}\,\mathcal{D}_\ell \,, \tag{14}$$

$$\mathcal{D}_\ell = \sum_m \Big(\sum_{m_1\, m_2} \langle 2m_1\; 2m_2 | \ell m \rangle\, \beta_{m_1}\, \beta_{m_2} \Big)^2 \quad .$$

Using eq.(10), one has

$$\mathcal{D}_0 = \frac{1}{5}\,\beta^4 \quad , \quad \mathcal{D}_2 = \frac{2}{7}\,\beta^4 \quad , \quad \mathcal{D}_4 = \frac{18}{35}\,\beta^4 \quad ,$$

$$\mathcal{D}_1 = \mathcal{D}_3 = 0 \quad . \tag{15}$$

5. $$\langle (d^\dagger \tilde{d}\,)^{(\ell)} \cdot (d^\dagger \tilde{d}\,)^{(\ell)} \rangle = \frac{2\ell+1}{5}\,\frac{n\beta^2}{1+\beta^2} + \frac{n(n-1)}{(1+\beta^2)^2}\,\mathcal{D}_\ell \quad . \tag{16}$$

6. $$\langle Q\cdot Q \rangle = N_n^{-1} \sum_m \langle B^n | (d_m^\dagger s + (-1)^m s^\dagger d_{-m} + \chi \sum_{m_1 m_2} \langle 2m_1\; 2m_2 | 2m \rangle d_{m_1}^\dagger (-1)^{m_2} d_{-m_2})$$
$$(s^\dagger d_m + (-1)^m d_{-m}^\dagger s + \chi \sum_{m_1' m_2'} \langle 2m_1'\; 2m_2' | 2m \rangle (-1)^{m_2'} d_{-m_2'}^\dagger d_{m_1'}) | B^{\dagger n} \rangle$$

$$= N_n^{-1} \sum_m n^2 \langle B^{n-1} | (\beta_m s + (-1)^m d_{-m} + \chi \sum_{m_1 m_2} \langle 2m_1\; 2m_2 | 2m \rangle \beta_{m_1} (-1)^{m_2} d_{-m_2})$$
$$(\beta_m s^\dagger + (-1)^m d_{-m}^\dagger + \chi \sum_{m_1' m_2'} \langle 2m_1'\; 2m_2' | 2m \rangle \beta_{m_1'} (-1)^{m_2'} d_{-m_2'}^\dagger) | B^{\dagger n-1} \rangle$$

$$= N_n^{-1} \sum_m n^2 \langle B^{n-1} | (\beta_m \frac{\partial}{\partial s^\dagger} + (-1)^m \frac{\partial}{\partial d_{-m}^\dagger} +$$
$$+ \chi \sum_{m_1 m_2} \langle 2m_1\; 2m_2 | 2m \rangle \beta_{m_1} (-1)^{m_2} \frac{\partial}{\partial d_{-m_2}^\dagger})$$

$$(\beta_m s^\dagger+(-1)^m d^\dagger_{-m}+\chi \sum_{m_1'm_2'} \langle 2m_1'\ 2m_2'|2m\rangle \beta_{m_1'}(-1)^{m_2'} d^\dagger_{-m_2'})|B^{\dagger n-1}\rangle$$

$$= \frac{n}{1+\beta^2}\left(5+(1+\chi^2)\beta^2\right) + \frac{n(n-1)}{(1+\beta^2)^2}\left(\frac{2}{7}\chi^2\ \beta^4-4\chi\frac{\sqrt{2}}{\sqrt{7}}\beta^3\cos 3\gamma+4\beta^2\right). \quad (17)$$

7. $$\langle S_+\ S_-\rangle = n(n-1)\ \frac{\beta^4}{(1+\beta^2)^2}\ , \quad (18)$$

with $S_+ = \sum_m (-1)^m\ d^\dagger_m\ d^\dagger_{-m}$.

8. $$\langle S'_+\ S'_-\rangle = n(n-1)\left[\frac{\xi+\eta\beta^2}{1+\beta^2}\right]^2\ , \quad (19)$$

with $S'_+ = \xi s^\dagger s^\dagger + \eta \sum_m (-1)^m\ d^\dagger_m\ d^\dagger_{-m}$.

9. $$\langle L\cdot L\rangle = -10\sqrt{3}\ \langle[(d^\dagger\tilde{d}\ )^{(1)}\ (d^\dagger\tilde{d}\ )^{(1)}]^{(0)}\rangle = \frac{6n\beta^2}{1+\beta^2}\ . \quad (20)$$

For the most general Hamiltonian¶

$$\begin{aligned} H = {} & \varepsilon_s n_s + \varepsilon_d n_d + \sum_\ell c_\ell\ (2\ell+1)^{1/2}\ [(d^\dagger d^\dagger)^{(\ell)}\ (\tilde{d}\ \tilde{d}\ )^{(\ell)}]^{(0)} \\ & + v_2[(d^\dagger d^\dagger)^{(2)}\ (\tilde{d}\ s\ )^{(2)} + \text{h.c.}]^{(0)} \\ & + v_0[(d^\dagger d^\dagger)^{(0)}\ (s\ s\ )^{(0)} + \text{h.c.}]^{(0)} \\ & + u_2[(d^\dagger s^\dagger)^{(2)}\ (\tilde{d}\ s\ )^{(2)}]^{(0)} + u_0[(s^\dagger s^\dagger)(s\ s\ )]\ , \end{aligned} \quad (21)$$

one has

$$\langle H\rangle = \frac{n}{1+\beta^2}\ (\varepsilon_s+\varepsilon_d\beta^2) + \frac{n(n-1)}{(1+\beta^2)^2}\ (a_1\beta^4+a_2\beta^3\cos 3\gamma+a_3\beta^2+u_0)\ , \quad (22a)$$

with†

¶ The parameters we used are slightly different from those in ref. 2.

† Remark that in the general Hamiltonian of eq.(21) there are 9 parameters, while in the expression of its classical limit of eq.(22a), there are only six independent parameters.

$$a_1 = \frac{1}{5}c_0 + \frac{2}{7}c_2 + \frac{18}{35}c_4 \quad ,$$

$$a_2 = -\sqrt{\frac{8}{35}}\, \upsilon_2 \ , \tag{22b}$$

$$a_3 = \frac{2}{\sqrt{5}}\upsilon_0 + \frac{1}{\sqrt{5}}u_2 \ .$$

By equating the parameters in eq. (22) to the special values corresponding to the three limiting cases of the IBM, we can obtain the three classical limits. However, it is more straightforward to use the group chain reduction in the three limiting cases. In this way, using the expressions (11)-(20) for the different Casimir operators, one obtains:

(I) SU(6) → SU(5) → SO(5) → SO(3)

$$H^{(I)} = \varepsilon_d n_d + \frac{a}{2}n_d(n_d-1) + bS_+S_- + c(L\cdot L-6n_d) \quad . \tag{23a}$$

$$\langle H^{(I)}\rangle \equiv E^{(I)}(\beta,\gamma) = \varepsilon_d \frac{n\beta^2}{1+\beta^2} + \kappa \frac{n(n-1)}{(1+\beta^2)^2}\beta^4 \quad , \tag{23b}$$

with $\kappa = \frac{a}{2} + b$.

(II) SU(6) → SU(3) → SO(3)

$$H^{(II)} = -\kappa_1 Q\cdot Q - \kappa_2 L\cdot L \quad . \tag{24a}$$

Putting $\chi = -\frac{\sqrt{7}}{2}$ in eq.(17), one obtains from eqs.(17) and (20)

$$E^{(II)}(\beta,\gamma) = -\kappa_1\left[\frac{n}{1+\beta^2}\left(5+\frac{11}{4}\beta^2\right)+\frac{n(n-1)}{(1+\beta^2)^2}\left(\frac{\beta^4}{2}+2\sqrt{2}\beta^3\cos3\gamma+4\beta^2\right)\right]$$
$$-\kappa_2\frac{6n\beta^2}{1+\beta^2} \quad . \tag{24b}$$

(III) SU(6) → SO(6) → SO(5) → SO(3)

$$H^{(III)} = AS_+S_- + \frac{B}{3}C_5 + \frac{C}{10}L\cdot L \quad , \tag{25a}$$

where

$$S_+ = \frac{1}{2}\sum_m (-1)^m d_m^\dagger d_{-m}^\dagger - \frac{1}{2}s^\dagger s^\dagger \quad ,$$

$$C_5 = -\sqrt{7}[(d^\dagger\tilde{d})^{(3)}(d^\dagger\tilde{d})^{(3)}]^{(0)} - \sqrt{3}[(d^\dagger\tilde{d})^{(1)}(d^\dagger\tilde{d})^{(1)}]^{(0)}.$$

From eqs.(15),(16),(19) and (20) one has

$$E^{(III)}(\beta,\gamma) = \kappa_3 \frac{n\beta^2}{1+\beta^2} + \kappa_4 n(n-1)\left(\frac{1-\beta^2}{1+\beta^2}\right)^2 , \tag{25b}$$

with

$$\kappa_3 = \frac{2}{3}B + \frac{3}{5}C \quad , \quad \kappa_4 = \frac{A}{4} \quad .$$

From eqs.(23b),(24b) and (25b) we know that there are two independent parameters for each limiting case and that $E^{(III)}$ as well as $E^{(I)}$ are γ-independent. For the limiting case (I) the equilibrium shapes of the nuclei are always spherical ($\beta = 0$). For the limiting cases (II) and (III), there are two competing factors: one favours spherical form and another favours deformation.

We will give an illustration of this phenomenon in both the limiting cases (II) and (III). In the O(6) limit we obtain, by minimizing $E^{(III)}(\beta,\gamma)$ with respect to β, a minimum at $\beta = 0$ when $\kappa_3 > 4(n-1)\kappa_4$ or a minimum at $\beta = \left(\frac{4\kappa_4(n-1)-\kappa_3}{4\kappa_4(n-1)+\kappa_3}\right)^{1/2}$ when $\kappa_3 < 4(n-1)\kappa_4$. We can illustrate this further in the case of the nucleus $^{196}_{78}Pt_{118}$, for which the empirical parameters are[4] A = 171 keV, B = 300 keV and C = 100 keV. Thus, $\kappa_3 = 266$ keV and $\kappa_4 \simeq 43$ keV, which shows that the equilibrium shape of $^{196}Pt_{118}$ is deformed (β=.7263).

In the SU(3) limit, the equilibrium shape can be found by minimizing $E^{(II)}(\beta,\gamma)$ of eq.(24b) with respect to β and γ. First of all, we remark that $E^{(II)}(\beta,\gamma)$ is invariant under the substitutions $(\beta,\gamma) \to (\beta,-\gamma)$, $(\beta,\gamma) \to (\beta,\gamma+120°)$ and $(\beta,\gamma) \to (-\beta,\gamma+60°)$ as should be,and consequently, we may restrict ourselves to the region $\beta > 0$ and $0° \leq \gamma \leq 60°$. From eq.(24b) it can be seen that for positive values of κ_1, $E^{(II)}(\beta,\gamma)$ reaches a minimum for $\gamma = 0°$. Furthermore, it can be shown that the extremum values of $E^{(II)}(\beta,\gamma=0°)$ with respect to β, are given by $\beta = 0$ and the roots of the equation

$$\beta^3 - D_n\beta^2 - 3\beta - D_n\frac{7}{\sqrt{2}} = 0 \quad , \tag{26}$$

where

$$D_n = \frac{1}{(n-1)\sqrt{2}}(6\kappa - 3n + 3/4) \quad , \quad \kappa = \kappa_2/\kappa_1 \quad .$$

Again we illustrate this in the case of the nucleus $^{156}Gd_{92}$, where $\kappa_1 = 7.6$ keV and $\kappa_2 = -7.7$ keV^3. With these values, we obtain one minimum at $\beta = 1.2426$, only slightly different from the value $\beta = \sqrt{2}$, quoted in ref. 1.

ACKNOWLEDGEMENT

We wish to thank Prof. F. Iachello for stimulating our interest in this subject and for interesting discussions. One of the authors (PVI) acknowledges the NFWO for financial support.

REFERENCES

1. A.E.L. Dieperink, O.Scholten and F.Iachello, Classical Limit of the Interacting Boson Model, to be published
2. A.Arima and F.Iachello, Ann. Phys. 99 (1976) 253
3. A.Arima and F.Iachello, Ann. Phys. 111 (1978) 201
4. A.Arima and F.Iachello, Ann. Phys. 123 (1979) 468
5. R.Gilmore and D.H.Feng, Phys. Lett 76B (1978) 26; Nucl. Phys. A301 (1978) 189

ARE BOSONS NUCLEON PAIRS?

K. Allaart

Natuurkundig Laboratorium
Vrije Universiteit
1081 HV Amsterdam

Since it has become clear that the Interacting Boson Model can successfully describe collective states in nuclei, especially of quadrupole type, the question has been raised whether the s- and d-bosons can simply be understood as pairs of nucleons coupled to angular momentum $J=0$ (the S-pair) and $J=2$ (the D-pair)[1-5].

A possible suggestion is that the correlated S-pair corresponds to the Cooper pair in BCS theory of the nucleus. It is a superposition of pair states in the many valence shells with quantum numbers $(n_a \ell_a j_a m_a) \equiv \alpha$

$$S^\dagger = \sum_{n_a \ell_a j_a} \varphi_a \frac{1}{2} \hat{a} \sum_{m_a} (j_a m_a j_a - m_a/00) \, a^\dagger_\alpha a^\dagger_{\bar{\alpha}} \tag{1}$$

with $\bar{\alpha} \equiv (n_a \ell_a j_a, -m_a)$, and φ_a is a coefficient. If many orbits occur in (1) with roughly equal coefficients, one expects that the precise microscopic structure of (1) does not show up in a pronounced way and the pair can be considered as an s-boson. Similarly one can argue that the D-pair, created by

$$D^\dagger_\mu = \sum_{(ab)} \chi_{ab} \sum_{m_a m_b} (j_a m_a j_b m_b/2\mu) \, a^\dagger_\alpha a^\dagger_\beta , \tag{2}$$

will play the role of a d-boson. Many studies have been devoted to the question whether one can find Shell Model Hamiltonians which, within a Shell Model space essentially composed by acting with the operators (1) and (2), reproduce the features of boson models[6-9]. In this note I shall try to add a small contribution to this discussion from the viewpoint of the number-conserving BCS[10] or broken-pair model [11]. The line of thought is much the same as in refs.[12].

THE S-PAIR STATE

In the Broken-Pair model one assumes that all but a few nucleons occur in pairs created by the S-pair operator (1). The ground state of even nuclei should then be given by the p-pair state

$$|\,\psi_{2p}\rangle = \left\{N^{(2p)}\right\}^{-\frac{1}{2}} \left\{\frac{1}{p!}\right\}^{\frac{1}{2}} (S^{\dagger})^{p}\,|\,0\,\rangle. \tag{3}$$

For single-closed-shell (SCS) nuclei, such as Sn isotopes or isotones with $N = 50$ or 82, this is an excellent approximation to a seniority $v = 0$ shell model wave function [13,14], provided that the coefficients φ_a in (1) are optimized for each particle number 2p separately. We shall forget for a moment that in SCS nuclei the ground state may contain about ten percent [15] admixture of two-broken-pair configurations.

We shall consider two questions, which may be of some relevance to the interpretation of the boson model. The first is whether the microscopic structure of the S-pair operator, as given by the coefficients φ_a in (1), is independent of the number of pairs p. This would be a nice feature, because in boson models one does not want to change the properties of the bosons when considering a series of isotopes, moving for example from a vibrational to a rotational domain[16]. The other question is how strongly the normalization constant $N^{(2p)}$ in (3) depends on p.

The answer to the first of these two questions can already be taken from studies in the beginning of the seventies by Talmi [17] and by Lorazo [14,18]. This answer is that the coefficients φ_a in the p-pair wave function (3) will be independent of p if the matrix elements of the shell model Hamiltonian [10,11,13]

$$H_{SM} = \sum_{\alpha} \varepsilon_a\, a^{\dagger}_{\alpha} a_{\alpha} + \frac{1}{4} \sum_{abcdJM} G(a\,b\,c\,d\,J)\, A^{\dagger}_{JM}(ab)\, A_{JM}(cd) \tag{4}$$

satisfy the relation:

$$-G(a\,a\,b\,b\,0)\left\{\frac{\varphi_a}{\varphi_b} + \frac{\varphi_b}{\varphi_a}\right\} + 4\sum_{J} \hat{J}^2 G(ab\,ab\,J)\hat{a}^{-1}\hat{b}^{-1} = V(\hat{a}\,\hat{b} - \tfrac{1}{2}\delta_{ab}), \tag{5}$$

where V is a constant, which can be interpreted as the interaction between the S-pairs. This one may notice from the result which a modified pairing force

$$G^{MPF}(a\,b\,c\,d\,J) = -G\,\delta_{ab}\,\delta_{cd}\,\delta_{J0}\,\hat{a}\,\hat{c}\,2\left\{\frac{\varphi_a}{\varphi_c} + \frac{\varphi_c}{\varphi_a}\right\}^{-1} \tag{6}$$

yields, if the coefficients φ in (6) are the same as in the S-pair operators (1):

$$H(S^{\dagger})^{p} \mid 0> = \{p\, E_{S} - \tfrac{1}{2}p(p-1)\, V\}(S^{\dagger})^{p} \mid 0> \tag{7}$$

with $V = 2G$ in this case. For a surface-delta interaction (SDI) [19] one obtains the same results, with $V = 0$, if its pairing matrix elements are modified according to (6). The expression for E_S is simply

$$E_S = 2 \sum_a \hat{a}^2\, \varepsilon_a \sum_a \hat{a}^2 - \tfrac{1}{2} G \sum_a \hat{a}^2 . \tag{8}$$

A reasonable value is to take E_S about 2 MeV lower than the lowest of the ε_a; this means that the breaking of a pair costs about 2 MeV (a BCS gap of about 1 MeV). Given a set of single-particle energies ε_a and a chosen value of E_S one computes the strength G from (8) and subsequently solves the equations for the φ_a:

$$\varphi_a = G \,/\, (2\varepsilon_a - E_S) \sum_b \hat{b}^2\, \varphi_b \left(\frac{\varphi_a}{\varphi_b} + \frac{\varphi_b}{\varphi_a} \right)^{-1} \tag{9}$$

numerically. With this solution the interaction (6), with the desired property that the structure of the S-pair does not depend on p, is obtained. The value V of the S-pair interaction in (7) can always be changed afterwards by adding to the interaction a term proportional to the nucleon number squared. How realistic is this artificial Hamiltonian? For realistic shell-model calculations within one major shell, the ratio between the largest coefficient φ_a and the smallest one is typically a factor five or so; this means that the modified pairing matrix elements (9) which scatter pairs between low and high orbits are considerably reduced compared to those of a pairing force or SDI. This is reasonable, since the pairing force or SDI overestimate these matrix elements as they ignore the radial part of the single-particle wave function. In fig.1 it is shown that the SDI, of which the pairing part has been modified according to (6), yields quite acceptable results, comparable to those of forces which are currently used in shell-model and broken-pair-model calculations. So the assumption that the microscopic structure of the S-pair is the same for a whole series of nuclei is a reasonable one; one may consider this to be in favour of the interpretation of s-bosons as S-pairs. This idea is not supported by the normalization factors of the S-pair state (3) however. The numbers in table 1, for six and nine S-pairs, deviate strongly from the boson-state normalization $N^{(2p)} = 1$. This result can easily be understood. As the coefficient φ for the $g_{7/2}$ orbit is the largest one (cf.fig.1) the largest term in the six-pair state (3) would be the one with six pairs in this $g_{7/2}$ orbit, if not the Pauli principle

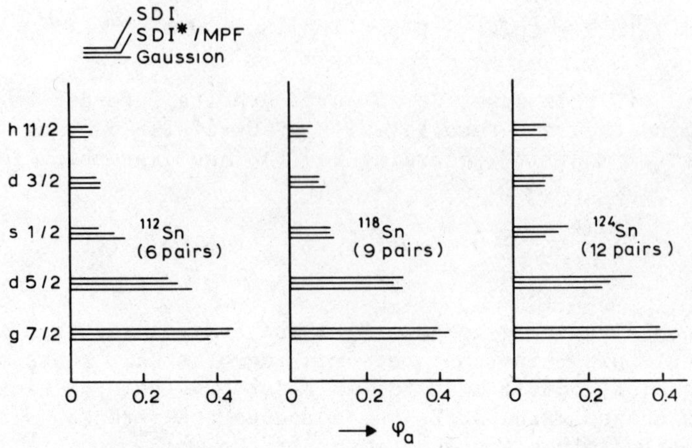

Fig.1. Amplitudes of the S-pair operator (1) in the zero-broken-pair ground state of three Sn isotopes, calculated with three different interactions. These are the SDI [19], the modified version given in the text and a Gaussian force which is frequently used [15] in broken-pair model calculations. For the first two interactions the single-particle energies were 0.0, 0.6, 2.0, 2.4 and 2.7 MeV for $g_{7/2}$, $d_{5/2}$, $s_{1/2}$, $d_{3/2}$ and $h_{11/2}$ orbits respectively; the strength $G = 0.31875$ which yields $E_S = -2$ MeV. For the Gaussian force the parameters ε_a vary smoothly with particle number to fit the data on odd nuclei.

Table 1: Normalization factors for p S-pairs, which have the structure displayed in fig.1.

	p = 1	p = 6	p = 9
SDI	1.0	0.040	0.00042
SDI*/MPF	1.0	0.055	0.00038
Gaussian	1.0	0.158	0.00045

would rule this out. So the Pauli principle cancels most of these dominant terms with too many pairs in the lowest shells, thereby lowering $N^{(2p)}$. This effect is more drastic when the lowest shell, here $g_{7/2}$, dominates more strongly in the S-pair structure. The empirical fact that still the boson model works well, without inclusion of extra "Pauli-normalization" factors of this type, may perhaps be understood if one redefines the S-pair operator to include a normalization factor in it:

$$\left[S^{\dagger}_{(p)}\right]' = \eta_{(p)}\, S^{\dagger} \tag{1'}$$

with $(\eta_{(p)})^{2p} = N^{(2p)}$. The factor $\eta_{(p)}$ varies then not too much with p.

THE D-PAIR

Along the same line of thought one may investigate whether the lowest, collective, 2^+ state in the one-broken-pair model, which is of the form

$$|\,2^+;2p> = \left\{N^{(2p,1)}\right\}^{-\frac{1}{2}} \left\{\frac{1}{(p-1)!}\right\}^{\frac{1}{2}} (S^{\dagger})^{p-1}\, D^{\dagger}_{\mu}\,|\,0>, \tag{10}$$

may correspond to a state with p - 1 s-bosons and one d boson. At first sight one would perhaps not expect that the microscopic structure of the D-pair, given by the coefficients χ_{ab} in (2), is independent of the number of pairs p. For in a nucleus with few pairs the two nucleons which couple to angular momentum 2 will mainly be found in the lowest orbits, while in a nucleus with many pairs these nucleons will be forced to stay in higher orbits. For example in light Sn nuclei the unpaired particles will predominantly occur in the $g_{7/2}$ and $d_{5/2}$ shells; in heavy Sn isotopes they will be in the $h_{11/2}$, $s_{1/2}$ and $d_{3/2}$ orbits. This is shown in the left half of fig.2, where the coefficients C_{ab} of the lowest 2^+ state are given for a realistic calculation which reproduces the spectra reasonably well[15]. These coefficients C_{ab} are <u>not</u> the same as χ_{ab} in (2), but they are defined by

$$|2^+,2p> = \sum_{(ab)} C_{ab} \left\{N^{(2p,ab)}\right\}^{-\frac{1}{2}} A^{\dagger}_{2\mu}(ab) \left\{\frac{1}{(p-1)!}\right\}^{\frac{1}{2}} (S^{\dagger})^{p-1}\,|\,0>; \tag{11}$$

that is they are the coefficients of <u>normalized</u> components. In the BCS model the coefficients C_{ab} are those of the normalized two-quasiparticle components, whereas $\chi_{ab} = C_{ab}/(u_a u_b)$. The coefficients χ_{ab} for the same case are shown in the right half of fig.2. As the broken-pair model, with which the results of figs.2 were obtained, is a very good approximation to the seniority v = 2 shell model, one may conclude that the assumption that the D-pair microscopic structure is independent of the total number of pairs is not an unreasonable

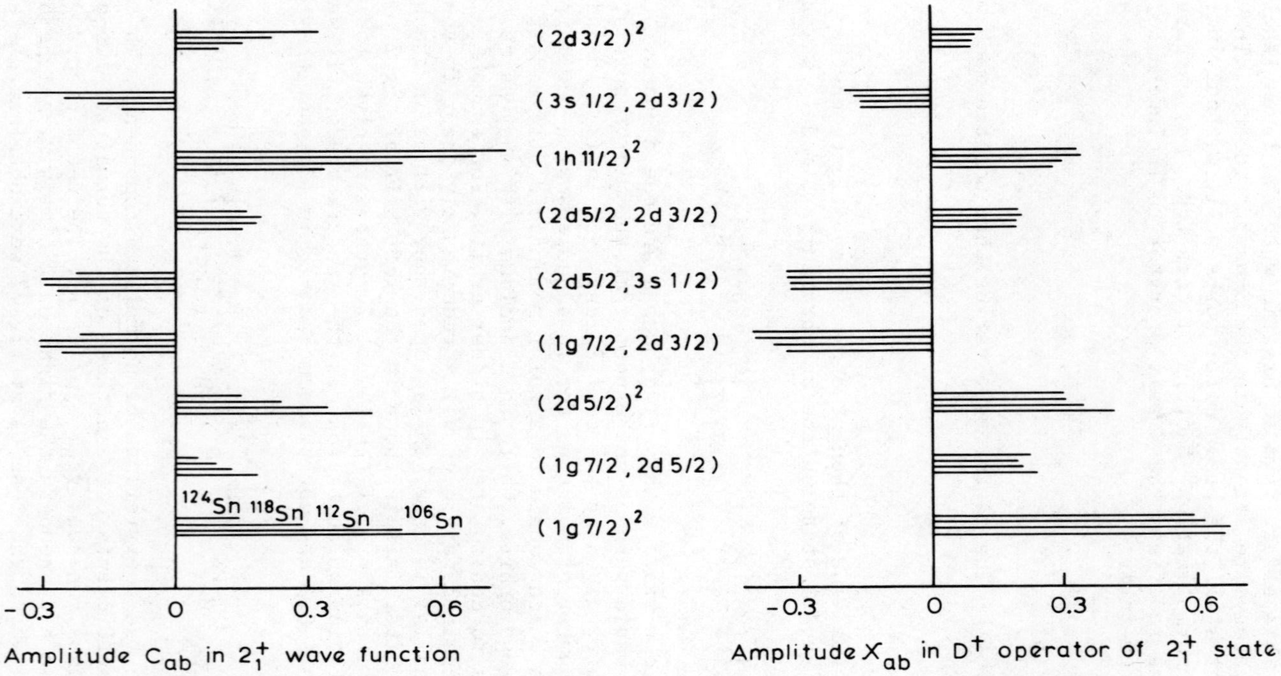

Fig. 2. Amplitudes of the lowest 2^+ state, obtained in a description of Sn isotopes using the Gaussian interaction of refs.[10,15]. In the left half of the figure the amplitude of the normalized components is indicated, in the right half the coefficients of the corresponding D-pair operator are displayed.

one. In other words it does not make much difference whether one first puts the D-pair into the valence space and next the (p-1) S-pairs or first the S-pairs and then the D-pair.

BOSONS OR PHONONS?

Paar has shown that the SU(6) interacting boson model is mathematically equivalent to the truncated quadrupole phonon model[22]. So one may raise the question whether the states of the broken-pair model may also be represented as, for example,

$$| 2^+, 2p > = Q^\dagger | \psi_{2p} > \tag{12}$$

with $Q^\dagger$ the collective particle-hole or phonon creation operator

$$Q^\dagger_\mu = \sum_{(ab)} f_{ab} \, (a_a^\dagger \tilde{a}_b)_{2,\mu}. \tag{13}$$

From the identity

$$(a_a^\dagger \tilde{a}_b)_{2\mu} (S^\dagger)^p | 0 > = -p \, \varphi_b \left(a_a^\dagger a_b^\dagger \right)_{2\mu} (S^\dagger)^{p-1} | 0 > \tag{14}$$

one may immediately conclude that for this case the phonon operator (13) is equivalent to the boson operator (2) if

$$\chi_{ab} = f_{ab} \, \varphi_b. \tag{15}$$

So the same conclusions which hold for the bosons also hold for the phonon operator.

CONCLUSIONS

One may conclude that also in realistic cases, where the spacing of single-particle energies is not small compared to the collective correlation energy, i.e. for φ_a and C_{ab} not approximately equal, the S-pair and D-pair microscopic structure still remain roughly constant with varying number of nucleon pairs. This supports the treatment of series of isotopes or isotones in the IBA model without changing the boson properties. Only the normalization factors of table 1 may cast some doubt on the interpretation of the bosons as nucleon pairs. Their interpretation as (particle-hole) phonons does not yield anything new as far as one-broken-pair states are concerned.

REFERENCES

1) A.Arima and F.Iachello, Ann.of Phys. 99 (1976) 253
2) A.Arima and F.Iachello, Ann.of Phys. 111 (1978) 201

[3] A.Arima and F.Iachello, Phys.Rev.Lett. 40 (1978)40
[4] T.Otsuka, A.Arima, F.Iachello and I.Talmi, Phys.Lett. 66B (1977) 205
[5] T.Otsuka, A.Arima and F.Iachello, Nucl.Phys. A309 (1978) 1
[6] J.N.Ginocchio, Phys.Lett. 79B (1978) 173
[7] J.N.Ginocchio, Phys.Lett. 85B (1979) 9
[8] J.N.Ginocchio, Ann.of Phys. (1980) to be published
[9] J.N.Ginocchio and I.Talmi, Nuci.Phys. A337 (1980) 431
[10] P.L.Ottaviani and M.Savoia, Phys.Rev. 178 (1969) 1594; Nuovo Cim. 47A (1970) 630
[11] Y.K.Gambhir, A.Rimini and T.Weber, Phys.Rev. 188 (1969) 1573
[12] M.Berard and N.de Takacsy, Phys.Rev. C20 (1979) 2439
N.de Takacsy, Nucl.Phys. A339 (1980) 54
[13] K.Allaart and E.Boeker, Nucl.Phys. A168 (1971) 630
[14] B.Lorazo, Nucl.Phys. A153 (1970) 255
[15] G.Bonsignori and M.Savoia, Nuovo Cim. 44A (1978) 121
[16] O.Scholten, F.Iachello and A.Arima, Ann.of Phys. 115 (1978)325
[17] I.Talmi, Nucl.Phys. A172(1971) 1
[18] B.Lorazo, Ann.of Phys. 92 (1975) 95
[19] A.Plastino, R.Arvieu and S.A.Moszkowski, Phys.Rev. 145 (1966) 837
[20] W.F.van Gunsteren, K.Allaart and P.Hofstra, Z.Physik A288 (1978) 49
[21] W.F.van Gunsteren and K.Allaart, Z.Physik A276 (1976) 1
[22] V.Paar, in Interacting Bosons in Nuclear Physics, ed. F.Iachello, Plenum Press (1979) p.163

NUCLEAR FIELD THEORY: EXACT THEORY OR SCIENCE FICTION?*

Da Hsuan Feng and Cheng-Li Wu†

Department of Physics and Atmospheric Science
Drexel University, Philadelphia, Pennsylvania 19104

INTRODUCTION

To understand the boson behavior of a fermion system is of fundamental importance in many fields of physics[1]. In nuclear physics, this problem too has attracted enormous interest because of the successes of the Interacting Boson Model (IBM)[2]. One possible way to carry out such studies is the nuclear field theory (NFT)[3] which was first proposed in 1968 by Mottelson in his attempt to treat the coupling between the so-called elementary excitation modes in nuclei; it may be convenient for this purpose because the bosonic degrees of freedom is contained inherently. Obviously, the nuclear system is a fermion system. Therefore, any theory which professes to describe the system must be equivalent to the fully fermionic theory. From this point of view, it is unfortunate that a solid foundation for the NFT has not been provided for thus resulting either in a limited acceptability of the theory or merely regarded by many as just a model.

There are two main reasons for this "state-of-affair".
(I) The NFT introduces extra bosonic degrees of freedom into the theory and then set up empirical rules[4] to correct it. It is not obvious that these rules are correct and sufficent. Although there have been work done[4] to demonstrate the equivalence between the NFT diagrams and the standard Feynman diagrams, it is still unclear whether the NFT perturbation series can converge, and if so, whether it can converge to the correct results.

*Work supported in part by the National Science Foundation under grant # Phy-7908402. and # Phy-8018613.
†On leave-of-absence from The Department of Physics, Jilin University, The People's Republic of China.

(II) Calculations performed up to now[4] take into account only the lowest order term which we shall call LNFT. The validity of the LNFT has not been established.

In this talk, we shall first present results of calculations for the "exact" NFT (NFT, for short, from now on) and the shell model and show that they are identical. A comparison is also made between the results of LNFT and the shell model; it shows that the suspicion of the LNFT is not groundless. This is followed by a development of the Boson-Fermion Hybrid (BFH) representations which constituted the foundations of the NFT.

OVERVIEW OF THE NFT

To facilitate later discussions, we shall present an overview of the NFT. The basic philosophy of the NFT is to treat a fermion system as an interacting Boson-Fermion system. To this end, it is crucial to introduce the concept of a "free" bosonic field. It is well known that when a group of even number fermions whose internal coordinates are "frozen", then such a group is termed as a "boson". Therefore, the boson in the NFT can simply be regarded as a pair of fermions which has the correlated two-fermion structure $\psi_{\alpha F}$ which satisfies the usual two-body equation

$$(H_F^o + V_F)\psi_{\alpha F} = \omega_\alpha \psi_{\alpha F} \quad ; \quad H_F^o = \sum_k \varepsilon_k a_k^\dagger a_k \tag{1}$$

In (1), ε_k is the single particle energies, ω_α the energy, V_F the interaction between the fermions and k labels all the single particle quantum number. The free bosonic field H_B^o and the boson-fermion vertices $Z_\alpha(k_1k_2)$ are just

$$H_B^o = \sum_\alpha \omega_\alpha B_\alpha^\dagger B_\alpha \quad ; \quad Z_\alpha(k_1k_2) = \langle k_1k_2 \, ; J_\alpha | V_F | \psi_{\alpha F} \rangle \tag{2}$$

It must be stressed here that the boson introduced here are "real" bosons, i.e., not only $[B_\alpha, B_\beta^\dagger]=\delta_{\alpha\beta}$ is satisfied, but $[B_\alpha, a_k^\dagger]=0$ is also satisfied exactly.

The Hamiltonian for the NFT is

$$H_{NFT} = H_B^o + H_F^o + V \, ; \, V = V_F + V_{FB} + V_{BF} \tag{3}$$

where

$$V_{FB} = V_{BF}^\dagger = \sum_{\alpha k_1 k_2} Z_\alpha(k_1k_2) A_{J_\alpha}^\dagger(k_1k_2) B_\alpha \tag{4}$$

$$A_{J_\alpha}^\dagger(k_1k_2) = [a_{k_1}^\dagger \times a_{k_2}^\dagger]_{J_\alpha} / (1 + \delta_{k_1k_2})^{1/2} \tag{5}$$

The corresponding Schrodinger equation $H_{NFT}\Psi=E\Psi$ can be solved within the framework of the perturbation theory. If one denotes ψ_p as the wavefunction in the (P) model space, then according to the Brilloiun-Wigner perturbation scheme, we have

$$(H_B^o + V_{eff})\psi_p = E\psi_p \tag{6a}$$

$$V_{eff} = V + VDV + VDVDV + VDVDVDV + \ldots \tag{6b}$$

$$D = (E - H_B^o - H_F^o)^{-1}\hat{Q} \tag{6c}$$

where $\hat{Q}$ is the Q space (intermediate space) projection operator. Finally, the wavefunction Ψ is just

$$\Psi = \psi_p + DV_{eff}\psi_p \tag{7}$$

Eventhough the NFT now regards the bosons and fermions as "equal", so-to-speak, it must nevertheless be remembered that the bosonic degrees of freedom originated from the fermionic degrees of freedom. Therefore, any chosen basis of this theory must have the inherent problem of overcompleteness as well as the violation of the Pauli principle. Recognizing these difficulties, the NFT introduces three correction procedures. They are (i) allow only boson states in the P space; (ii) all diagrams in (6b) which have a bubble (or bubbles) must be discarded and (iii) only states which are normalizable are regarded as physical ones, all others are regarded as spurious and discarded.

We would like to elaborate on the correction procedure (iii). The normalization of Ψ of (7) is just (by assuming that ψ_p is normalized)

$$< \Psi|\Psi > = 1 + < \psi_p|V_{eff}^\dagger D^\dagger D\, V_{eff}|\psi_p > \tag{7a}$$

Normally, there is no question of normalizability. However, the special correction requirement (ii) of NFT will allow certain terms for the 2nd term on the r.h.s. of (7a) be discarded, thus allowing the possibility that the remaining terms be less than -1. When that happens, the state thus becomes unnormalizable. This is a very crucial point about the NFT. This completes our overview of the NFT.

RESULTS OF THE NFT CALCULATIONS

Obviously the question of exactness comes in naturally, i.e. are the criteriors (i), (ii) and (iii) introduced into the NFT sufficient or even proper to correct for the haunting problems of the violation of the Pauli principle and overcompleteness. A step towards answering this question would be to give an example of a general nature to demonstrate the exactness. To this end, we will treat the problem of four identical nucleons moving in a single j level as such an example. We are aided by a successful summing method of Wu[5] which would now allow us to sum up the NFT diagrams exactly without any special requirements, thus setting the stage for the comparison between the NFT and the Shell model results. It must also be pointed out that with this summing method, we can now calculate $\hat{V}_{eff}$ exactly. Since ψ_p must be in the boson space, V_{eff} must constitute the precise definition of the boson-boson interaction; the work has been done and the results are given in ref.6. It is clear from eq.(7) that we can now calculate the NFT wavefunction $\bar{\Psi}$. As far as we are aware of, this has not been presented in the literature before. In ref.6, we presented the first discussion on the NFT wavefunctions.

We have done both the shell model and NFT calculations for the aforementioned example for j between 3/2 to 15/2, with arbitrary two-body interactions. In Fig. 1, we show the results for the j=15/2 case. The various multipole strengths V_o, V_2 and V_4 are chosen to be -1.0, -0.5 and -0.1 respectively. These numbers have arbitrary units. The quantity V_λ is defined as $<jj;\lambda|V_F|jj;\lambda>$. The energy levels labeled by EXACT are the shell model (calculated via the C.F.P. of ref. 7) and the NFT results; they agree completely. Finally, using the NFT wavefunctions of (7), along with requirements (i), (ii), and (iii), we were able to demonstrate that all the interesting physical observables are in complete agreement with the shell model predictions. Furthermore, the normalizability of these states removes completely all the spurious states from the NFT. The energy levels labeled by LNFT correspond to the NFT calculations using the Rayleigh-Schrodinger perturbation scheme[8] but only take the lowest order term into account; thus these results have the same "input" for all the other calculations presented in papers on the NFT in this volume. What is disturbing about the LNFT are the deviations of the normalizations from the exact results. In Table 1 we listed the normalizations of some of the low lying states for both the LNFT and the NFT and the differences are uncomfortably large. In fact, if one follows the criterior of picking out the physical states, then the 2^+ and the 4^+ states will have to be discarded by the LNFT normalizations. We should add further that any physical quantities which are associated with the normalizations (e.g. the two particle transfer form factors) are also to be suspected in view of our results. Finally, we should also point out that although the energy spectrum calculated by LNFT may be tolerable, those for the lower j cases are actually quite bad! It is clear that one must be very careful in interpreting the LNFT results since the deviations from the exact results are enormous even for such a large j.

From the above results, we reach the following conclusions
(a) All the exact results are reproduced by the NFT without exception
(b) The LNFT gives normalization constants which are quite different from the exact results thereby the removal of spurious states becomes a difficult task. This means that the LNFT is not a good approximation in general.

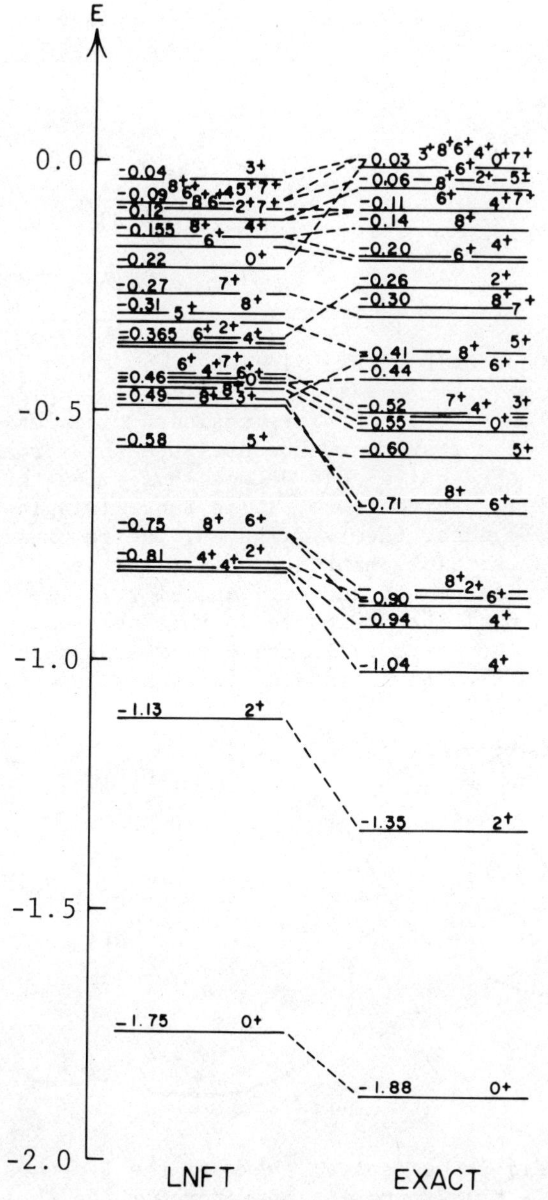

Fig. 1. Comparison between exact and LNFT results.

Table 1. Normalizations for LNFT and NFT

J^{π}	LNFT norm	NFT(EXACT) norm
0^{+}	0.38	0.79
2^{+}	-0.50	0.60
4^{+}	-4.30	0.30
6^{+}	0.75	1.11

THE BOSON-FERMION HYBRID REPRESENTATION

Since the "input" to the aforementioned examples is very general, therefore, the results must constitute the strongest demonstration yet for the equivalence between a fermion theory (shell model) and the NFT. Furthermore, it is tantalizingly suggestive that the NFT is an exact theory. Indeed, the results presented in the last section certainly hint strongly that the NFT must be just a special representation of the fermion theory. This suggestion is now realized and shall be discussed in this section. For obvious reasons, we have chosen to call such a representation the Boson-Fermion Hybrid (BFH) representation. It's basic ideas are given succinctly in Fig.2.

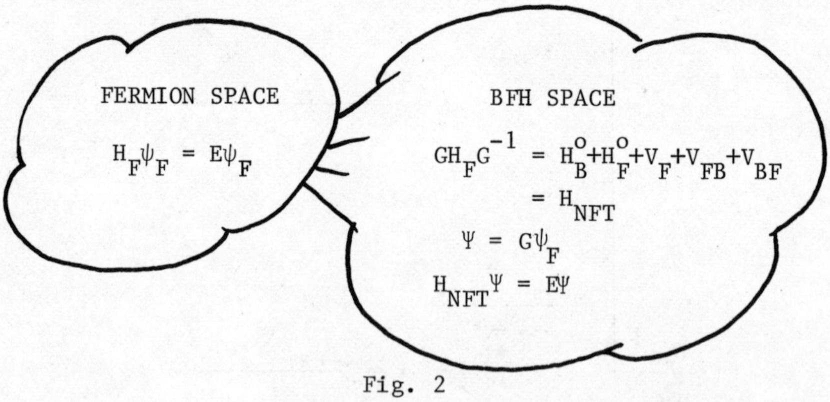

Fig. 2

We see that in Fig.2 the central theme of the BFH theory is the introduction of a transformation operator G whose actions are depicted in the r.h. "cloud" of Fig.2. However, by enlarging the space to include the bosonic degrees of freedom, the BFH Schrodinger equation may give solutions which are non-physical. Thus to ensure the "physicalness" of the solutions, we need to introduce an operator R such that

$$\psi_{\text{physical}} = R\Psi \;;\; H_{\text{physical}} = RH_{NFT}R^{-1} \tag{8a}$$

and $$<\psi^{(\alpha)}_{\text{physical}}|\hat{O}_{\text{physical}}|\psi^{(\beta)}_{\text{physical}}> = <\psi_{\alpha F}|\hat{O}_F|\psi_{\beta F}> \tag{8b}$$

Now the crucial question comes in. What is R?

In the simplest system where there is only a boson (or two fermions), it can be vividly demonstrated what R must be. Again, we begin by recognizing a boson as an interacting fermion pair $\psi_{\alpha F}$ with boson "energy" ω_α. The BFH basis consists of $\psi_{\alpha B}$ and $|k_1k_2;j_\alpha>$ where $\psi_{\alpha B}$ is the boson wavefunction ($H^o_B\psi_{\alpha B} = \omega_\alpha\psi_{\alpha B}$). The wavefunction is

$$\Psi_\alpha = \psi_{\alpha B} + \sum_{k_1k_2} C_{k_1k_2\alpha}|k_1k_2;J_\alpha> \tag{9}$$

Following eq.(6a), we have

$$(H^o_B + V_{eff})\psi_{\alpha B} = E\psi_{\alpha B} \tag{10}$$

The energy E of equation (10) can be shown to be

$$E = \omega_\alpha + <\psi_{\alpha B}|V_{eff}|\psi_{\alpha B}> \tag{11}$$

The second term on the r.h.s. of eq.(11) must vanish if the correct energy $E=\omega_\alpha$ is to be obtained. It is not difficult to visualize that all the terms for the matrix element of (11) involve only bubble diagrams. Therefore, by choosing the operator R to be such that its action is to remove all the bubble (or bubbles), then of course, the term will vanish identically, and we will recover the correct energy for the system. Furthermore, we would like to point out that $R\Psi_\alpha$ can be demonstrated to be the physical wavefunction. It can also be demonstrated that if one were to use the fermion states as the P states, then the results would amount to choosing R as an operator whose action is to remove all the diagrams with boson lines. Thus the problem is reduced to the fermion case again. This is why the NFT introduces the condition (ii). From this brief discussion of the simple system of one boson, we may now summarize the meaning of R as follows:

$$R = R_F = \text{removal of all boson lines; } \psi^{(\alpha)}_{\text{physical}} = \psi_{\alpha F}(\hat{O}_F) \tag{12}$$

$$R = R_n = \text{removal of all bubbles; } \psi^{(\alpha)}_{\text{physical}} = \psi_\alpha(\tilde{O}) \tag{13}$$

In (13), ψ_α is just the wavefunction of eqn.(7) (with conditions (i), (ii) and (iii)) in an α state. The quantity $\tilde{O}$ is any BFH operator and "$\sim$" means the removal of all the bubbles diagrams.

For the single-j four-fermion system, eq.(12) and (13) cannot easily, and vividly be demonstrated. However, numerical results do show that indeed (8b) is satisfied!

Thus it is now very clear that the NFT is merely a special representation with a transformation operator U, where U=RG. The NFT Hamiltonian H_{NFT} comes from the transformation G, while the forbiddenness of the bubble diagrams as well as the restriction of the P states to be boson states come from R. The normalizability of the physical states is, of course, just a very natural requirement. Hence, all the NFT rules can emerge naturally from the BFH representation. Thus we feel confident to say that the NFT is indeed an exact theory, and that it can be used to explore the boson behavior of a fermion system.

ACKNOWLEDGEMENTS

We express our gratitude to Professor Aage Bohr for his hospitality during our stay at the Niels Bohr Institute, where part of this work was carried out. Very useful discussions with Professors B.R. Mottelson, W.T. Pinkston and P. Siemens on various occasions are acknowledged. We also thank Professors F.K. Davis, W.W. Eidson and J.H. Hamilton for their support. Finally, we are grateful to Mr. Antony Joseph who prepared this manuscript.

REFERENCES

1. For example, J.Freed, Is Spin-Aligned Hydrogen A Bose Gas? J. Chem. Phys. 72:1414(1980) and references therein.

2. For example, A.Arima and F.Iachello, New Symmetry in the sd Boson Model of Nuclei: The Group o(6), Phys. Rev. Lett. 40: 385(1978).

3. B.R. Mottelson, J. Phys. Soc. Japan, Supple 24:87(1968).
4. P.F. Bortignon et.al. Nuclear Field Theory, Phys. Rep. 30C: 305(1977).

5. C.L.Wu, The Summing Method of the Two-Boson Diagrams in the Nuclear Field Theory, Nucl. Phys. A (in Press).

6. C.L.Wy and D.H.Feng, The Boson-Fermion Hybrid Representation (I) Formulation, Ann. of Phys. (to be published); C.L.Wu and D.H.Feng, The Boson-Fermion Hybrid Wavefunction and the Probabilities of Boson States in Nuclei, Phys. Lett. (to be published).

7. B.F.Bayman and A.Lande, Nucl. Phys. 77: 1(1966).

8. P.F.Bortignon, R.A.Broglia and D.R.Bes, Phys. Lett. 76B: 153(1978).

APPLICATIONS OF THE NUCLEAR FIELD THEORY TREATMENT

OF THE PAIR ALIGNED MODEL

E. Maglione

The Niels Bohr Institute, University of Copenhagen
DK-2100 Copenhagen Ø, Denmark, and
Istituto Nazionale di Fisica Nucleare-Laboratori
Nazionali di Legnaro (Padova), Italy

1. INTRODUCTION

It has been shown [1], [2] that the aligned wavefunction, and the wavefunction obtained by restricting pairs of particles to be coupled to angular momentum zero and two as assumed by the Quadrupole Phonon Model (QPM) and by the Interacting Boson Model (IBM) (cf. refs. [3]-[5] and references therein) are, for strongly deformed systems, rather different. They become identical in the vibrational limit and display a different degree of similarity for intermediate (anharmonic) situations [6], [9]. To which extent this difference reveals itself in the predicted properties of the low-energy nuclear spectrum is an open question. In an attempt to clarify this point we have calculated the spectrum and the electromagnetic and two-nucleon transfer probabilities for some of the Kr- and Sm-isotopes*) in the framework of the NFT [10]-[12] and in a basis of monopole and quadrupole pairing modes.

2. THE MODEL

The particles are assumed to move in a single j-shell and interact through a multipole $(\lambda = 0, 2)$ pairing force and through a quadrupole particle-hole interaction. The multipole pairing

* It is noted that the spectrum of the Kr- and of the Sm-isotopes and associated electromagnetic and two-nucleon transfer probabilities have already been calculated utilizing the IBM and are reported in the literature [13], [14]. The calculations reported here are however microscopic.

Hamiltonian is defined as in Ref. [10] and reads

$$H_P = -\sum_{\lambda,\mu} G_\lambda (2\lambda+1) P^+_{\lambda\mu} P_{\lambda\mu} \quad , \tag{1}$$

where, for a single j-shell,

$$P^+_{\lambda\mu} = \left(\frac{2\pi}{2\lambda+1}\right)^{1/2} \langle j \| T_\lambda \| j \rangle \, [c^+_j c^+_j]^\lambda_\mu / \sqrt{2} \, . \tag{2}$$

From systematical analysis of the nuclear spectrum utilizing $T_{\lambda\mu} = Y_{\lambda\mu}$ it has been found (cf. e.g. Ref. [15]) that $G \sim 27/A$ MeV, independent of λ ($\lambda = 0,2,$ and 4).

The s- and d-boson of the IBM are here interpreted as monopole and quadrupole pairing vibrations. For a single j-shell, the RPA relation which determines the energy W_λ of the multipole pairing vibration reduces to

$$\varepsilon - W_\lambda = 2\pi G_\lambda |\langle j \| T_\lambda \| j \rangle|^2 . \tag{3}$$

The quantity

$$Z_\lambda = \varepsilon - W_\lambda \quad , \tag{4}$$

is the correlation energy associated with the mode λ, ε being the energy of the j-shell where the particles correlate. From systematics, $Z_s \sim 1.5$ MeV and $Z_d \sim 0.5$ MeV.

The RPA amplitude associated with the mode λ is

$$d = 1 = \frac{\Lambda_\lambda}{\sqrt{2}} \; \frac{\langle j \| T_\lambda \| j \rangle}{\varepsilon - W_\lambda} \, . \tag{5}$$

The particle-vibration coupling strength is given by

$$\sqrt{2}\, Z_\lambda = \Lambda_\lambda \langle j \| T_\lambda \| j \rangle , \tag{6}$$

and measures the strength with which the particle couples to the pairing modes.

The monopole and quadrupole pairing vibrations interact through a quadrupole particle-hole force

$$H_Q = -K \sum_\mu (-1)^\mu Q_{2\mu} Q_{2-\mu} \, . \tag{7}$$

3. THE Kr-ISOTOPES

The calculations have been carried out utilizing both proton and neutron pairing modes. They were obtained by allowing the valence particles to move in a $j = 21/2$ shell, corresponding to the degeneracy between the closed shells at nucleon number 28 and 50.

The basis states contain four proton boson, and from 0 to 4 neutron-hole bosons. Already with such a small subspace, the matrices to be constructed are of the order of 200×200 (for ^{78}Kr). The values of the parameters utilized in the calculation are displayed in Fig. 1. The spectrum associated with the different

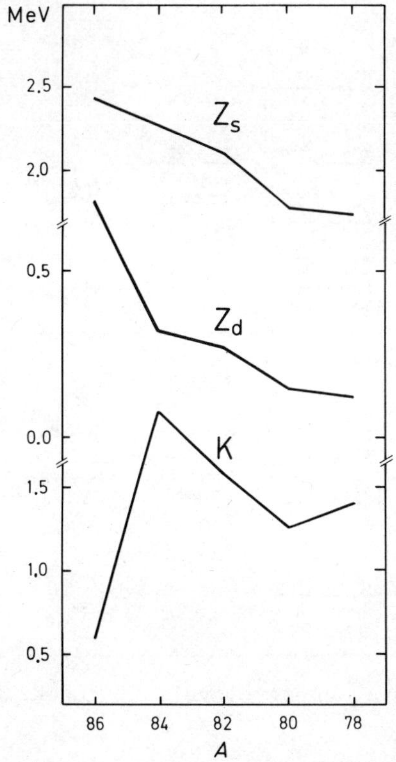

Fig. 1. Value of the parameters utilized in the calculation for the Kr isotopes. The quantities Z_s and Z_d are the correlation energy of a pair of particles coupled to angular momentum zero and two, and interacting through a monopole and quadrupole pairing force respectively. The parameter $K \equiv \frac{1}{5} \kappa \langle j \| Q_2 \| j \rangle^2$ is proportional to the strength of the quadrupole particle-hole interaction.

isotopes is shown in Fig. 2, in comparison with the experimental data. The grouping of levels in bands is done on the basis of the wavefunctions. Examples are displayed in Fig. 3.

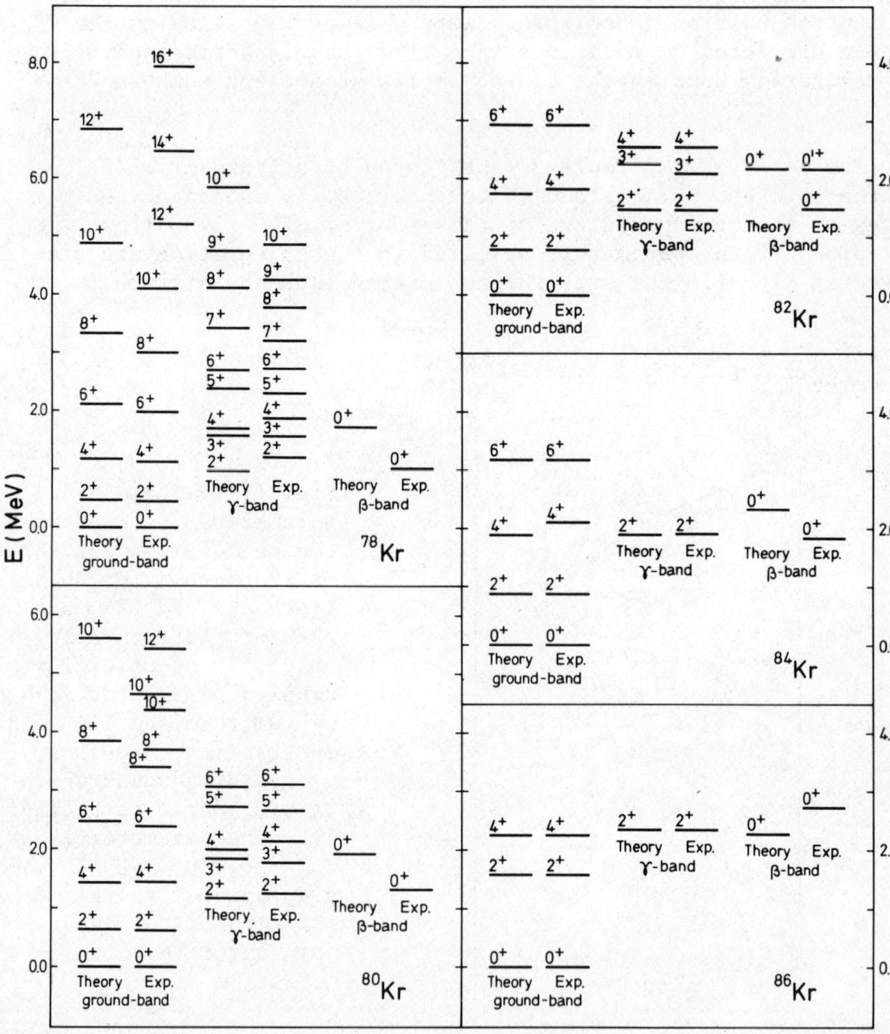

Fig. 2. Calculated and experimental energy levels in the Kr isotopes.

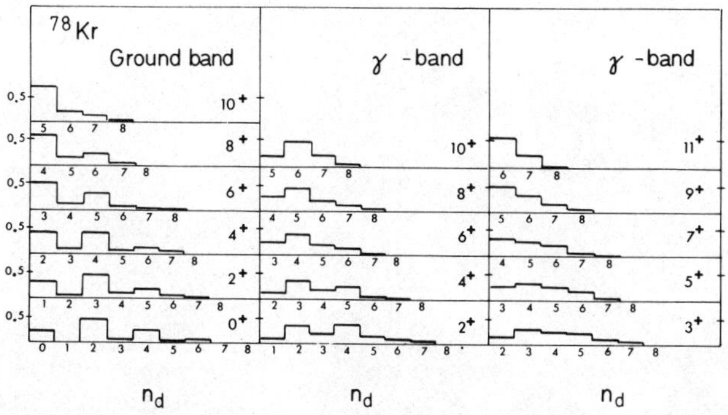

Fig. 3. Wave functions of the ground- and of the γ-band of ^{78}Kr.

The model provides with a good fitting to the ground state rotational band. Concerning the γ-band the splitting between even and odd members (i.e. $3^+ - 4^+$, $5^+ - 6^+$, etc.) are predicted too small. It is noted that these states are degenerated in the vibrational limit. The model thus predicts a quasi γ-band which is too close to the spherical limit. Concerning the β-band, the model predicts a state which, in general, is much too high in energy.

The calculated electromagnetic transition amplitudes for the first 2^+ and the branching ratio for the second 2^+ states are shown in Fig. 4, for the various isotopes. The experimental patterns are fairly well reproduced. The $R = B(E2; I \rightarrow I-2)/B(E2; 2 \rightarrow gs)$ ratios for the members of the yrast line are shown in Fig. 5 for ^{78}Kr and ^{80}Kr. Because of the finite model subspace, there is a rather strong limitation of the maximum angular momentum the nucleus can have, as compared to the shell model prediction. For example, $I_{max} = 16$ for ^{78}Kr, which corresponds to the total alignment of 8 d-bosons. Because of the artificial cut-off, predictions past beyond $I = 8$ are affected by the imposed limitations. The decrease of the ratio R reflects in fact this limitation.

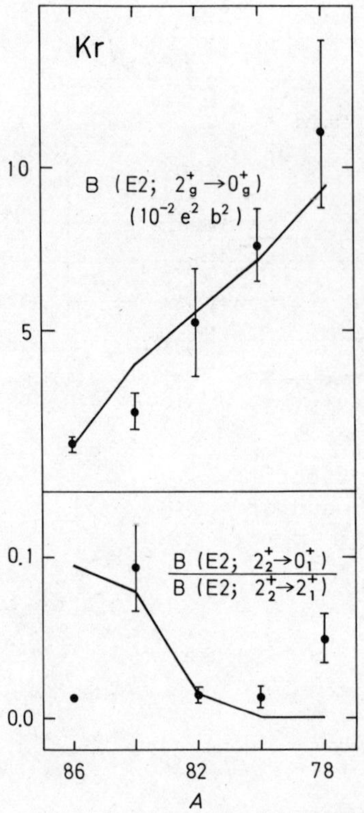

Fig. 4. Calculated and experimental electromagnetic transition probabilities for the first 2^+ state and branching ratio for the second 2^+ states of the Kr isotopes.

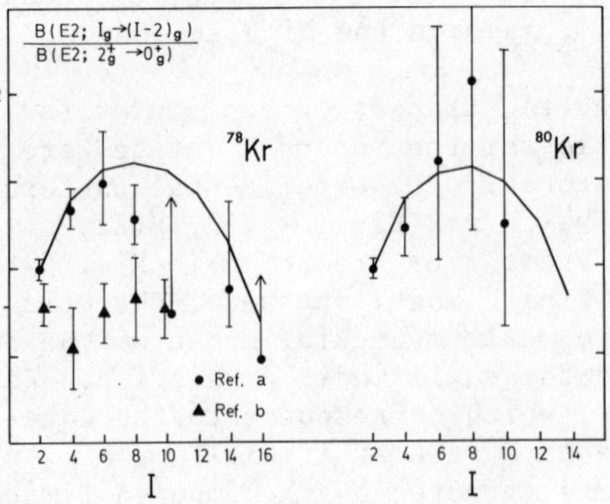

Fig. 5. Calculated and experimental B(E2;I I-2)/B(E2;2g 0g) ratios for the members of the yrast band of ^{78}Kr and ^{80}Kr. The experimental points for ^{78}Kr are taken from: H.P. Hellmeister et al., Nucl. Phys. A332 (1979) 241, and R.L. Robinson et al., Phys. Rev. 21C (1980) 603.

The Sm-isotopes

In this calculation no distinction is made between neutrons and protons. It is thus essentially the same as that reported in Ref. [14]. All particles outside the $Z = 50$ and $N = 82$ closed shells are explicitly considered. They move in a single j-shell with degeneracy $\Omega = 36$. The parameters of the calculation resulting from a fitting of the data are shown in Fig. 6. Would we

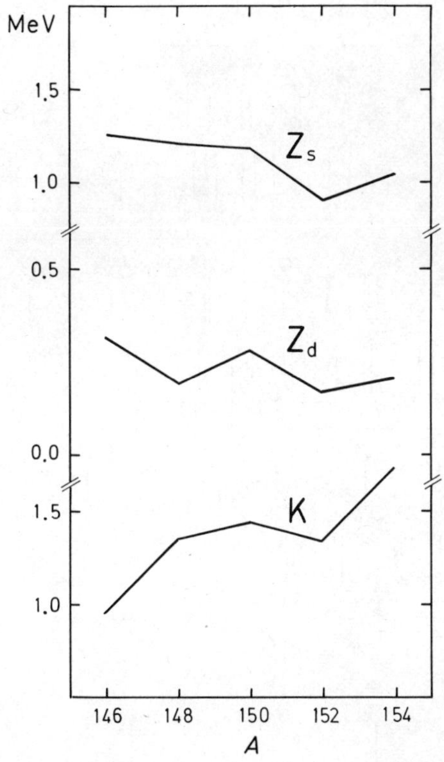

Fig. 6. Same as fig. 1) but for the Sm isotopes.

have taken constant parameters the quality of the fitting would not have been altered in any essential way. The predicted spectra in comparison with the experimental data is shown in Fig. 7.

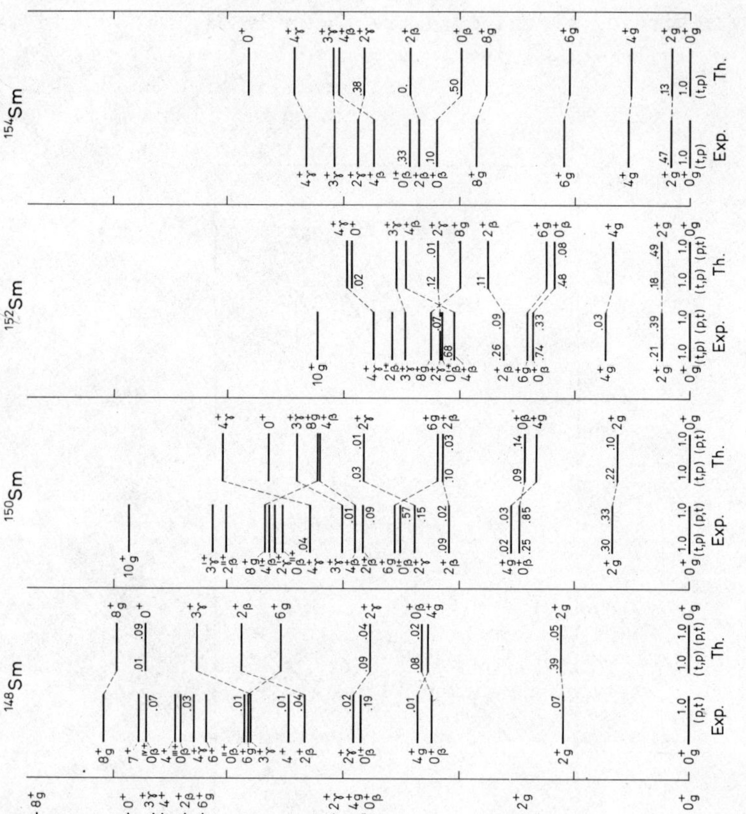

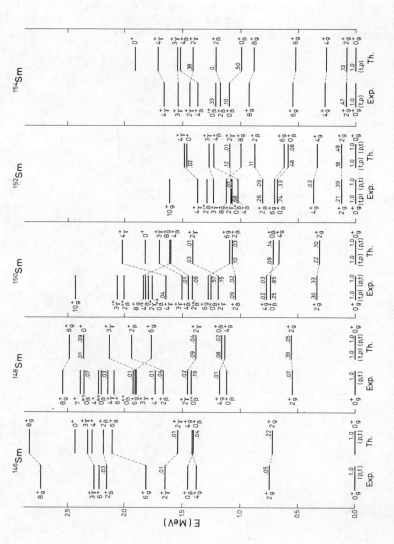

Fig. 7. Calculated and experimental energy levels and two-nucleon transfer probabilities for the Sm Isotopes.

The phase transtion is accounted for. However, the moment of inertia of both the quasi-beta and quasi-gamma bands are, for ^{152}Sm and ^{154}Sm, about 50% smaller than the experimental values, the predicted bands being still too vibrational. The change in the coupling scheme taking place through the Sm-isotopes is further evidenced by the behaviour of the wavefunctions describing the different states. Examples are given in Fig. 8.

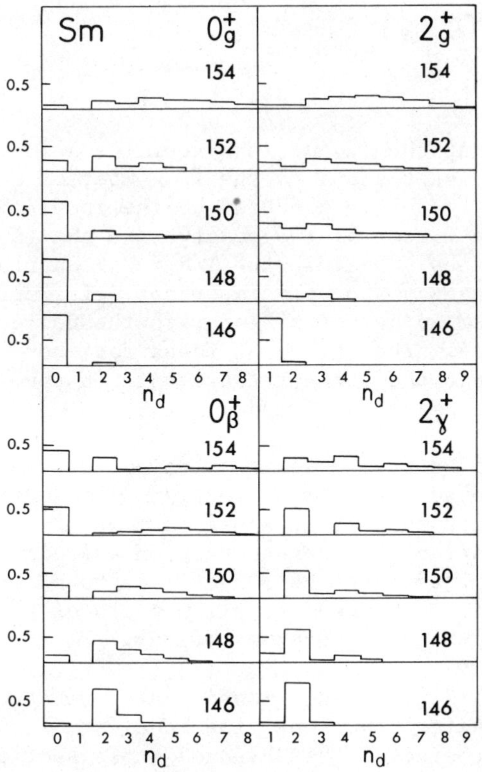

Fig. 8. Wave functions for the 0_g^+, 2_g^+, 0_β^+, 2_γ^+ states of the different Sm-isotopes.

In Fig. 7 we also display the two-nucleon transfer probabilities predicted by the model in comparison with the experimental data. It has been shown [16] that two-nucleon transfer reactions to the different members of the ground state rotational band probe, in deformed nuclei, the multipole pairing deformations, that is, the deformations of the Nilsson orbitals close to the Fermi surface. In fact

$$\sigma(\lambda) \approx \left(\sum_{\nu} U_{\nu} V_{\nu} q_{\nu}(\lambda)\right)^2 = \left(\frac{\Delta(\lambda)}{G_{\lambda}}\right)^2 \tag{8}$$

where $q_{\nu}(\lambda) \sim \langle \nu \| r^{\lambda} Y_{\lambda} \| \nu \rangle$ and where U_{ν}, V_{ν} are the BCS occupation parameters. The quantity G_{λ} is the multipole pairing force (~ 27/A MeV, cf. [15]), and $\Delta(\lambda)$ is the multipole pairing gap. For $\lambda = 0$ one obtains the well known relation valid for superfluid systems

$$\sigma(gs \rightarrow gs) = (\Delta/G)^2, \tag{9}$$

the quantity $\Delta = \Delta(0)$ being the monopole pairing gap.

From Fig. 7 it is seen that the (t,p) cross sections associated with transition between ground states and to the 2^+ member of the ground state rotational band are well reproduced by the model. Note however the deviations up to a factor of two observed in the (p,t) cross sections to the 2_g^+ states. Because the $\Delta N = 2$ contributions to the quadrupole pairing gap are very important (cf. Ref. [17]), these discrepancies seem to be associated with the neglect of the deformations of the core. The agreement found for the ground state transitions reflect the correct treatment, in the model, of the monopole pairing degree of freedom.

The transition to the 4_g^+ state, observed with a cross section which is ~ 3-4% the ground state cross section, lies outside the model. For nuclei for which the hexadecapole term in the Nilsson potential becomes important, the cross section associated with this transition can become as large as 30% of the ground state cross section [18]. In such case the need to allow pairs of particles to couple to $\lambda = 4$ (g-bosons) becomes obvious.

Of the 0^+ states which aside from the ground state are strongly excited in two nucleon transfer reactions, the model predicts only one. Even for the one predicted, the observed cross section is about a factor of two larger than predicted. This is because the model does not contain the degree of freedom associated with pairing vibrations [19]. In this case, the large change of the shell structure which seems to give rise to two minima in the potential energy surface, leads to large fluctuations in the gap and in the shape degrees of freedom, and the two modes mix strongly. It is noted that the pairing vibration is not an intruder state (cf. Ref. [20]) in a system in which pairing correlations are important, but the expression of the quantal fluctuations of the average pairing field. The breaking of gauge invariance (i.e. violation of the number of particles) is at the basis of the collective properties of this degree of freedom, in a similar way as the breaking of rotational invariance (i.e. non-conservation of the angular momentum), is at the basis of the rotational spectra in normal space.

Conclusions

A description of the nuclear spectrum in terms of pairs of fermions coupled to 0^+ and to 2^+ and moving in a single j-shell seems to display the general properties experimentally observed. It however leads to γ-bands which are too close to the spherical, vibrational pattern. This seems also to be the case for the β-vibrational bands. In fact, the associated moment of inertia is about half of the ground state band moment of inertia.

The intensity with which the β-vibrations are excited in two-nucleon transfer reactions is smaller than experimentally observed. This is because the model does not contain the pairing vibrational degrees of freedom. Pairing vibrations, which play an important role in the low-energy nuclear spectrum, can mix with the β-vibrations and renormalize their energy and two-nucleon transfer strength in a major way. Extra s'-bosons are to be introduced to take this degree of freedom into account.

Concerning the ground state band, departures of about a factor of two are observed in the (p,t) cross sections associated with the 2^+_g , while the cross section associated with the 4^+_g is zero. Because of the central role played by the hexadecapole deformation in two-nucleon transfer processes, a g-boson seems to be needed in many cases, not only to account for the 4^+_g cross section, but because of the renormalization effect it has in the 2^+_g cross section.

The cut-off displayed by the model due to the finite space is a limitation which has to be removed if a meaningful comparison with the high-spin states $(I > 8^+)$ of the yrast line is to be carried out.

References

[1] A. Bohr and B. R. Mottelson, Preprint NORDITA-80/19 (1980)
[2] R. A. Broglia, Invited talk to the Workshop on "Interacting Bose-Fermi Systems in Nuclei", Erice, June 12-19, 1980
[3] D. Janssen, R. V. Jolos and F. Dönau, Nucl. Phys. A224 (1974) 93
[4] Interacting Bosons in Nuclear Physics, Ed. F. Iachello, Plenum Press, N.Y. 1979
[5] V. Paar, Interacting Bosons in Nuclear Physics, Ed. F. Iachello, G. Kyrchev, Preprint of the JINR E4-80-178, Dubna, 1979
[6] T. Otsuka, A. Arima, F. Iachello and I. Talmi, Phys. Lett. 76B (1978) 139; 66B (1977) 205

[7] T. Otsuka, Interacting Bosons in Nuclear Physics, Ed. F. Iachello, Plenum Press, N.Y. 1979, p. 93; J. McGrory, ibid. p. 121
[8] F. Sakata, private communication
[9] T. Otsuka, A. Arima and F. Iachello, Nucl. Phys. A309 (1978) 1
[10] P. F. Bortignon, R. A. Broglia, D. R. Bes and R. Liotta, Phys. Rep. 30C (1977) 305
[11] D.R. Bes and R. A. Broglia, Interacting Bosons in Nuclear Physics, Ed. F. Iachello, Plenum Press, N.Y. 1979 p. 143
[12] R. A. Broglia, K. Matsuyanagi, H. Sofia and A. Vitturi, Nucl. Phys. (in press)
[13] A. Gelberg and U. Kaup, Interacting Bosons in Nuclear Physics, Ed. F. Iachello, Plenum Press, 1979, p. 59; Z. Physik A293 (1979) 311
[14] O. Scholten, Interacting Bosons in Nuclear Physics, Ed. F. Iachello, Plenum Press, 1979, p. 17
O. Scholten, F. Iachello and A. Arima, Ann. of Phys. 115 (1978) 325
A. Saha, O. Scholten, D. C. J. M. Hageman and H. T. Fortune, Phys. Lett. 85B (1979) 215
[15] D. R. Bes, R. A. Broglia and B. Nilsson, Phys. Lett. 50B (1974) 213
[16] R. A. Broglia, C. Riedel and T. Udagawa, Nucl. Phys. A135 (1969) 561
[17] I. Ragnarsson and R. A. Broglia, Nucl. Phys. A263 (1976) 315
[18] K. Kubo, R. A. Broglia, C. Riedel and T. Udagawa, Phys. Lett. 32B (1970) 29
[19] D. R. Bes and R. A. Broglia, Nucl. Phys. 80 (1966) 289
[20] KVI Report 1979, p. 101

DYSON'S FINITE BOSON EXPANSION AND THE INTERACTING BOSON MODEL

T. S. Yang and L. M. Yang

Department of Physics, Peking University

Peking, The People's Republic of China

1. INTRODUCTION

In this note a microscopic description of the IBM is presented. Dyson's finite Boson expansion[1,2] is used to map the shell model subspace concerned into a Boson subspace. Each $s^\dagger$- or $d^\dagger$-operator is expressed as a linear combination of creation operators of ideal Bosons. A complete set of equations is given for determining these operators and calculating all the coefficients appearing in the IBA1 Hamiltonian. Furthermore, according to our method, the difficulties associated with violations of the Pauli principle can be overcome by using the so-called modified Jancovici-Schiff substitution[2] which transforms the Boson wave functions back into fermion wave functions. After the transformation the s- or d-Bosons in the IBM wave functions become correlated nucleon pairs.

Fig. 1 shows an outline of our method.

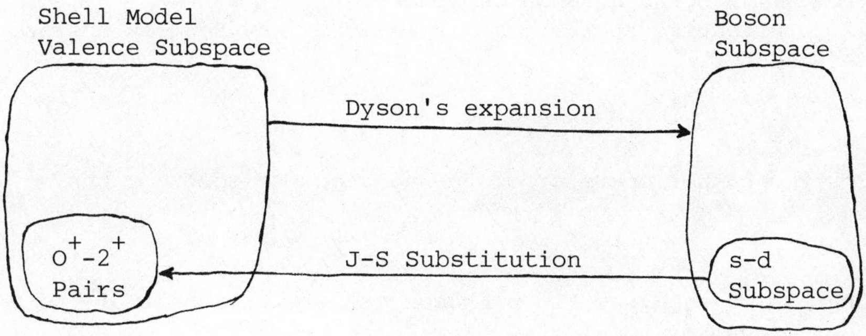

Fig. 1.

2. THE EFFECTIVE HAMILTONIAN IN DYSON'S EXPANSION

We assume that the shell model valence subspace (SF) concerned is spanned by the state vectors belonging to the configurations

$$(j_1)^{k} \;\ldots\ldots\; (j_s)^{k_s} \tag{2.1}$$

$$(k_1 + k_2 + \ldots + k_s = \text{even}) \;.$$

The four quantum numbers of the one-particle states, nljm, will often be denoted by a single symbol μ or α . Let $\xi_\mu^\dagger$, ξ_ν be the creation and annihilation operators of the valence nucleons, and let $|0>$ be the state vector of the inert core. Thus

$$\xi_\nu |0> = 0 \;. \tag{2.2}$$

An arbitrary state vector $|\Psi>$ in (SF) can be expressed as :

$$|\Psi> = \Sigma C \, \xi_{\mu_1} \xi_{\mu_2} \cdots |0> \tag{2.3}$$

where each μ has a j belonging to $(j_1, j_2, \ldots j_s)$.

According to Dyson's Boson expansion, an arbitrary Fermion state vector $|\Psi>$ in (SF) is mapped into a Boson vector $\S\Psi)$:

$$\S\Psi) = U|\Psi> \tag{2.4}$$

where

$$U = <0| \, e^{\frac{1}{2}\sum_{\mu\nu} b^\dagger_{\mu\nu} \xi_\nu \xi_\mu} |0) \;. \tag{2.5}$$

$\{b^\dagger_{\mu\nu}\}$ are the so-called ideal Boson creation operators $(b^\dagger_{\mu\nu} = -b^\dagger_{\nu\mu})$ and $|0)$ is the ideal Boson vacuum :

$$b_{\mu\nu} |0) = 0 \;. \tag{2.6}$$

When $|\Psi>$ runs through the subspace (SF), the set of $\S\Psi)$ obtained from (2.4) spans a Boson subspace (SB).

The effective operator g_B of an arbitrary Fermion operator g is defined by

$$Ug = g_B U \;. \tag{2.7}$$

For the number operator of the valence nucleons one has

$$n = \sum_\mu (\xi_\mu^\dagger \xi_\mu)_B = 2 \sum_{\mu<\nu} b^\dagger_{\mu\nu} b_{\mu\nu} \;. \tag{2.8}$$

That is to say, each ideal Boson carries two units of the valence nucleon number.

Let the Fermion Hamiltonian H_f acting on (SF) take the form :

$$H = \sum_{\alpha} E_\alpha \, \xi^\dagger_\alpha \xi_\alpha + \sum_{\alpha\beta\gamma\delta} P_{\alpha\beta\gamma\delta} \, \xi^\dagger_\alpha \xi^\dagger_\beta \xi_\gamma \xi_\delta \ . \tag{2.9}$$

According to (2.7), we can choose a hermitian expression for the effective Hamiltonian :

$$H_B = H^{(1)} + H^{(2)} \tag{2.10}$$

$$H_B^{(1)} = \sum_{\alpha} E_\alpha \, b^\dagger_{\alpha\nu} b_{\alpha\nu} + \sum_{\alpha\beta\gamma\delta} P_{\alpha\beta\gamma\delta} b^\dagger_{\alpha\beta} b_{\gamma\delta} \tag{2.11}$$

$$H_B^{(2)} = -\sum_{\alpha\beta\gamma\delta} P_{\alpha\beta\gamma\delta} \sum_{\lambda\lambda'} b^\dagger_{\delta\lambda'} b^\dagger_{\gamma\lambda} b_{\beta\lambda'} b_{\alpha\lambda} \ . \tag{2.12}$$

The effective Schrodinger equation takes the form

$$H_B \, \Phi(b^\dagger)|0) = \mathcal{E} \, \Phi(b^\dagger)|0) \ . \tag{2.13}$$

This equation has a number of solutions which do not belong to (SB). However, with the help of the so-called modified Jancovici-Schiff substitution[2)] each eigen-function of H_B can be transformed into an eigen function of H with the same energy (see sect. 4).

3. THE DETERMINATIONS OF $s^\dagger$, $d^\dagger$ AND THE IBA1 HAMILTONIAN

The most general form of the IBA1 Hamiltonian is[3,4)]

$$\begin{aligned} h = \; & \epsilon_s s^\dagger s + \epsilon_d \sum_m d^\dagger_m d_m + \sum_I \tfrac{1}{2}(2I+1)^{\frac{1}{2}} \, C_I[(d^\dagger d^\dagger)_I (\tilde{d}\tilde{d})_I]_0 \\ & + (\tfrac{1}{2})^{\frac{1}{2}} V_1 \{ (d^\dagger d^\dagger)_0 ss + s^\dagger s^\dagger (\tilde{d}\tilde{d})_0 \} \\ & + (\tfrac{5}{2})^{\frac{1}{2}} V_2 \{ [(d^\dagger d^\dagger)_2 \tilde{d}]_0 \, s + s^\dagger [d^\dagger (\tilde{d}\tilde{d})_2]_0 \} \\ & + V_3 s^\dagger s^\dagger ss + (5)^{\frac{1}{2}} V_4 \, s^\dagger (d^\dagger \tilde{d})_0 \, s \end{aligned} \tag{3.1}$$

where

$$\tilde{d}_m = (-1)^m d_{-m} \ . \tag{3.2}$$

Owing to the fact that a s or d-Boson like an ideal Boson carries two units of the valence nucleon number, one has

$$s|0) = d_m|0) = 0 \tag{3.3}$$

$$s^\dagger|0) = \sum_{\mu<\nu} \chi^{(s)}_{\mu\nu} b^\dagger_{\mu\nu}|0) \tag{3.4}$$

$$d^\dagger_m|0) = \sum_{\mu<\nu} \chi^{(d)}_{(m)\mu\nu} b^\dagger_{\mu\nu}|0) \ . \tag{3.5}$$

In the cases that the IBM is actually valid, an eigen-function of h is an approximate eigen-function of H . Thus

$$H_B \, s^\dagger \, 0) = \epsilon_s \, s^\dagger \, 0) \tag{3.6}$$

$$H_B d^\dagger_m |0) = \epsilon_d d^\dagger_m |0) \quad . \tag{3.7}$$

Using

$$H^{(2)}_B s^\dagger |0) = H^{(2)}_B d^\dagger_m |0) = 0$$

one has

$$H^{(1)}_B s^\dagger |0) = \epsilon_s s^\dagger |0) \tag{3.8}$$

$$H^{(1)}_B d^\dagger_m |0) = \epsilon_d d^\dagger_m |0) \quad . \tag{3.9}$$

Now the s- or d-Bosons are supposed to be the lowest elementary excitation with $n = 2$ and $J\Pi = 0^+$ or 2^+. With the help of these supplementary conditions we can determine $\chi^{(s)}_{\mu\nu}$, $\chi^{(d)}_{\mu\nu}(m)$, ϵ_s and ϵ_d from equations (3.8) and (3.9).

Leaving aside the other possibilities, we assume here that the operator $s^\dagger$ or $d^\dagger_m$ is a linear combination of $\{b^\dagger\}$. This means that

$$s^\dagger = \sum_{\mu<\nu} \chi^{(s)}_{\mu\nu} b^\dagger_{\mu\nu} \tag{3.10}$$

$$d^\dagger_m = \sum_{\mu<\nu} \chi^{(d)}_{\mu\nu}(m) b^\dagger_{\mu\nu} \quad . \tag{3.11}$$

Secondly, using all the one-Boson eigen functions of $H^{(1)}_B$,

$$\sum_{\mu<\nu} \chi^{\gamma\Pi J}_{\mu\nu}(M)\, b^\dagger_{\mu\nu} |0) \tag{3.12}$$

and constructing a complete set of creation operators of Bosons with $n = 2$:

$$Q^\dagger_{\gamma\Pi JM} = \sum_{\mu<\nu} \chi^{\gamma\Pi J}_{\mu\nu}(M)\, b^\dagger_{\mu\nu} \tag{3.13}$$

one can expressed H as

$$H^{(1)}_B = \sum_{\mu<\nu} \epsilon_{\gamma\Pi J}\, Q^\dagger_{\gamma\Pi JM} Q_{\gamma\Pi JM} \tag{3.14}$$

where γ is used to distinguish between states with the same (Π JM). Let the s- and d-Bosons correspond to $\gamma = 0$. Thus

$$Q^\dagger_{0+00} = s^\dagger \tag{3.15}$$

$$Q^\dagger_{0+2m} = d^\dagger_m \tag{3.16}$$

$$\epsilon_{0+0} = \epsilon_s \tag{3.17}$$

$$\epsilon_{0+2} = \epsilon_d \quad . \tag{3.18}$$

From the Boson commutation relations satisfied by the Q- and b-operators, one obtains

$$b^{\dagger}_{\mu\nu} = \sum_{\gamma\Pi JM} \chi^{\gamma\Pi J}_{\mu\nu}(M)\, Q^{\dagger}_{\gamma\Pi JM} \qquad (\mu<\nu) \tag{3.19}$$

$$b_{\mu\nu} = \sum_{\gamma\Pi JM} \chi^{\gamma\Pi J}_{\mu\nu}(M)\, Q_{\gamma\Pi JM} \qquad (\mu<\nu) \quad . \tag{3.20}$$

For $\mu>\nu$ one can use

$$b^{\dagger}_{\mu\nu} = -b^{\dagger}_{\nu\mu} \tag{3.21}$$

$$b_{\mu\nu} = -b_{\nu\mu} \quad . \tag{3.22}$$

Using (3.19)-(3.22) to (2.12) and extracting the pure s-d parts, one can determine the coefficients of the interaction terms in (3.1). These coefficients can be expressed as[5] :

$$C_I = \frac{1}{2}(0|\ (\tilde{d}\tilde{d})_{I0}\ H^{(2)}_B\ (d^{\dagger} d^{\dagger})_{I0}\ |0) \tag{3.23}$$

$$v_1 = (\tfrac{1}{2})\ (0\ |\ ss\ H^{(2)}_B\ (d^{\dagger} d^{\dagger})_0\ \ |0) \tag{3.24}$$

$$v_2 = (\tfrac{1}{2})\ (0\ |\ d_0 s\ H^{(2)}_B\ (d^{\dagger} d^{\dagger})_{20}\ |0) \tag{3.25}$$

$$v_3 = (0|\ ss\ H^{(2)}_B\ s^{\dagger} s^{\dagger}|0) \tag{3.26}$$

$$v_4 = (0|\ d_0 s\ H^{(2)}_B\ s^{\dagger} d^{\dagger}_0|0) \quad . \tag{3.27}$$

Equation (3.8)-(3.11) and (3.23)-(3.27) represent a complete set of equations for determining the operators s^{+}, d^{+} and h.

We note that it is enough to start with a smaller subspace of (SF) chosen according to the J-Π conditions of the s- and d-Bosons.

4. TRANSFORMING THE BOSON WAVE FUNCTIONS INTO FERMION WAVE FUNCTIONS

Let $\Phi(s^{\dagger},d^{\dagger})\ |0)$ be an eigen function of h :

$$h\ \Phi(s^{\dagger},d^{\dagger})\ |0) = \mathcal{E}\ \Phi(s^{\dagger},d^{\dagger})\ |\ 0) \quad . \tag{4.1}$$

As has been pointed out in sect. 3, when the IBM is actually valid, $\Phi(s^{\dagger},d^{\dagger})\ |0)$ is an approximate eigen-function of H :

$$H_B\ \Phi(s^{\dagger},d^{\dagger})\ |0) = \mathcal{E}\ \Phi(s^{\dagger},d^{\dagger})\ |0) \quad . \tag{4.2}$$

Using the so-called modified Jancovici-Schiff substitution[2) , one has

$$H_f\ |\Phi_f> = \mathcal{E}\ |\Phi_f> \tag{4.3}$$

where

$$|\Phi_f> = \Phi\ (\ s^{+} \rightarrow \sum \chi^{(s)}_{\mu\nu} \xi^{\dagger}_{\mu} \xi^{\dagger}_{\nu}\ ,\ d^{+}_{m} \rightarrow \sum \chi^{(d)}_{\mu\nu}(m)\ \xi^{\dagger}_{\mu} \xi^{\dagger}_{\nu}\ )|0> . \tag{4.4}$$

In this way every IBM eigen function is transformed into a Fermion wave function which, if different from zero, is an approximate eigen-function of H_f with the same eigenvalue. We note from (4.4) that after the transformation, the s- or d-Bosons in the IBM wave functions become correlated nucleon pairs.

If we had used the usual Boson expansion in place of the Dyson expansion, we would still have obtained (2.10)-(2.12) and all the results given in sect. 3. However, such an approach is not convenient to overcome the difficulties that face the phenomenological IMB due to its violation of the Pauli principle. This is the underlying reason we use the Dyson expansion in this work.

Dyson's finite Boson expansion and the modified Jancovici-Schiff substitution have been generalized to include both Boson and Fermion veriables[6], and therefore the method of the present paper can be generalized and used in the interacting Bose-Fermi systems of nuclei.

One of us (T. S. Yang) is grateful to Dr. R. Hilton and Dr. Cynthia Mather for reading the manuscript.

REFERENCES

1) D. Janssen, F. Donau, S. Frauendorf and R.V. Jolos, Nucl. Phys. A172(1971) 145
2) T. S. Yang and H. J. Mang, (to be published)
3) A. Arima and F. Iachello, Phys. Rev. Lett. 35(1975) 1069
4) A. Arima and F. Iachello, Ann. (N. Y.) 123(1979) 468
5) T. S. Yang and L. M. Yang, (to be published)
6) T. S. Yang, (to be published)

THE PARTICLE-VIBRATION COUPLING MECHANISM

P.F. Bortignon

Istituto di Fisica dell'Università, Padova - Italy*
INFN, Lab. Nazionali di Legnaro - Italy
The Niels Bohr Institute, University of Copenhagen, Denmark

In my talk, I would like to discuss how a number of physical phenomena within the field of nuclear structure can be related and interpreted on the basis of the particle-vibration coupling mechanism[1].

The framework utilized is that of the nuclear field theory (NFT, cf. e.g. ref. 2 and ref. therein), that is a field theory in which the nuclear elementary modes of excitation play the role of the fermionic (quasi-particles) and of the bosonic (shape vibrations, pairing vibrations etc.) free fields.

Because the nuclear vibrational fields are built out of the quasi-particle degrees of freedom, which already exhaust all the nuclear degrees of freedom, it is an essential feature of the product basis used in the NFT to be overcomplete and to violate the Pauli principle. On the other hand, this basis is directly related to the observables of the system and the different experiments (inelastic scattering, one and two particle transfer, etc.) project out only one or two of its components.

The NFT has been shown[2] to provide a graphical perturbative approach to obtain the exact solutions in the product basis, eliminating the above discussed spurious contributions to the different physical quantities to the order of perturbation in which the calculation is carried out.

The rules for evaluating the effect of the couplings between

* Permanent address.

fermions and bosons involve a number of restrictions as compared with the usual rules of perturbation theory, because a significant part of the original interaction acting in the fermion space has already been included in generating the vibrational modes (phonons). The NFT rules read as follows[2]:

1) The energies of the uncoupled particle and vibrational fields are to be calculated by utilizing the Hartree-Fock approximation and the Randon Phase Approximation, respectively. The RPA equations determine the particle-vibration coupling strength, which is not a free parameter.

2) The particle-vibration couplings as well as the interactions acting among the fermions have to be used in all combinations to generate the diagrams appropriate to any given order of perturbation theory.

3) The usual rules of perturbation theory yield the value of each graph, but one must omit bubbles diagrams, i.e. diagrams in which a fermion pair is created and subsequently annihilated having propagated freely between these two times.

4) In initial and final states, proper diagrams involve vibrational and particle modes, but no fermion configuration which can be replaced by a product of vibrational modes.
The external fields acting on the system are allowed to create any state to generate the different diagrams of perturbation theory. The corresponding matrix elements should be weighted with the amplitude of the component through which the final state is excited.

I reminded you of the basic features of the NFT as an introduction to the new developments and applications which will be presented at this workshop by A.Vitturi and E.Maglione[3].

In ref. 3, the vibrational fields are mainly pairing vibrations interacting with a fermion and among themselves. Therefore, in what follows, I will confine myself to discuss some effects connected to the excitation of shape vibrations (see also, e.g., ref. 4 and 5). In this way the different aspects of the recent applications of the NFT will be covered.

The particle-hole excitations correspond, in a classical picture, to oscillations of the nuclear surface about the equilibrium shape[1]. They give rise to collective low-frequency and high-frequency modes, strongly excited in inelastic scattering experiments. Examples are the low-lying quadrupole and octupole collective states, well known, I believe, to each of us. The number of experimental evidences about the high-lying collective states (the so-called giant resonances) of different multipolarity is growing very fastly (see, e.g. ref. 6). They appear at an excitation energy of 10-30 MeV, exhausting an appreciable fraction of the appropriate energy weighted sum rule and show a full width half maximum Γ of 2-8 MeV.

Microscopically, in spherical nuclei the low-frequency and the high-frequency modes are successfully described in the Random Phase Approximation[1,7], that is as a linear combination of 1 particle-1 hole (1p-1h) excitations on a correlated ground state. In the RPA calculations the residual (p-h) interaction has to be the functional derivative of the Hartree Fock hamiltonian with respect to density [7].

In fig. 1 are collected few NFT diagrams describing in lowest order the effects of the particle-vibration coupling I will discuss now*.

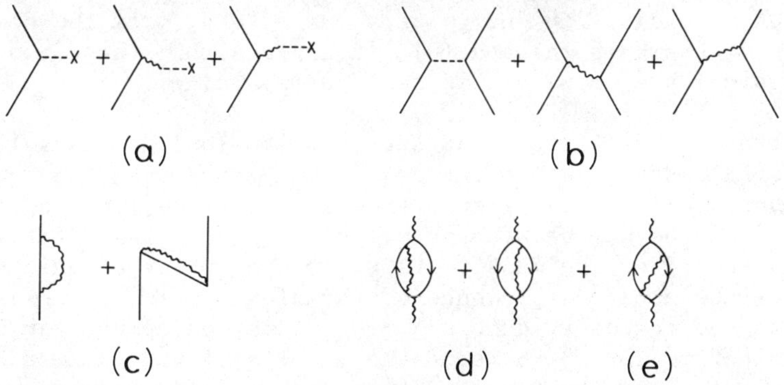

Fig. 1.

I already noticed that the matrix elements of this coupling are completely determined by the HF and RPA solutions of the original fermion hamiltonian.

* I will not discuss the anharmonicity problem in the even-even nuclei and the particle-phonon interaction energy in the odd-A nuclei, because of the lack of time (space) to cover adequately so wide a subject (see, e.g., 1-5).

The three diagrams (a) in fig. 1 describe how a transition between two single-particle states is modified by the coupling to a vibration (wavy line), giving rise, for example, to an effective charge, which can be explicitly calculated and compared with the experimental values. For quadrupole and octupole transitions, e.g., a good agreement is found[1] for values which are often several times larger than the bare charge (cf. also ref. 5). As an indication, the experimental value of (6±1)e for the octupole transition in ^{209}Bi between the ground state and the $1i_{13/2}$ state at 1.61 MeV is well reproduced, the calculated[1] value being 5.4e. In the same way, a value of (0.5-1)e is calculated[1,5] for the neutron effective charge for quadrupole transitions in the lead-isotopes, in agreement with the experimental values.

The exchange of a phonon between two particles gives rise to an effective interaction, as shown by the diagrams in (b), fig. 1. The effect of the coupling can be very large[1,5], the polarization interaction resulting considerably stronger than the bare force. In both the above discussed cases, the contributions of the high-lying isoscalar and isovector modes have to be included. A correct treatment of the coupling to these modes has allowed, e.g., to reproduced in ref. 5 the splitting of 250 keV between two excited 2^+ states in ^{208}Pb at an excitation energy of about 5.7MeV. From the comparison of the diagrams in (a) and (b), it appears that these two different physical phenomena are in fact correlated[1,5].

The graphs in (c) represent the contributions to the self-energy of single-particle (-hole) states generated from the coupling to phonons. For states around the Fermi surface, this coupling acts to reduce the gap between occupied and unoccupied HF orbits, increasing the effective mass to values of the order of unity[8,9] (because of the energy dependence of the diagrams in (c)), in agreement with the empirical evidences[10]. The reduction of the gap calculated in ref. 8 for ^{208}Pb is of the order of few MeV. More extensive calculations about this effect are probably necessary. For deeply-bound hole states and high-lying particles (scattering states) the contribution of the graphs (c) gives rise to a spreading width[11,12,13], to be compared with the experimental measured fragmentations of hole-states[14] or with the matrix elements of the imaginary part of the optical potential. The resulting full width half maximum are typically of the order of 4-5 MeV, in encouraging agreement with the experimental values.

The spreading width of the single-particle and single-hole state will induce a spreading width for the high-lying collective state (giant resonances) too, as shown by the diagrams (d) in fig. 1[11,15]. The process displayed by the graph (e) tends to cancel the contribution of the two in (d), (in the case of isoscalar normal parity vibrations in the intermediate states) reducing the spreading width of the giant modes. We are performing the calculation of this damping effect in different nuclei (^{16}O, ^{40}Ca, ^{90}Zr,

^{120}Sn, ^{208}Pb) and for the giant resonances of different multipolarity and parity (2^+, 3^-, 1^-, 0^+, etc.)[13]. The resulting values are of the order of 2-6 MeV, in promising agreement with the experiments.

The possibility of a dynamical description of the interplay between the shell structure and the collective degrees of freedom in an extremely rich variety of related physical phenomena is well indicated, I hope, by the diagrams of fig. 1 and by the above discussed results.

At the end, let me remind you that the very recent extensions[16] of the NFT to deal with deformed systems open new perspectives for the treatment of strongly anharmonic nuclei.

REFERENCES

1. A.Bohr and B.R.Mottelson, Nuclear Structure, Vol.II, Benjamin, New York (1975).
2. R.A.Broglia, B.R.Mottelson, D.R.Bes, R.J.Liotta and H.Sofia, Phys. Lett. 64B: 29(1976);
   P.F.Bortignon, R.A.Broglia, D.R.Bes and R.Liotta, Phys. Rev. 30C: 305(1977).
3. A.Vitturi, in these proceedings;
   E.Maglione, in these proceedings.
4. I.Hamamoto, Phys. Rep. 10C: 63 (1974).
5. D.R.Bes, R.A.Broglia and B.Nilsson, Phys. Rep.16C: 1(1975).
6. F.E.Bertrand, Ann. Rev. Nucl.Sci. 26: 457 (1976).
7. G.F.Bertsch and S.F.Tsai, Phys. Rep. 18C: 126 (1975).
8. I.Hamamoto and P.Siemens, Nucl. Phys. A269: 199 (1976);
   V.Bernard and N.Van Giai, Orsay preprint IPNO/TH 79-55;
   R.P.J.Perazzo, S.L.Reich and H.M.Sofia, Nucl. Phys. A339: 23 (1980).
9. Cf. also e.g. R.Sartor and C.Mahaux, Phys. Rev. 21C:2613(1980) and references therein.
10. Cf. e.g. K.Bear and P.E.Hodgson, J.Phys. G4: L 287 (1978);
    M.M.Giannini, G.Ricco and A.Zucchiati, Ann.Phys. 124:208(1980).
11. G.F.Bertsch, P.F.Bortignon, R.A.Broglia and C.H.Dasso, Phys. Lett. 80B: 161 (1979).
12. Cf. also, e.g., V.G.Soloviev, Ch.Stoyanov, A.I.Vdovin, Nucl. Phys. A342: 261 (1980);
    N.Van Giai, invited talk at the International Symposium on Highly Excited States in Nuclear Reaction, May 1980, Osaka, to be published and references therein.
13 P.F.Bortignon and R.A.Broglia, Nucl. Phys., to be published.
14. Cf. e.g. J.Guillot et al., Phys. Rev. 21C: 879 (1980) and references therein;
    G.J.Wagner, invited talk at the International Symposium on Highly Excited States in Nuclear Reaction, May 1980, Osaka, to be published.

15. Cf. also e.g. J.S.Dehesa, S.Krewald, J.Speth and A.Faessler, Phys. Rev. 15C: 1858 (1977);
V.G.Soloviev, Ch. Stoyanov, A.I.Vdovin, Nucl. Phys. A288:504 (1977);
G.E.Brown, J.S.Dehesa and J.Speth, Nucl. Phys. A330: 290 (1979).
16. Cf. e.g. V.Alessandrini, D.R.Bes, B.Machet, Nucl. Phys. B142: 489 (1978).

NUCLEAR STRUCTURE INFERRED FROM HEAVY ION COULOMB EXCITATION

Douglas Cline

Nuclear Structure Research Laboratory*
University of Rochester
Rochester, N.Y. 14627

INTRODUCTION

Low-lying nuclear spectra in all nuclei exhibit to some extent, features characteristic of quadrupole collective modes of motion. Thus it is interesting to ascertain to what extent it is possible to describe nuclear properties using phenomenological collective models such as the Interacting Boson Model[1] or geometrical models since they can give a clear insight into the structure of a complex many body system. Nuclear E2 properties are a direct and unambiguous measure of the collective shape parameters for quadrupole collectivity. Unfortunately, this sensitivity could not be exploited fully in the past due to the sparseness of the available E2 data. Consequently most collective models have emphasized fits to level energies because of the extensive body of data available. However, the level energies of the collective states are sensitive to both microscopic and collective parameters. As a result the collective parameters extracted from level energies can be ambiguous. Coulomb excitation selectively excites collective states and the excitation cross sections are a direct measure of the strong E2 matrix elements. It is just these enhanced E2 matrix elements that collective models should predict the best. Recent advances in the field of Coulomb excitation makes it feasible, for the first time, to measure essentially all the E2 matrix elements for the low lying nuclear levels. These data should prove to be a stringent test of collective models. In particular it is now practical to use a simple model independent method for extracting collective parameters from E2 data. Application of these techniques to study heavy

*Supported by the National Science Foundation.

rotational and shape transitional nuclei and the model implications are discussed in this and the adjoining paper[2].

RECENT ADVANCES IN COULOMB EXCITATION

Heavy ion beams such as ^{208}Pb which recently become available from the accelerators at GSI and Berkeley have produced a renaissance in the field of Coulomb excitation. Now it is possible for the first time to exploit fully the considerable potential of Coulomb excitation for studying nuclear spectroscopy. Coulomb excitation data are most sensitive to E2 matrix elements involving the highest states excited. Consequently it is necessary to vary the strength of the interaction in order to measure accurately the E2 matrix elements for all the states excited. This can be achieved by varying either (1) the projectile Zp, (2) the bombarding energy or (3) the scattering angle. Both Zp and scattering angle are varied in our work.

Gamma-ray spectroscopy using Ge gamma detectors is the only viable experimental technique for observing multiple Coulomb excitation of high spin states when heavy ions are utilized. Unfortunately, the recoil velocities of the excited target nuclei can be large, i.e. 5% to 12% of the velocity of light. Consequently, the Doppler shift and more importantly the Doppler broadening in the gamma detectors are large. The problem of Doppler broadening can be overcome by using thin targets so that the excited nuclei recoil into vacuum, and by observing the de-excitation gamma-rays in coincidence with scattered ions detected at known scattering angles, in order to specify the recoil direction and velocity. Thus the individual detected gamma ray signals can be corrected for the Doppler shift on an event by event basis. The remaining large contribution to the Doppler broadening is due to the finite size of the gamma detectors and this is minimized by placing the gamma detectors in the recoil direction where the Doppler shift is a maximum and the Doppler broadening is a minimum. The resulting gamma energy resolution for 1 MeV typically is better than 1%. Large area position sensitive parallel plate proportional detectors or silicon detector arrays are used to detect the scattered ions over a wide angular range. The decay scheme and spin values are deduced from (1) the magnitude of the γ-ray yields, (2) the scattering angle dependence, (3) the excitation functions, (4) the Zp dependence, (5) measurements of the γ-ray multiplicity using an array of NaI detection in coincidence with the Ge detection and (6) from the gamma-ray angular distributions.

Groups at both GSI and Berkeley enthusiastically started using the ^{208}Pb beam for multiple Coulomb excitation when these beams became available about 3 years ago. The analysis of these data proved to be extremely difficult because of the dramatic increase in the number of unknown E2 matrix elements involved when many levels are excited. For shape transitional nuclei there can be $\sim$100 unknown E2 matrix elements. This forced both the Rochester[3] and the GSI groups to develop least squares Coulomb excited codes to fit the unknown E2 matrix elements to the measured γ-ray yields. All analyses of heavy-ion multiple Coulomb excitation data in the past have made model dependent assumptions to account for the influence of the static quadrupole moments and the Eλ matrix elements involving side bands in order to extract the ground band B(E2) values from the data. Fortunately, for strongly deformed nuclei these model dependent assumptions are reasonable since the influence of the static moments and side bands on the ground band excitation is relatively small. A model dependent analysis of multiple Coulomb excitation data for shape transitional nuclei is inadequate because (1) the coupling between the various collective bands is strong and the static moments produce large changes in the cross-sections and (2) the various collective models give widely different predictions for the static moments and matrix elements involving the side bands. We have found that we can extract all the diagonal and off-diagonal E2 matrix elements including the relative signs of the matrix elements, in a model independent analysis of multiple Coulomb excitation data recorded over a wide range of scattering angles using a wide range of projectile Z values. The new least squares fitting code we have developed at Rochester[3] is crucial for such a model independent analysis.

I believe that an important breakthrough has been made in the field of Coulomb excitation. Now for the first time we have the necessary heavy ion beams and have developed the experimental and analysis techniques required to measure the complete set of E2 matrix elements for the low lying collective states in nuclei. That is, both static moments and the relative signs and magnitudes of off-diagonal matrix elements can be measured.

EXTRACTION OF COLLECTIVE PARAMETERS FROM E2 MATRIX ELEMENTS

Collectivity produces strong correlations of the E2 matrix elements and there are far fewer significant collective variables than there are data. The usual method of comparing a list of the experimental E2 matrix elements with the model values exhibits neither the uniqueness nor the sensitivity of the data to the collective model parameters. In addition such comparisons do not show whether the discrepancies between the experimental and theoretical values reflect a fundamental failure of the model or just deficiencies in the collective model parameters used. Considerably more insight is obtained by projecting the collective degrees of freedom from both the data and the model calculations for comparison. These projected collective model parameters more clearly show which collective parameters are determined by the data and the goodness of the collective model description is more clearly exhibited. Since the E2 properties are primarily sensitive to the quadrupole shape parameters the natural collective parameters are the E2 moments in the instantaneous principal axis frame. If the collectivity is strong then the E2 parameters in the principal axis frame will be similar for all states belonging to a given collective band. It is difficult to use model calculations to perform the projection because the complete Bohr collective Hamiltonian for quadrupole deformation involves 6 inertial and 1 potential energy terms all of which are complicated functions of Bohr's quadrupole shape parameters β and γ. It is possible to interpret experimental electromagnetic matrix elements directly in terms of collective degrees of freedom for any state without recourse to a model. The method uses rotational invariant products of multipole operators to directly relate properties in the principal axis frame to those in the laboratory frame[4,5,6].

The electromagnetic multipole operators are spherical tensors and thus zero coupled products of such operators can be formed which are rotationally invariant. That is, these products are identical in the instantaneous intrinsic frame and the laboratory frame. Consider the particular case of the electric quadrupole operator. The electric quadrupole tensor can be rotated into an instantaneous principal axis frame which has only two non-zero quadrupole moments. By analogy with Bohr's parameters ($\beta\gamma$) we can express the principal axis frame electric moments in terms of two parameters Q, δ, where $E(2,0)\equiv Q\cos\delta$ and $E(2,+2)=E(2,-2)\equiv Q/\sqrt{2}\sin\delta$ and $E(2,1)=E(2,-1)=0$. Zero coupled products of E2 operators can be evaluated in the principal axis frame, e.g.

$$\{E2 \times E2\}^0 = 1/\sqrt{5}\, Q^2$$

$$\{[E2 \times E2]^2 \times E2\}^0 = -\sqrt{2/35}\, Q^3 \cos 3\delta$$

$$\{[E2 \times E2]^0\, [E2 \times E2]^0\}^0 = 1/5\, Q^4$$

$$\{[(E2 \times E2)^2 \times E2]^0\, [E2 \times (E2 \times E2)^2]^0\}^0 = 2/35\, Q^6 \cos^2 3\delta .$$

Similar rotationally invariant products of the shape deformation tensor α^2_μ can be expressed in terms of the intrinsic frame parameters ($\beta\gamma$). However, in the case of the electric quadrupole tensor, it is possible to evaluate expectation values of the E2 invariants in the laboratory frame using experimental E2 matrix elements. The expectation values of the invariants can be written as sums of products of E2 reduced matrix elements by making intermediate state expansions. For example

$$\langle s|[E2\times E2]^0|s\rangle = \frac{(-)^{2s}}{\sqrt{2s+1}} \sum_r \langle s||E2||r\rangle\langle r||E2||s\rangle \begin{Bmatrix} 2 & 2 & 0 \\ s & s & r \end{Bmatrix} .$$

These sum rules can be evaluated using experimental data, it the relative signs and magnitudes of the E2 matrix elements are available. Thus, in principle, the expectation values, for a state s, of all the rotationally invariant products of E2 operator can be evaluated directly using sum rules of the type shown. These directly determine the distribution of the intrinsic frame electric quadrupole moments, parametrized by Q and δ. The intrinsic frame electric quadrupole moments distribution can be plotted as contours on a Q-δ plot analogous to the usual β-γ plot as illustrated in Figure 1. Knowing the values of the various invariants, for a state s, directly determines the centroids, $\langle s|Q^2|s\rangle$ and $\langle s|Q^3\mathrm{Cos}3\delta|s\rangle$, the variances, $\sigma(Q^2)^2 = \langle Q^4\rangle - \langle Q^2\rangle^2$ and

$$\sigma(\mathrm{Cos}3\delta)^2 = \frac{\langle Q^6\mathrm{Cos}^2 3\delta\rangle}{\langle Q^6\rangle} - \left(\frac{\langle Q^3\mathrm{Cos}3\delta\rangle}{\langle Q^3\rangle}\right)^2,$$

skewness, cross correlation coefficients etc., describing the distribution in the Q-δ plane of the expectation value of the electric moments for the state s.

Although the present monopole sum rule technique has been discussed in the context of its application to the collective model, the method is completely model independent and is applicable to any spherical tensor operator. These invariants can be evaluated exactly for model calculations and approximately using experimental data and are equivalent to observables. The significance and usefulness of presenting the experimental data in the form of model independent invariants depends on the degree to which the nuclear properties can be described in terms of a few collective degrees of freedom. That is, if collective effects are important then the invariants for various states will have reasonable values and be correlated indicating that a few collective degrees of freedom can

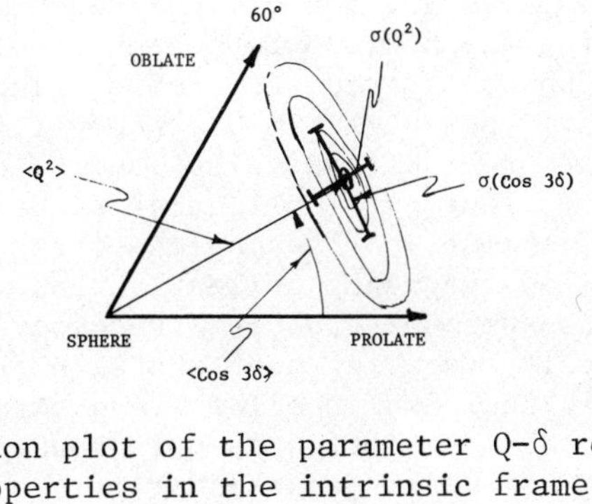

Fig. 1. Distribution plot of the parameter Q-δ required to define the E2 properties in the intrinsic frame. All possible E2 moments are defined by the region $Q \geqslant 0$ and $0° \leqslant \delta \leqslant 60°$.

effectively describe a wealth of data. On the other hand, if collective effects are absent then the invariants will be uncorrelated and as such would be a useless representation of the data. This model independent method directly determines the distribution of the intrinsic frame electric moments. Model assumptions can be used to relate the Q-δ E2 distributions to the β-γ shape distribution if so desired.

The practical problems associated with evaluating sum rules have severely limited the use of sum rule techniques for analyzing experimental data. The sensitivity of sum rules to missing strength, due to incomplete summation, usually increases markedly with the order and complexity of the sum rule. Fortunately zero-coupled products of four or more operators can be formed with various spins for the intermediate couplings. These different intermediate spin couplings lead to sum rules involving summations over different sets of data. Identities relate these various intermediate spin zero coupled products thus allowing a self consistency check of the related sum rules. For example, there are three independent sum rules associated with determining $\sigma(Q^2)$ and eight independent sum rules associated with determining $\sigma(\text{Cos}3\delta)$. These identities allow a direct test of completeness and convergence of the higher order sum rules. These identities have been used to study the convergence properties of the sum rules in conjunction with E2 matrix elements predicted by various model calculations. It was found that for E2 the centroids and variances for low-lying states in collective nuclei can be reproduced using the E2 properties measurable by Coulomb excitation. This sum rule method is useful for any spherical tensor operator for which the strength is localized to a region amenable to investigation.

The low-lying levels contain only $\approx$10% of the E2 energy weighted sum rule strength in all nuclei. The giant E2 resonance, which we have ignored so far, could contain the other 90% of this strength. Fortunately we are considering non-energy weighted sum rules. About 90 to 95% of the non-energy weighted E2 strength is in the lowest few states for the strongly deformed nuclei and greater than 70% for any nucleus for which a collective model description would be considered reasonable. The high lying E2 strength is obviously non-negligible. However, the separation of the E2 strength into two quite distinct and separate collective modes is important. Thus on a collective picture we have a high frequency collective mode, corresponding to the giant resonance, superimposed in a low frequency collective mode. Most collective models implicitly assume only the low frequency mode. The same result of effectively averaging over the high frequency mode may be achieved by only including the low lying strength in our sum rules. Of course this assumption is only reasonable if the two modes are distinct and weakly coupled or if the high frequency oscillation about the low frequency shape is relatively small.

I have tried to show that electromagnetic data can be expressed directly in a form which is ideal for studying the extent to which collective phenomena are important in nuclei. The power of the sum rule technique is that it expresses the data in a form which clearly exhibits the degree of collectivity without using model calculations. That is, a wealth of data can be transformed into a form which clearly shows the extent to which the data are correlated due to collectivity and which features are a manifestation of microscopic structure. The recent developments in the field of Coulomb excitation makes it possible to determine all the E2 matrix element required to apply this sum rule technique to the low lying levels in nuclei. This model independent method is equally useful for projecting the collective degrees of freedom out of model calculations, where it is possible to evaluate the sum rules exactly. For example, it provides an alternative[7] method for projecting collective shape parameters out of the Interacting Boson Model[1].

STRONGLY DEFORMED NUCLEI

The termination of rotational bands and the large retardation of the B(E2) values near this terminator are two general features discriminating the Interacting Boson Model from most other collective models. There is now considerable evidence, from Coulomb excitation and lifetime measurements that neither the termination nor the retardation of B(E2) values are observed in the ground band of strongly deformed nuclei. An early demonstration[8] of this is shown in Figure 2. In this and many other examples[9,10] the ground band B(E2) values obey the rigid spheroidal rotor relation in conflict with the I.B.A. predictions. These data also demonstrate that the spin dependence of the moment of inertia is not due to centrifugal stretching implying that Coriolis antipairing is causing the behavior. The failure of the Interacting Boson Model at high spin values implies that either the number of active boson must be increased or the addition of higher ℓ boson are required.

THE Os-Pt SHAPE TRANSITIONAL NUCLEI

The geometrical collective model is remarkably successful in strongly deformed nuclei. Thus it is interesting to ascertain the extent to which the geometrical collective model is applicable to shape transitional nuclei such as the Os-Pt isotopes. The B(E2) values in the Os-Pt nuclei are enhanced; i.e. 50 to 100 single particle units. Thus it is plausible that collective models are still useful for correlating the nuclear properties of the low-lying levels in these nuclei. It has been known for many years that the smooth transition in the energy level spectra through this mass region is characteristic of a prolate to oblate shape transition. The experimental evidence for non-axial collective motion

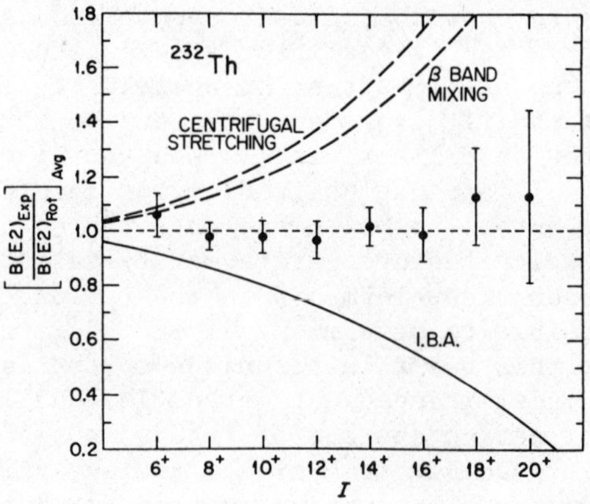

Fig. 2. The measured B(E2;I→I-2) normalized to the axially symmetric rigid rotor values for the ground band of ^{232}Th. The I.B.A. calculations assume 12 bosons outside a ^{208}Pb core and the SU(3) limit. The centrifugal stretching model values are those that result from attributing deviations of the ground-band moment of inertia to centrifugal stretching.

in these nuclei is quite convincing. The static quadrupole moments[11,12] shown in Figure 3 exhibit just such a prolate to oblate shape transition. The E2 data imply an average value of the E2 asymmetry parameter δ which changes smoothly for 12° in ^{182}W to 35° in the Pt isotopes. These are similar to the γ values derived using the triaxial rigid rotor model and the energies of the 2_1^+, 2_2^+ and 4_1^+ states for the W-Os isotopes. For example, using these γ values and the $B(E2:0_1^+\to 2_1^+)$ reproduces $Q_{2_1^+}$ for these nuclei. The Kumar-Baranger[14], I.B.M.[15] and B.E.T[16] calculations all predict appreciable γ-softness yet they produce agreement comparable with the triaxial rigid rotor model for the low-lying states because the data are sensitive to only the centroids of the deformation parameters and not the softness. Note that the I.B.M. predictions of $Q_{2_1^+}$ deviate markedly from the data from $A \geqslant 190$.

Recently[17,18] we made the first measurements of B(E2) values sensitive to γ-softness in order to differentiate between these contradictory collective models. The enhanced B(E2) values, for which collective model predictions should be reliable, which are sensitive to γ softness are the $\Delta I=2$ transitions in the γ band, the $\Delta I=0$ ground to gamma band transitions and the transitions involving the 0^+ double γ-phonon state. The even mass isotopes of Os and Pt were Coulomb excited[17,18] using a ^{136}Xe beam from the Berkeley SuperHILAC. Figure 4 illustrates that the Coulomb excitation γ-ray

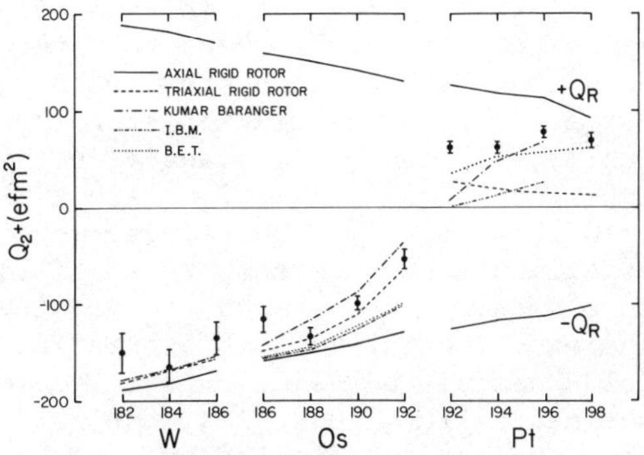

Fig. 3. Static electric quadrupole moments of the 2_1^+ states in even W, Os and Pt isotopes[11,12]. The solid lines correspond to values derived from the $B(E2:0_1^+\rightarrow 2_1^+)$ assuming the rigid spheroidal rotor model. Negative Q_2+ corresponds to prolate intrinsic shapes and positive Q_2+ to oblate intrinsic shapes. The rigid triaxial rotor model values were derived from the $B(E2:0_1^+\rightarrow 2_1^+)$ using a γ value determined by the relative energies of the 2_1^+, 4_1^+ and 2_2^+ states.

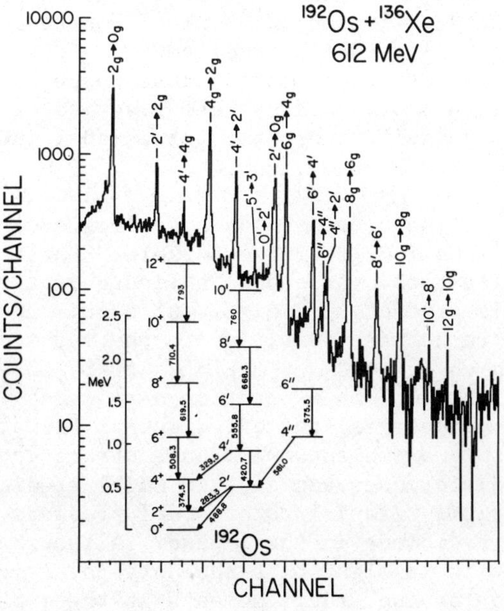

Fig. 4. Typical coincidence γ-ray spectrum and decay scheme derived from Coulomb excitation of ^{192}Os using 612 MeV ^{136}Xe ions.

spectra look quite different from what is seen for strongly deformed rotational nuclei. The most notable features are: (1) the ground state, γ-band and 4^+ band states connected to the ground state by the same number of enhanced E2 excitation steps are populated approximately equally; (2) the γ-ray transition energies of the $4'\rightarrow2'$, $6'\rightarrow4'$, $8'\rightarrow6'$ and $10'\rightarrow8'$ transitions are similar to the $6\rightarrow4$, $8\rightarrow6$, $10\rightarrow8$ and $12\rightarrow10$ respectively. There were more unknowns than data for these experiments because each γ-ray yield depends on several E2 matrix elements. Consequently a model dependent analysis was made where the experimental Coulomb excitation γ-yields were compared with yields calculated using extreme geometric collective models. The experimental level energies, not calculated ones, were used in the Coulomb excitation calculations. Both the rigid triaxial model and an empirical γ-soft collective Hamiltonian were used to calculate the E2 matrix elements for the Coulomb excitation calculations. The centroid of the γ-value was fixed to the value given by the experimental E2 data. All these calculations reproduced the experimental γ-ray yields for the ground band but there were factors of 5-10 difference between the predictions of the models for the yields for transitions involving the γ-band and 4^+ band. The experimental yields were in good agreement with the rigid triaxial rotor model and in clear contradiction with the predictions of γ-soft geometric collective models. Moreover, the experimental yields for the 0^+ states in $^{194,196}Pt$ were only 1/4 of that predicted by the γ-soft model. The conclusion drawn[17,18] from this model dependent analysis was that the E2 properties obeyed the rigid triaxial rotor relation in complete contradiction with several collective models all of which predicted γ-soft shapes. This unexpected result plus the complicated dependence of the γ-ray yields on E2 matrix elements provoked us into a series of experiments designed to provide sufficient data for a model independent analysis to be made.

The Os and Pt isotopes have been Coulomb excited using ^{208}Pb, ^{58}Ni and ^{40}Ca ions from the Berkeley, Rochester, and Brookhaven accelerators. As shown in Figure 5, the different collective models predict very different static E2 moments and the extracted B(E2) values are sensitive to the assumed static moments. We[19] are now in the process of making a truly model independent analysis of all the Coulomb excitation data which will determine both the static moments and the off-diagonal E2 matrix elements. The extracted diagonal moments will have errors appreciably smaller than the large differences between the model predictions shown in Figure 5. The rigid triaxial rotor model predicts dramatic changes in the diagonal E2 moment in the γ-band. Although the analysis is still incomplete it is clear that the data for ^{194}Pt imply a less dramatic behavior of the static moments in the γ-band. These measured static moments result in B(E2) values involving the γ-band that do not agree as well with the rigid triaxial rotor model as was obtained in a analysis assuming model values for the moments[18].

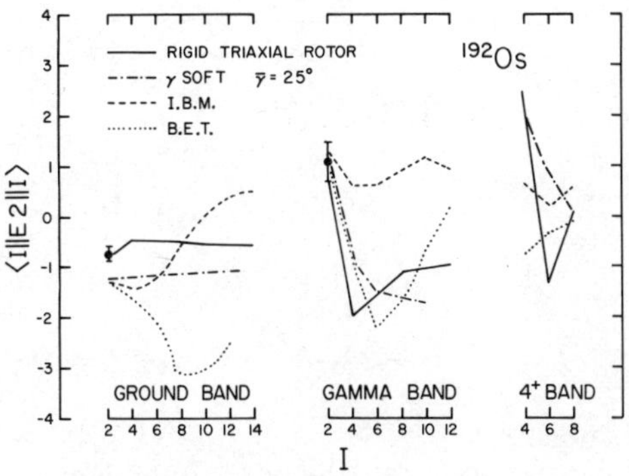

Fig. 5. The diagonal E2 matrix elements for ^{192}Os.

The experimental B(E2) values for ^{194}Pt extracted from the Coulomb excitation data[19] are illustrated in Figure 6. Only the most enhanced inter and intra band transitions are compared with the model predictions. The ground band B(E2) values are in good agreement with the values from G.S.I.[10] The B(E2: $0_2^+ \to 2_2^+$) is about one order of magnitude weaker than predicted by the I.B.M[15], B.E.T.[16] and γ-unstable models. This B(E2) is zero in the rigid triaxial rotor model. Both enhanced B(E2) values deexciting the 4_γ^+ state are sensitive to γsoftness and suggest considerable γ rigidity while the B(E2) values involving the 6_γ^+ are closer to the γ-soft model predictions. The B(E2) values for both ^{192}Pt and ^{196}Pt are closer to the rigid triaxial rotor model than the γ-soft models[20]. It appears that an additional term is required in the I.B.M. Hamiltonian for the Pt nuclei to introduce the amount of γ-rigidity required by these E2 data. In addition, although it is not apparent in Figure 6, the G.S.I. and the Berkeley data for the ground state band are not consistent with the retardation of the high spin B(E2) values predicted by the I.B.M. calculations[15].

The B(E2) values for ^{192}Os shown in Figure 7. The I.B.M. calculations are a factor of two too low for the $2_\gamma^+ \to 2_g^+$ transition and 50% too high for the $2_\gamma^+ \to 4_\gamma^+$ transition. These results are typical of the situation for the other Os isotopes. No model reproduces all the measured B(E2) values.

Although the E2 matrix element shown are not the best values it is interesting to make sum rule analyses of the current set of experimental E2 matrix elements and those predicted using the models. The experimental values of $\langle I|Q^2|I\rangle$ and $\sigma(Q^2)$ for the ground and γ-bands of ^{192}Os and ^{194}Pt are similar to the values extracted using the I.B.M. and B.E.T. E2 matrix elements and are in clear conflict with the harmonic vibrator model. The experimental ground band

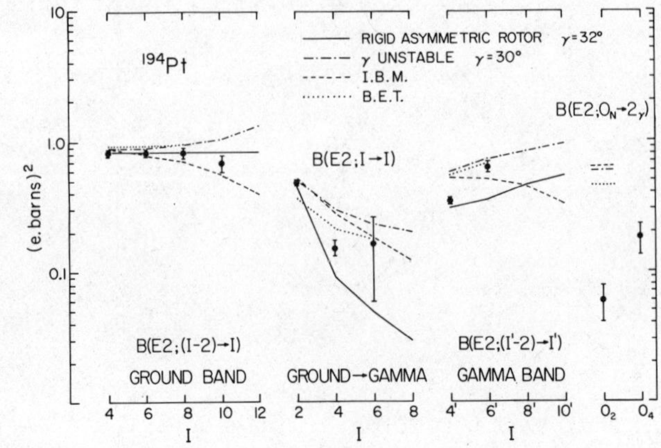

Fig. 6. B(E2) values for ^{194}Pt derived from the present data[19].

centroids of the asymmetry parameter, Cos3δ, deviate markedly from the predictions of both the I.B.M. and B.E.T. predictions for ^{192}Os. The present data imply ground band values of σ(Cos3δ) of around 0.4, i.e. a moderate amount of γ-softness. This is in agreement with the I.B.M. and B.E.T. predictions for ^{192}Os but not as γ-soft as these models predict for ^{194}Pt. The dependence of the E2 matrix elements on σ(Cos3δ) is very non linear. Many of the enhanced B(E2) values are relatively close to the rigid triaxial rotor model values for σ(Cos3δ) $\leqslant$0.4. Consequently it is possible to have E2

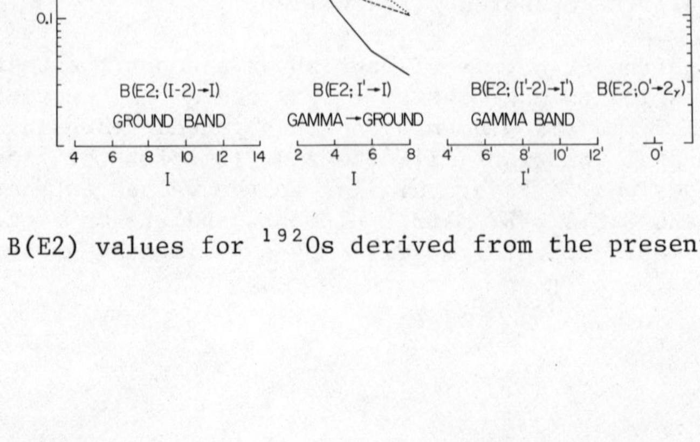

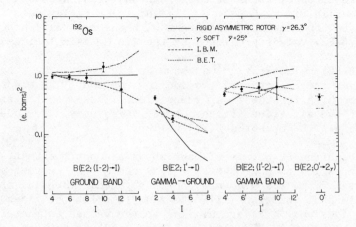

Fig. 7. B(E2) values for ^{192}Os derived from the present data[19].

properties suggestive of rigid triaxial rotation with a reasonable amount of γ softness. The slightly larger value of $\sigma(\text{Cos}3\delta)$ predicted by the I.B.M. calculations[15] for the Pt nuclei is sufficient to give E2 matrix elements which more closely resemble the γ-soft rather then the γ-rigid predictions of the geometric collective models.

The results presented here were derived from a subset of our data. They are inadequate to differentiate clearly between the different models in that all the models appear to be deficient. The analysis now in progress of all of our Coulomb excitation data using the new least squares computer code should give appreciably more E2 information and more accurate values. In particular, values of the static moments will be obtained which should serve to differentiate between the various collective models as illustrated in Figure 5. In spite of the deficiencies in the present data it is apparent that the collective parameters predicted by the I.B.M. calculations have too large a value of $\sigma(\text{Cos}3\delta)$ for the Pt isotopes and too prolate an asymmetry centroid for ^{192}Os.

CONCLUSIONS

Even though the field of Coulomb excitation is 1/4 century old it is only now that this powerful techniques can be exploited to the fullest extent. The availability of a wide range of heavy ion beams from the newest accelerators, recent developments in experimental techniques and especially the new breakthrough in the development of powerful analysis programs makes it possible, for the first time, to measure essentially the complete set of E2 matrix elements for the low lying states in nuclei. That is, static moments, B(E2) values and the signs of the many interference terms involving E2 matrix elements. These data will provide stringent tests of nuclear models.

These developments in the field of Coulomb excitation makes it possible, for the first time, to effectively exploit the powerful model independent sum rule method developed several years ago[4,5,6]. The power of the sum rule technique is that it allows a wealth of data to be transformed into a form which clearly shows the extent to which the data are correlated due to collectivity and which features are a manifestation of microscopic structure. Moreover, the method is equally useful for projecting out the significant collective shape parameters from theoretical calculations.

Although the analysis of many of the Coulomb excitation experiments is incomplete, certain implications can be drawn regarding the Interacting Boson Model. The first is that all the evidence contradicts the cut-off of the ground band predicted by the I.B.M. model for both strongly deformed nuclei and the Os-Pt nuclei. The second is that the I.B.M. predictions for the Pt nuclei correspond to geometrical shape parameters that are slightly too γ-soft while the magnitude of the asymmetry is incorrect for ^{192}Os. These results imply that the I.B.M. in its present form fails at high spin values. It will be especially valuable if the I.B.M. Hamiltonian or configuration space could be modified in order to extend to high spin the usefulness of this simple and elegant theory.

ACKNOWLEDGEMENTS

The author wishes to thank his many collaborators involved in this work[11,12,17,19]. Especially valuable contributions were made by T. Czosnyka in developing the Coulomb excitation code, L. Hasselgren in proving that a model independent analysis of Coulomb excitation data was possible and C. Y. Wu for analyzing the Os-Pt data.

REFERENCES

1. F. Iachello, ed., "Interacting Bosons in Nuclear Physics," Plenum Press, New York (1979).
2. L. Hasselgren and D. Cline, "Interacting Bose-Fermi Systems in Nuclei," F. Iachello, ed., Plenum Press, New York (1980).
3. T. Czosnyka, C. Y. Wu and D. Cline, unpublished.
4. K. Kumar, Phys. Rev. Lett. 28:249 (1972).
5. D. Cline, "Proc. of the Orsay Colloquium on Intermediate Nuclei," Foucher, Perrin and Veneroni, ed., p. 4 (1971).
6. D. Cline and C. Flaum, "Proc. of the Int. Conf. on Nucl. Struct. Studies Using Electron Scattering and Photoreaction, Sendai 1972," K. Shoda and H. Ui, ed., Vol. 5, p. 61, Tohoku University, Sendai (1972).
7. J. N. Ginocchio and M. W. Kirson, Phys. Rev. Lett. 44:1744 (1980).
8. M. W. Guidry, P. A. Butler, P. Colombani, I. Y. Lee, D. Ward, R. M. Diamond, F. S. Stephens, E. Eichler, N. R. Johnson and R. Sturm, Nucl. Phys. A266:228 (1976).
9. A. Gelberg, "Interacting Bose-Fermi Systems in Nuclei," F. Iachello, ed., Plenum Press, New York (1980).
10. J. Stachel, ibid.
11. P. Russo, D. Cline, J. K. Sprinkle, R. P. Scharenberg and P. B. Vold, Bull. Am. Phys. Soc. 22:545 (1978).
12. J. K. Sprinkle, D. Cline, P. Russo, R. P. Scharenberg and P. B. Vold, Bull. Am. Phys. Soc. 22:545 (1978).
13. A. S. Davydov and G. F. Filippov, Nucl. Phys. 8:237 (1958).
14. K. Kumar and M. Baranger, Nucl. Phys. A122:273 (1968).
15. R. Bijker, A. E. L. Dieperink, O. Scholten and R. Spanhoff, Preprint (1980).

16. K. J. Weeks and T. Tamura, Phys. Rev. C22:1323 (1980).
17. I. Y. Lee, D. Cline, P. A. Butler, R. M. Diamond, J. O. Newton, R. S. Simon and F. S. Stephens, Phys. Rev. Lett. 39:684 (1977).
18. D. Cline, "Lecture Notes in Physics 92, Nuclear Interaction," B. A. Robson, ed., p. 39, Springer Verlag (1979).
19. C. Y. Wu, D. Cline, T. Czosnyka, L. Hasselgren, P. Russo, J. K. Sprinkle, I. Y. Lee, P. A. Butler, R. M. Diamond, F. S. Stephens, C. Baktash and S. Steadman, Bull. Am. Phys. Soc. 25:605 (1980).
20. M. Vergnes, "Interacting Bose-Fermi Systems in Nuclei," F. Iachello, ed., Plenum Press, New York (1980).

A CONSIDERATION OF BAND STRUCTURE IN TERMS OF ENERGY EXPANSION COEFFICIENTS

Mitsuo Sakai

Institute for Nuclear Study
University of Tokyo
Tanashi-shi, Tokyo 188

1. INTRODUCTION - Retrospective -

The presence of band structures of the nuclei in the whole nuclear region was first discussed in detail in 1967.[1-3] The essential idea lies in coordinating the excited levels in the vertical way instead in the lateral way. The excited level system in so-called vibrational nuclei were usually discussed in the latter way at that time. Two figures of level structure presented in the original paper[3] are shown in figs. 1-A and 1-B together with the levels discovered since then. The survey of the level structure revealed that the band like structure in the vibrational and transitional regions which was called according to the analogy with the collective bands as quasi ground-state rotational band, quasi-beta band and quasi-gamma band tends smoothly to the collective bands in the deformed region. It is worth nothing that the original paper[3] already pointed out the presence of band like structure in the superstructure of excited level system in the single closed-shell nuclei.

The author remembers a contribution to the International Conference on Nuclear Structure at Tokyo in 1967 submitted by Thosar et al.. They claimed a presence of a band-like structure in the odd nucleus of ^{117}In.

Such a recognition of the appearance of band structure was ignored for a long time. However, nowadays the band structure is a well known fact accepted in the nuclear scientific community. It appears not only in the even-even nuclei but also in the even-odd nuclei situating in the nuclear region close to the single closed-shell nuclei and even in the single closed-shell nuclei. It proves the validity of the findings presented in the earlier papers.[3,4]

The present author is very pleased to participate in this dedicated workshop to the band structure because he was one of the workers who discussed the band structure in the very early days.

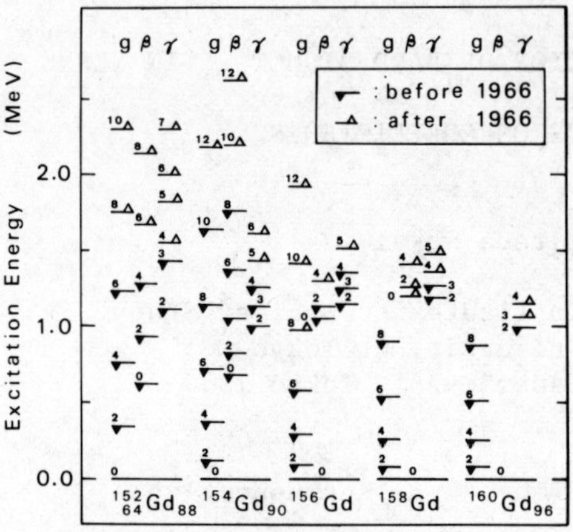

Fig. 1-A. Classification of the excited level structure in the Gd isotopes. The levels marked by ▾ are reported in the original paper and those marked by ▴ are those discovered since then.

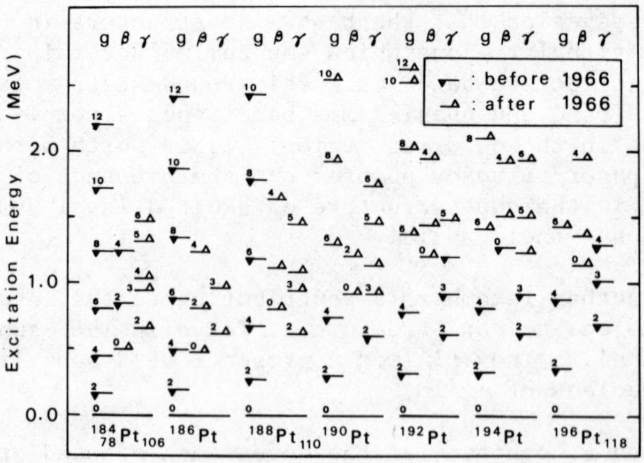

Fig. 1-B. Classification of the excited level structure in the Pt isotopes. Marks on the levels have the same meaning as in the fig. 1-A.

2. BEHAVIOR OF COEFFICIENTS OF ENERGY EXPANSION FOR THE QUASI GROUND-STATE ROTATIONAL BAND AND THE QUASI-GAMMA BAND

The energy expansion for bands has been discussed by many workers. The following two formulae for the quasi ground-state rotational bands (g.b.) and the quasi-gamma bands (γ.b.) are popular[5]:

$$E(I) = AI(I+1) + BI^2(I+1)^2 + \cdots \quad \text{for g.b.} \tag{1}$$

and

$$E(I) = E_2 + AI(I+1) + BI^2(I+1)^2 + C(-1)^I(I-1)(I+1)(I+2) + \cdots \quad \text{for } \gamma\text{.b..} \tag{2}$$

Of course there are many sophisticated formulae of which the most famous ones are those in the framework of the VMI model[6] and the IBM model.[7]

In the present paper we will introduce a new technique for investigating the band structure. So far the energy-expansion formulae like eqs. (1) and (2) were discussed by comparing the experimental level energies with the calculated energy values deduced from the formulae with the fixed expansion coefficients which are usually determined by using energies of several levels with low spin. However, it may be more natural to consider the coefficients as variable. Based on this consideration we investigated the variation of these coefficients as a function of spin value. The coefficients were calculated by using excitation energies of consecutive levels on the assumption of the constant coefficient in the neighborhood of the relevant level. In other words they are calculated differentially for each relevant level. This method may be regarded as a variation of the VMI model.

The eqs. (1) and (2) for the second order expansion of I(I+1) can be given as

$$E(I) = AI(I+1) + BI^2(I+1)^2 \quad \text{for g.b.} \tag{3}$$

and

$$E(I) = E_2 + AI(I+1) + BI^2(I+1)^2 + C(-)^I(I-1)(I+1)(I+2) \quad \text{for } \gamma\text{.b..} \tag{4}$$

First we discuss g.b. As eq. (3) has two parameters, they can be calculated by using two consecutive levels. Figs. 2 and 3 show the trends of A and B as a function of I, respectively. As A can be written as $\hbar^2/2J$ and J is proportional to $A_0^{5/3}$, the calcula-

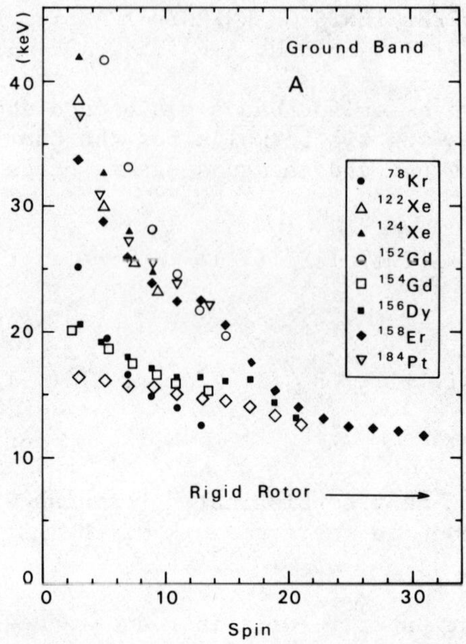

Fig. 2. Plott of coefficient A for g.b.. The values are multiplied by $(A/160)^{5/3}$. ◇ : ^{158}Dy

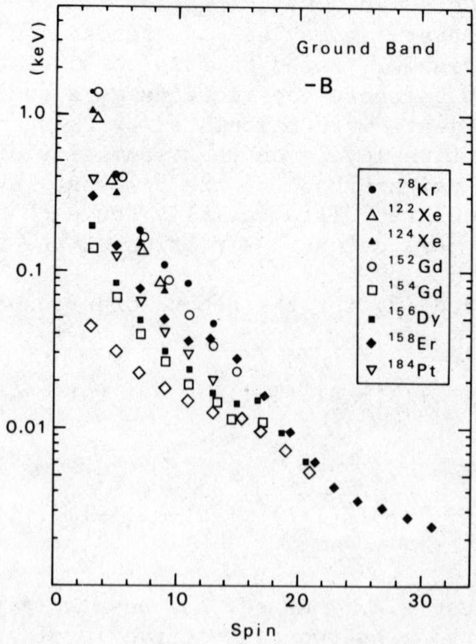

Fig. 3. Plott of coefficient B for g.b.. ◇ : ^{158}Dy

ted values multiplied by $(A_0/160)^{5/3}$ are plotted in fig. 2, where J stands for moment of inertia and A_0 for mass number. It is clear that this procedure is to normalize the value to that in case of $A_0 = 160$. In these figures we can notice several conspicuous features. In fig. 2 the normalization factor makes the curves converge in a narrow region. All the nuclei treated have the common sharply decreasing trend with increasing spin value. They look like to tend together to a final value which we might consider as the moment of inertia of the rigid rotor indicated in the figure by an allow. On the other hand the curves in fig. 3 seem decrease exponentially and approach to a common curve at the high spin limit. Notice that the B values are always negative. It is of interest to observe that the curve of ^{158}Er has a bump at the backbending in both curves for A and B.

Second we discuss γ.b.. Eq. (4) has four parameters and we have to use four levels to deduce the relevant coefficients. In fig. 4 shows the plot of the calculated A values which are normalized to $A_0 = 160$ by multiplying the factor $(A/160)^{5/3}$. We can observe two groups, of which the first one has a large zigzaging

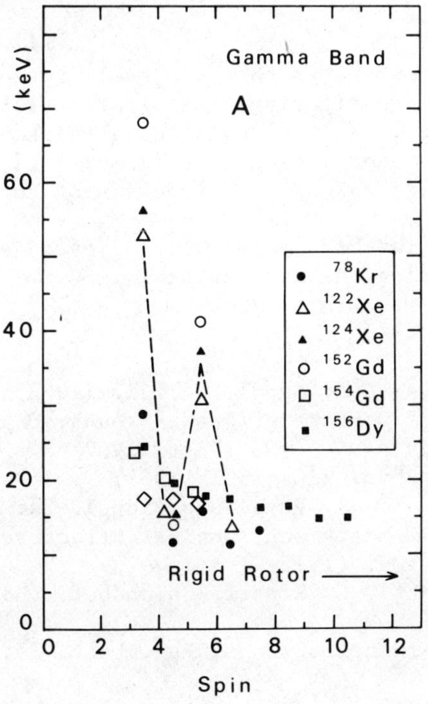

Fig. 4. Plott of coefficient A for γ. b.. The values are multiplied by $(A/160)^{5/3}$. ◇ : ^{158}Dy

of A and the second one has a smooth trend of A. The zigzagging demps to tend to a certain value at high spin value. The gross behaviour of curves looks like as if these curves tend to converge to about the same value as in the case of g.b..

3. DISCUSSION AND CONCLUSION

We have investigated the behaviour of energy-expansion coefficients, regarding them as variable as a function of spin value.

The A values tend to a certain value, maybe, to the rigid rotor value and the B values decrease exponentially and converge to a single curve at high spin limit. These gross features are common for all the cases treated in the present paper, being independent on the nuclear region. It may imply that the nuclei at low spin value behave with a way characteristic to the nuclear regions, while in the high spin limit they lose the special character and rotate with a moment of inertia of the rigid rotor and the B values are goint to disappear. The same argument seems to be applicable to the case of γ.b. It is worth noting that the curves of A for both of g.b. and γ.b. in ^{158}Er have a bump at the same spin value. These findings provide a strong support for the validity of the quasi-band concept.

This paper demonstrates that the newly introduced technique is very powerful for investigating gross features of band structures. Its essential point lies in calculating at each excited level the coefficients of the energy-expansion formula and in examining the gross trends of these values as a function of the spin value.

The author is indebted to Mr. T. Kurihara and Mrs. Y. Hirosawa for preparing the manuscript.

REFERENCES

1. M. Sakai, J. Phys. Soc. Suppl. 24: 577 (1968)
2. M. Sakai, List of Members of Quasi-Ground, Quasi-Beta and Quasi-Gamma Bands, INS-J-105 (Sept. 1967)
3. M. Sakai, Nucl. Phys. A104: 301 (1967)
4. B. V. Thosar et al., J. Phys. Soc. Suppl. 24: 175 (1968)
5. A. Bohr and B. R. Mottelson, Nuclear Structure Vol. II Ch.4, W. A. Benjamin, INC. (1975)
6. M. A. J. Mariscotti, G. Scharff-Goldhaber, and B. Buck, Phys. Rev. 178: 1864 (1969)
7. A. Arima and F. Iachello, Phys. Rev. Lett. 35: 1069 (1975)

ON THE PROBLEM OF "PHASE TRANSITIONS" IN NUCLEAR STRUCTURE*

G. Scharff-Goldhaber

Brookhaven National Laboratory, Upton, NY 11973 and Cornell University, Ithaca, NY 14850

In the setting of Erice, so evocative of Homeric adventures, it may be appropriate to state that the nuclear physicist is forced to steer perpetually between the Scylla of inductive and the Charybdis of deductive reasoning. The first approach consists in attempting to organize the empirical facts in as model independent a fashion as possible with the aim of describing them analytically in order to gain deeper insight. The second approach consists in model building, where the model, although usually based on vastly simplifying assumptions, is expected to yield completeness and accuracy in its predictions. The nuclear structure talks we have heard so far during this workshop have proven very effectively the success of an approach which belongs in the second category, namely that associated with the Interacting Boson Model (IBM), in classifying the complex band structure of even-even nuclei. On the other hand the inductive method yielded the Variable Moment of Inertia (VMI) equations,[1,2] which give a precise description of ground state band energies of non-magic nuclei in terms of the rotational energy expression:

$$E = \frac{C}{2}(\mathcal{J} - \mathcal{J}_o)^2 + \frac{J(J+1)}{2\mathcal{J}} \quad (1)$$

$$\left.\frac{\partial E}{\partial \mathcal{J}}\right|_J = 0 \quad (2)$$

From (1) and (2) follows the "equation of state" for the moment of inertia:

*Research supported by the U. S. Department of Energy under Contract Nos. DE-AC02-76CH00016 and DE-AC02-80ER10576.

$$\mathcal{J}^3 - \mathcal{J}_o\mathcal{J}^2 = \frac{J(J+1)}{2C} \qquad (3)$$

An impressive comparison of the VMI predictions with the predictions resulting from an anharmonic vibrator model (which are identical with those given by the three dynamical IBM symmetries) was published by Das, Dreizler, and Klein[3] (I, Fig. 6) ten years ago.

In a recent talk[2] I have described the evidence for a universal mechanism underlying these bands, as well as the developments preceding this finding. Let me briefly repeat the main points: Fig. 1 depicts the "heroic" stages which immediately followed the classical papers on the shell model: To the right we find the well known level scheme of the Bohr-Mottelson strongly deformed nucleus with its rotational ground state, beta and gamma bands. This model contains the assumption that the moment of inertia of the g.s. band as well as the "intrinsic" quadrupole moment are independent of the angular momentum J. To the left we see a near-harmonic level structure displayed by nuclei in the smaller shells, and at the beginning of the rare earths region. This pattern gave rise to the vibrational model as well as to the "gamma unstable" model. An abrupt change was found to exist between near-harmonic nuclei containing $\leq$ 88 neutrons and strongly deformed nuclei with $\geq$ 90 neutrons. Soon afterwards an almost as abrupt transition was found to exist at the beginning of the actinide region, between 86 and 88 protons.[4] The obvious question arose: is there no intermediate stage possible? It was soon answered when a gradual transition was discovered in level schemes of even Os (Z = 76) nuclei from rotational to vibrational. This transition which was shown to continue into the Pt (Z = 78) region appeared to be particularly simple and regular when instead of the level energies the ratios E_J/E_2 were presented. The reason for the increase in simplicity is due to the fact that the energy E_2 can be used as a "scale factor" which includes not only the nuclear size effect ($\propto A^{-2/3}$) but also the influence of residual forces. This procedure soon furnished evidence for the existence of "quasi-gamma bands" and "quasi-beta" bands, in addition to the g.s.bands, in all even-even nuclei.

At the same time that extensive level scheme studies suggested that nuclei in the Os-Pt transition region may not be axially symmetric, Edoardo Mallmann concentrated on the structure of ground state bands alone and discovered that if ratios E_J/E_2 (J = 6,8) are plotted vs. E_4/E_2, the points lie on two "universal curves." This implied that one and the same mechanism is responsible for the ground state bands of all non-magic nuclei (actually Mallmann included also the magic nuclei; the advent of the extended VMI model,[5] however, proved that this inclusion was incorrect). The mechanism, as I stated in the beginning, appears to be rotation in a harmonic vibrator potential. Apart from the strongly deformed, stable

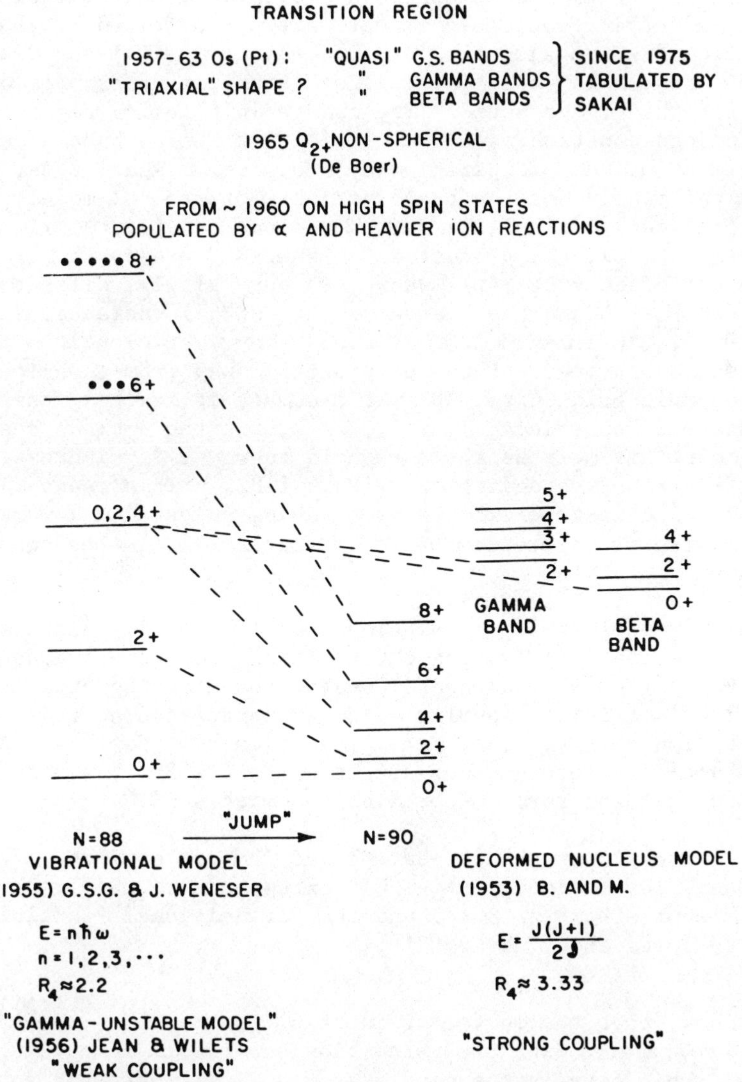

Fig. 1. Schematic graph depicting evolution of models of collective motions in e-e nuclei from the early 1950s on.

nuclei, it turned out that the nuclear moment of inertia is not constant, i.e. independent of J, but increases gradually from the "ground state moment of inertia" $\mathcal{J}_0$. Already in the middle sixties measurements of the static electric quadrupole moments of 2+ states of "spherical nuclei" had produced a contradiction to the assumption that electric quadrupole moments are not affected by the angular momentum.* Thus also the idea underlying the Nilsson diagrams that each nucleus has a definite deformation parameter β is only correct for the ground state. The change in thinking required by these findings constitutes a "revolution" in Thomas Kuhn's sense, a replacement of the old paradigm by a new one. What, then, is the physical meaning of the basic variable of the VMI model, E_4/E_2? At the rotational limit $E_4/E_2 = 10/3$; since $E \propto (J(J+1)$. As shown in I, Fig. 5, the product $\mathcal{J}_0 E_2$ (where $\mathcal{J}_0$ equals the ground state moment of inertia $\mathcal{J}(0)$) decreases very slowly, close to the point where $\mathcal{J}_0 = 0$ and the "hardness" $h(= 2C\mathcal{J}_0^3)$ vanishes, at $E_4/E_2 = 2.23$. Nuclei with $E_4/E_2 \leq 2.23$ possess $\leq$ two nucleon pairs of at least one type (neutrons or protons) away from a magic number. From this point until the magic number is reached, the nuclear ground state is spherical, i.e. $\mathcal{J}(0) = 0$, but the nucleus resists cranking more and more as the threshold energy $\frac{C}{2}\mathcal{J}_0^2$ increases until $\mathcal{J}_0 = -\infty$ corresponding to $E_4/E_2 = 1.82$. This means magic number nuclei cannot be excited by cranking unless one or two nucleon pairs are first promoted to a higher orbit, thus causing deformation.

The VMI model yields predictions for arbitrarily high angular momenta J, but in reality "phase transitions" due to various mechanisms produce (normally downward) deviations starting at a critical spin J_{cr}, or a critical angular velocity corresponding to 1 to 2 MeV excitation energy. (See Ref. 1, section 8.) However, in the deformed actinides energies of states up to J = 28 (∿ 4.5 MeV) agree with (2-parameter) VMI predictions within 1%.[6]

The correspondences between VMI and IBM are obvious: the "vibrational level scheme" ($\mathcal{J}_0 \sim 0$) corresponds to SU(5), the rotational level scheme to SU(3) and the "transitional"[8] nuclei ^{194,196}Pt nuclei and ^{120}Xe to O(6).

---

*We know now[7] that the moment of inertia $\mathcal{J}_{02}$ $\left(= \frac{\mathcal{J}(0) + \mathcal{J}(2)}{2}\right)$ is strongly correlated with the transition quadrupole moment Q_{02} (0 → 2). This correlation consists of a linear part for spherical nuclei which have small moments, and a quadratic correlation for transitional and strongly deformed nuclei. The linear part can be interpreted by a closed-shell, non-moving core with two clusters (on the average α-particles) at its perimeter which rotate (the "alpha dumbell model"), whereas the quadratic part is given by a two-fluid model (inertial fluid and superfluid). The rather sharp break corresponds precisely to $\mathcal{J}_0 = 0$.

Dr. Dieperink has given a beautiful proof for the occurrence of shape phase transitions of ground states between SU(5) and SU(3) (first order) and SU(5) and O(6) (second order), using an algorithm derived by Gilmore and Feng. No phase transitions between these three types occur within the framework of VMI (which is based solely on the structure of g.s. bands), but it is highly significant that one or the other of the VMI parameters takes on extreme values for the three special subgroups: SU(5) corresponds to $\mathcal{J}_o \sim 0$, SU(3) to maxima of $\mathcal{J}_o$ (Fig. 2) and O(6) to minima of the stiffness parameter (or spring constant) C (Fig. 3). The question whether phase transitions at $\mathcal{J}_o = 0$ and $\mathcal{J}_o = -\infty$ can be deduced from the VMI equations has recently been explored.[9] For this purpose it is instructive to discuss the VMI "equation of state" (3) in comparison with the cubic equations of state characterizing some systems of condensed matter. The systems we chose are a) the van der Waals equation

$$(p + \frac{a}{v^2})\ (v - b) = RT, \tag{4}$$

and b) the equation of state of a ferroelectric (perovskite) under the influence of an external electric field E

$$P^3 + \gamma(T - T_o)P - \alpha E = 0,\ P = \text{polarization}; \quad \gamma \text{ and } \alpha \text{ are constants.} \tag{5}$$

We showed that to $\mathcal{J}_o = 0$ corresponds the critical temperature T_c in the van der Waals equation, where

$$RT_c = \frac{8}{27}\frac{a}{b}, \tag{6}$$

and T_o in the ferroelectric case. For the latter system it was found empirically that the dielectric constant ε_o obeys the equation $\varepsilon_o = \frac{a}{\gamma(T - T_o)}$, where a is a constant. In these two many body systems two phases can exist below T_c, resp. T_o since the cubic equation has there three real roots, two minima and one (unphysical) maximum. However, in the VMI system only one phase exists for $\mathcal{J}_o < 0$, since, in order to be physically meaningful, the moment of inertia $\mathcal{J}$ has to be positive, corresponding to only one of the three real roots of (3). Hence, it follows that while the van der Waals gas undergoes a first order phase transition at constant pressure from gas to liquid below T_c, and the ferroelectric undergoes a phase transition as ε_o passes through infinity (which is second order for E = 0, first order for $E \neq 0$), no phase transition appears to occur at $\mathcal{J}_o = 0$.

Yet, as $\mathcal{J}_o \to -\infty$, a first order phase transition does occur. As mentioned before, at this point an external torque will not produce rotation: excitation of the nucleus can only take place due to a

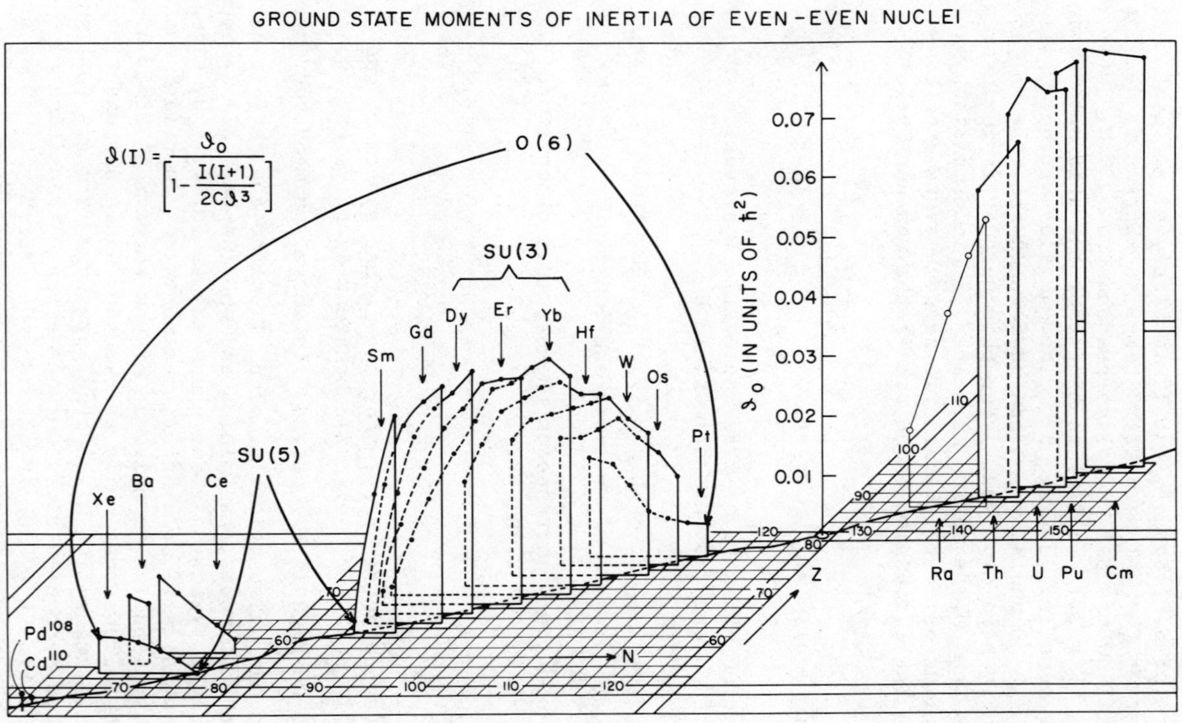

Fig. 2. Three dimensional diagram of $\mathcal{J}_0$ (computed from Eqs. 1 and 3) vs. N and Z. For nuclei corresponding to the group SU(5), $\mathcal{J}_0$ has minima, whereas maximum values occur for nuclei corresponding to SU(3). Small, but finite values characterize ^{120}Xe and $^{194,196}Pt$, the representatives of O(6).

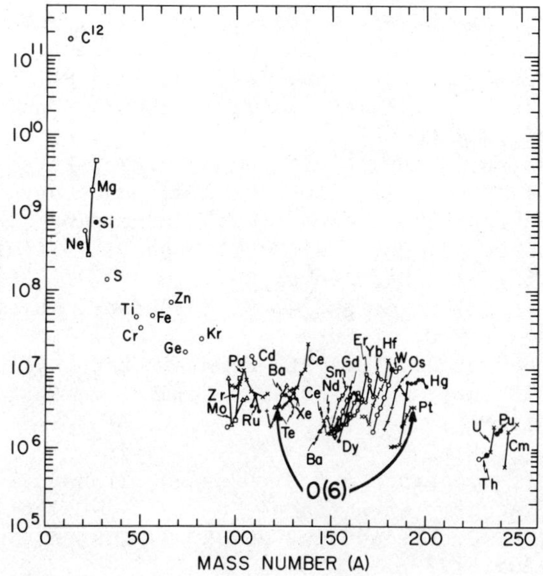

Fig. 3. Stiffness parameter C as a function of A. Points for isotopes of the same element are connected by solid lines, which peak at the most stable nuclei. For nuclei corresponding to O(6), C reaches a minimum.

rearrangement of nuclear matter. Similarly, in a ferroelectric, as $\varepsilon_0 \to 0$, no externally imposed field can produce an internal field in the solid. In the van der Waals case, if the attractive force vanishes, i.e. a = 0, the critical temperature $T_c = 0$. Hence it would take infinite pressure to liquefy the gas.

A discussion of phase transitions within ground state bands ("backbending") in comparison with phenomena in condensed matter systems has to be omitted because of lack of space. In all three systems the physical phenomena result from the competition between two opposing tendencies, which are exactly in balance at the critical point: In the van der Waals case these are the repulsive and attractive regions of the intermolecular forces, while the pressure acts as an external influence on the system. In the ferroelectric case these tendencies refer to the polarization which tends to pull the crystal apart, and the elastic forces tending to restore the original shape. In the VMI model the tendency to deform the nuclear shape vies with the tendency to preserve the spherical shape. Since the latter is caused by the analytically rather intractable Pauli principle, no microscopic derivation of the VMI model has yet succeeded.[10]

REFERENCES

1. G. Scharff-Goldhaber, C. Dover, and A. L. Goodman, Ann. Rev. Nucl. Sci. 26:239 (1976) and references therein.
2. G. Scharff-Goldhaber, Int. Conf. on Band Structure and Nuclear Dynamics, Nucl. Phys. A347:31 (1980). This reference will be referred to as I.
3. T. K. Das, R. M. Dreizler, and A. Klein, Phys. Rev. C 2:632 (1970). The deviations of g.s. band energies according to IBM from VMI increase with increasing J. See G. Scharff-Goldhaber, J. Phys. G: Nucl. Phys 5:L207 (1979), Fig. 1b.
4. G. Scharff-Goldhaber, Phys. Rev. 103:837 (1956).
5. G. Scharff-Goldhaber and A. S. Goldhaber, Phys. Rev. Lett. 24:1349 (1970).
6. L. K. Peker et al., private communication (BNL-NCS 28090).
7. A. S. Goldhaber and G. Scharff-Goldhaber, Phys. Rev. C 17:1171 (1978).
8. See Ref. 5, Fig. 1
9. G. Scharff-Goldhaber and M. Dresden, Transactions of the N. Y. Academy of Sciences, Ser. II, 40:166 (1980).
10. B. Buck, L. C. Biedenharn, and R. Y. Cusson, Nucl. Phys. A317:205 (1979).

Part II. THE INTERACTING BOSON-FERMION MODEL

PRESENT STATUS OF THE INTERACTING BOSON-FERMION MODEL

F. Iachello

Physics Department, Yale University, New Haven, Ct.06520
and
Kernfysisch Versneller Instituut, Groningen, The Netherlands

1. INTRODUCTION

Since its introduction in 1974, considerable effort has gone into the investigation of the properties of the interacting boson model, both from the theoretical and from the experimental point of view. The investigations carried out up to 1978 are summarized in the first book of this series[1], while the investigations carried out in the last two years are presented in the first part of the present volume. These investigations have provided a relatively clear view of the successes and limitations of the interacting boson model in the description of even-even nuclei.

However, in addition to the information available on even-even nuclei, there is a considerable amount of experimental information available on odd-A nuclei. Most of this information, especially that in the strongly deformed regions, can be (and has been) analyzed in terms of the Nilsson model[2]. But, there exists also information, especially in transitional nuclei, which is very difficult to analyze either in terms of this model, or of other models based on the geometric approach of Bohr and Mottelson[3]. For this reason, it is of interest to see whether or not another, algebraic, description of odd-A nuclei can be provided, capable of describing not only the simple situations treated by the Nilsson model, the particle-vibration model[4] and the triaxial rotor plus particle model[5], but also the other, more complex situations.

In an attempt to provide such description, a model has been developed in the last two years, called interacting boson fermion model (IBFA-1). The purpose of this lecture is to report on the present status of the model.

2. THE INTERACTING BOSON FERMION MODEL

In this model[6], one uses as building blocks both L=0, L=2 pairs (s,d bosons) and single particles (fermions with angular momenta j), Fig. 1.

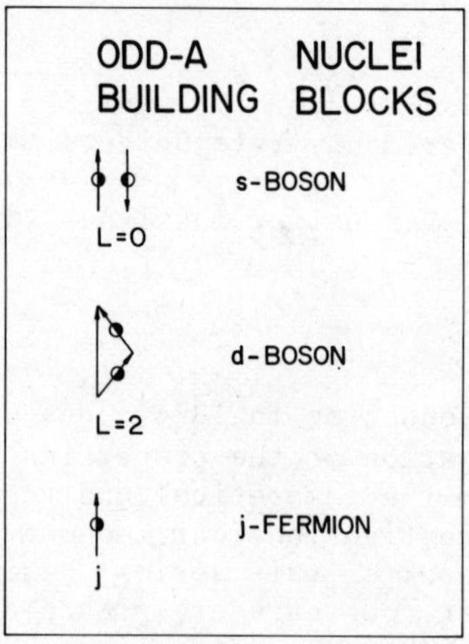

Fig. 1. Schematic representation of the building blocks of the interacting boson fermion model.

The Hamiltonian is then written as

$$H = H_B + H_F + V_{BF} \quad , \tag{2.1}$$

where H_B is the IBA Hamiltonian of Ref. 7, and

$$H_F = \sum_{j,m} \varepsilon_j \, (a^\dagger_{jm} a_{jm}) . \tag{2.2}$$

The boson-fermion interaction V_{BF} contains, in principle, many terms[8]. However, for a phenomenological analysis, it appears to be sufficient to take into account only three terms in V_{BF}, an effective monopole and quadrupole interaction, and an exchange term[6]

$$V_{BF} = \sum_j A_j \, [M_B^{(0)} \times M_{Fj}^{(0)}]^{(0)} \; +$$

$$+ \sum_{jj'} \Gamma_{jj'} \, [Q_B^{(2)} \times Q_{Fjj'}^{(2)}]^{(0)} \; + \tag{2.3}$$

$$+ \sum_{jj'j''} \Lambda_{jj'}^{j''} \; :[F_j^{\dagger(j'')} \times F_{j'}^{(j'')}]^{(0)}: \quad ,$$

where (:) denotes normal product, and

$$M_B^{(0)} = (d^\dagger \times \tilde{d})^{(0)} ,$$

$$M_{Fj}^{(0)} = (a_j^\dagger \times \tilde{a}_j)^{(0)} ,$$

$$Q_B^{(2)} = (s^\dagger \times \tilde{d} + d^\dagger \times \tilde{s})^{(2)} + \chi \, (d^\dagger \times \tilde{d})^{(2)} ,$$

$$Q_{Fjj'}^{(2)} = (a_j^\dagger \times \tilde{a}_{j'})^{(2)} , \tag{2.4}$$

$$F_j^{\dagger(j'')} = (a_j^\dagger \times \tilde{d})^{(j'')} ,$$

$$F_{j'}^{(j'')} = (d^\dagger \times \tilde{a}_{j'})^{(j'')} .$$

The coefficients A_j, $\Gamma_{jj'}$ and $\Lambda_{jj'}^{j''}$ depend on the labels j, j', j",.. of the single particle orbits which can be occupied by the odd fermion. If these coefficients are all taken as arbitrary parameters, a phenomenological study of situations in which the odd fermion can occupy several orbits becomes rather difficult. Several attempts have been made in order to estimate the j-dependence of the coefficients A_j, $\Gamma_{jj'}$ and $\Lambda_{jj'}^{j''}$. Two will be reported by Scholten[9] and Talmi[10]. The approach of Scholten gives

$$A_j = \sqrt{5}\,\sqrt{2j+1}\; A_0 ,$$

$$\Gamma_{jj'} = \sqrt{5}\; \Gamma_0 \; \gamma_{jj'} , \tag{2.5}$$

$$\Lambda_{jj'}^{j''} = \frac{-2\sqrt{5}\Lambda_0}{\sqrt{2j''+1}} \; \beta_{jj''} \; \beta_{j''j'} ,$$

where

$$\gamma_{jj'} = (u_j u_{j'} - v_j v_{j'}) Q_{jj'},$$

$$\beta_{jj'} = (u_j v_{j'} + v_j u_{j'}) Q_{jj'}, \qquad (2.6)$$

$$Q_{jj'} = \langle j \| Y_2 \| j' \rangle .$$

The coefficients u_j and v_j are given by

$$v_j = \alpha_j \sqrt{\frac{N}{\Omega_e}}, \qquad u_j = \sqrt{1-v_j^2}, \qquad (2.7)$$

where α_i and Ω_e are calculated using the procedure given by Arima and Otsuka[11]. The approach of Talmi leads to similar results. The formulas (2.5), (2.6) and (2.7) become much simpler when:

(i) the fermion can occupy only one single particle level j, for example the unique parity states $h_{11/2}$ and $i_{13/2}$ in the shells 50-82 and 82-126;
(ii) the single particle levels occupied by the fermion can be considered as degenerate, in which case the v_j's are independent of j.

In both cases, (i) and (ii), a calculation of the coefficient v_i's is not needed within the framework of a phenomenological analysis, since they can be absorbed in the strength parameters Γ_0 and Λ_0. Calculations for situations of this sort have been performed by

(i) Scholten and Blasi for the states originating from the $h_{11/2}$, $d_{5/2}$ and $g_{7/2}$ levels in the odd Eu isotopes;
(ii) Casten for the $h_{11/2}$ level in the odd Pd isotopes;
(iii) Kaup and Brentano for the $g_{9/2}$ level in the odd Rb isotopes;
(iv) Braga and Wood for the $h_{9/2}$ level in the odd Au isotopes;
(v) Bijker for the $h_{9/2}$ and $h_{11/2}$ levels in the odd Ir isotopes;
(vi) Cunningham for the $h_{11/2}$ level in the odd Xe isotopes.

Those of Scholten, Casten, Kaup and Wood are included in the Proceedings of this Workshop.

3. SYMMETRIES

In addition to performing numerical calculations, it is of interest to study the symmetries of the Hamiltonian (2.1). This is a difficult task since H in (2.1) contains both bosonic and fermionic degrees of freedom. However, some progress has been made in the last few months, and I will report in my contribution[12] on the latest developments. This particular problem, which bears relevance to the existence of supersymmetries in physics, will certainly experience a considerable amount of research activities in the immediate future.

4. LIMITING CASES

Some effort has gone into a comparison of the interacting boson fermion model with other existing models. This can be easily done in those cases in which the boson part of the Hamiltonian, H_B, has one of its three dynamical symmetries, SU(5), SU(3) and SO(6). Then, the spectra obtained as a solution of the interacting boson fermion model, share many features with those of

I) the particle-vibration model[4], when H_B has SU(5) symmetry;
II) the Nilsson model[2], when H_B has SU(3) symmetry;
III) the γ-unstable plus particle model[13], when H_B has SO(6) symmetry.

This is best illustrated by returning to (2.1), (2.2) and (2.3), and considering the case in which the odd particle can occupy only one level with j = 9/2. The energy levels of the odd-A nucleus then depend only on three parameters

$$A_{j=9/2} \equiv A \quad ,$$

$$\Gamma_{j=9/2,j'=9/2} \equiv \Gamma, \qquad (4.1)$$

$$\Lambda^{j''=9/2}_{j=9/2,j'=9/2} \equiv \Lambda.$$

If, for simplicity, A is set equal to zero, one can study the structure of the spectra when Γ and Λ vary, for any of the three limiting cases I), II) and III).

Case I. SU(5)

Fig. 2 shows a typical spectrum when $\Lambda = 0$, while Fig. 3 shows a typical spectrum when $\Gamma = 0$[14]. Particular attention should be given to the exchange term, since this term appears to be the dominant term in many situations. An example is shown in Fig. 4, where one observes the large depression of the state with angular momentum 9/2 = j-1 originating from the multiplet $2^+ \otimes j$, resulting from a large exchange term.

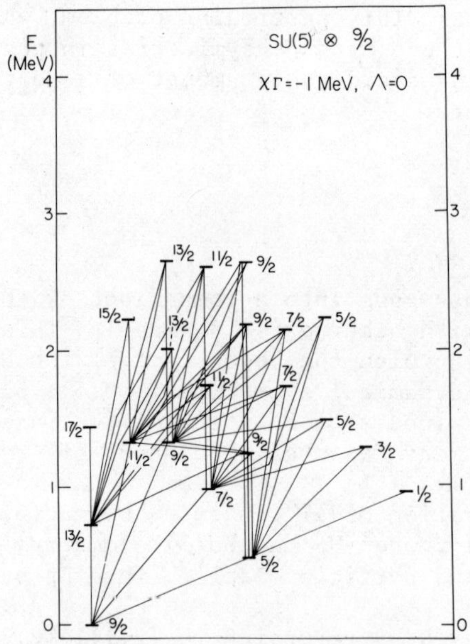

Fig. 2. Typical spectrum of an odd-A nucleus with j=9/2 in the SU(5) limit and Λ=0. Only levels up to n_d=2 are shown. The lines denote allowed E2 transitions.

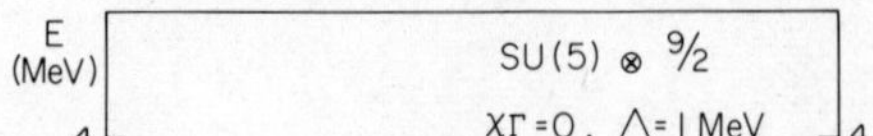

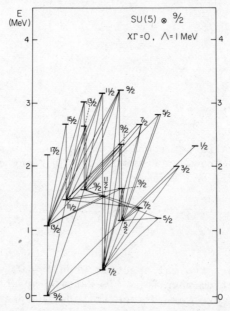

Fig. 3. Typical spectrum of an odd-A nucleus with j=9/2 in the SU(5) limit and Γ=0. Only levels up to n_d=2 are shown. The lines denote allowed transitions.

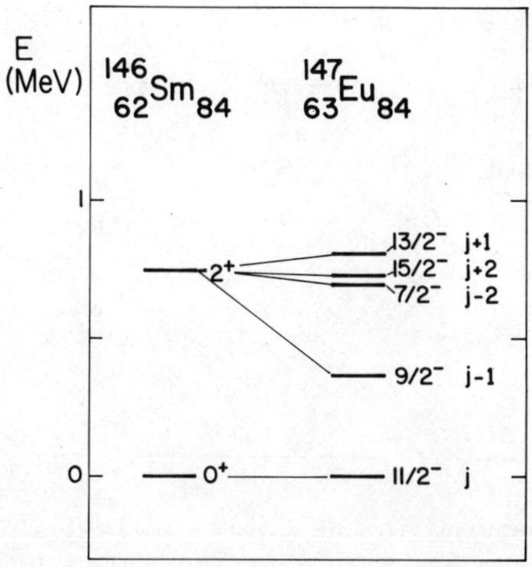

Fig. 4. The experimental spectrum[15] built on the $h_{11/2}$ proton configuration in $^{147}_{63}Eu_{84}$. In order to complete the multiplet $2^+ \otimes 11/2^-$ a state with $J^\pi=11/2^-$ should be found above the $13/2^-$ state.

Case II. SU(3)

The interesting feature of this limiting case[6,14] is the displacement of the various K_0 bands as Λ and Γ change. This displacement, shown in Figs. 5 and 6 is similar to the displacement of the various K_0 bands in the Nilsson model for fixed deformation, β, and varying Fermi energy, λ_F. Another point worth noting is that, while in the Nilsson scheme there is no relationship between the location of the β and γ bands and those of the bandheads for various K_0, such a relationship exists in the interacting boson fermion model. It would be interesting to see whether or not it is experimentally observed.

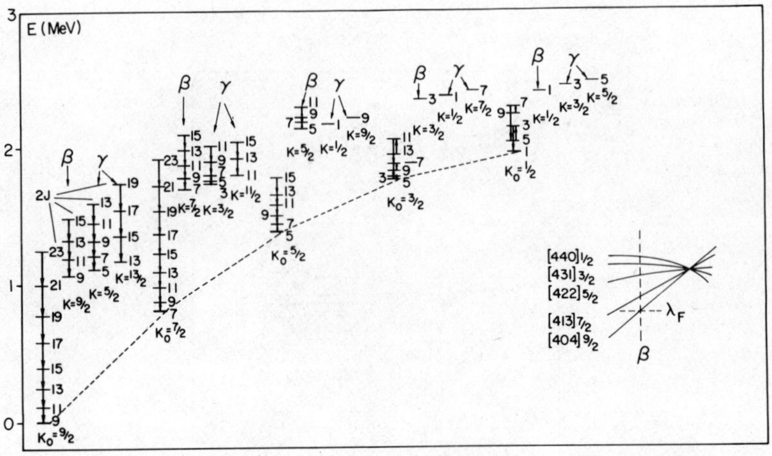

Fig. 5. A typical spectrum in the SU(3) limit of the interacting boson fermion model with Λ=0. The number of bosons is N=6 and j=9/2. Only a selected number of levels is shown. The levels are arranged into bands denoted by the lowest value of the angular momentum, K, contained in the band. This quantum number is only approximately equivalent to the quantum number K in the Nilsson model. In the inset, the corresponding situation in the Nilsson model is shown.

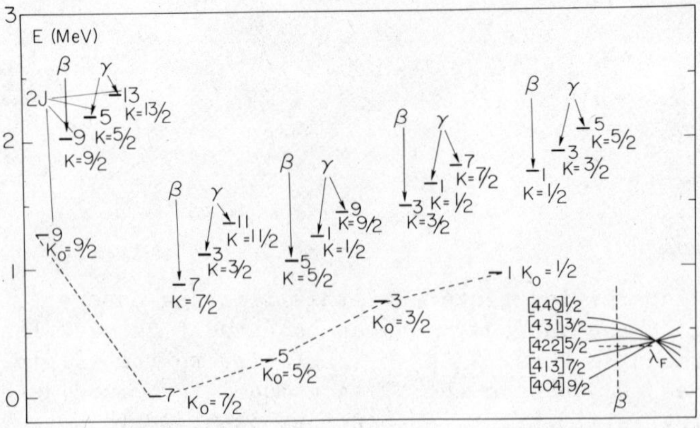

Fig. 6. The same as in Fig. 5 but with Λ≠0. Only the lowest state of each band is shown here.

Case III. SO(6)

This case is discussed in Refs. 14 and 16. Also here, the effect of the exchange term is that of displacing the relative location of the bandheads, as shown in Figs. 7 and 8.Of the three limiting cases, this is the most interesting one, because of the new and very intriguing triangular-like band structure displayed by its spectrum, Fig. 7.

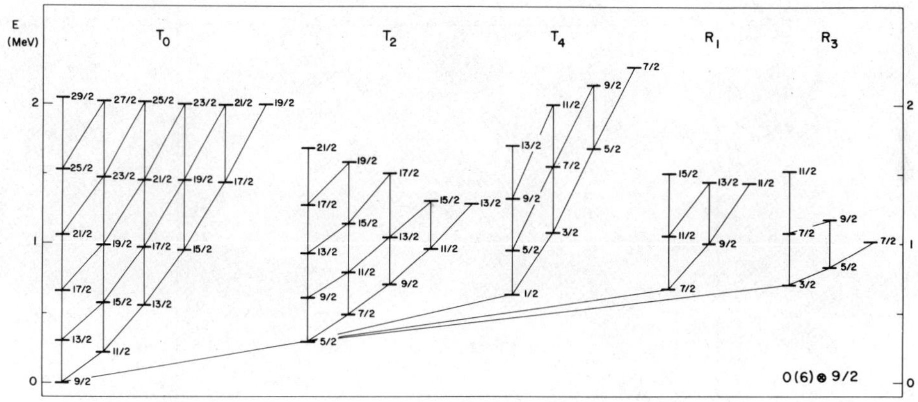

Fig. 7. Typical spectrum of an odd-A nucleus with j=9/2 in the SO(6) limit and Λ=0. The number of bosons is N=6. The lines connecting the levels denote large E2 transitions.

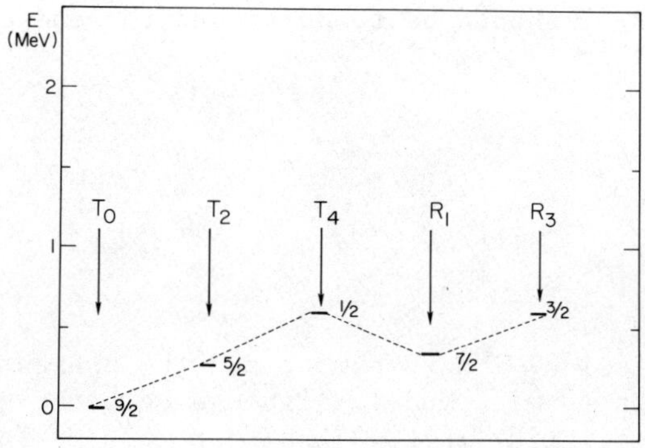

Fig. 8. The same as in Fig. 7 with $\Lambda \neq 0$. Only the lowest state of each band is shown.

Since the SO(6) limit corresponds, within certain approximations, to the γ-unstable model of Wilets and Jean, which, in turn, shares some properties with the triaxial rotor model, a similar pattern should also appear both in the γ-unstable[13] and triaxial rotor[5] plus particle calculations. One may remark that the particularly simple structure of Fig. 7 may be a consequence of some higher degree of symmetry of the Hamiltonian, for example of the Bose-Fermi symmetries discussed in Ref. 12. An experimental spectrum in which this triangular-like pattern is observed is shown in Fig. 9.

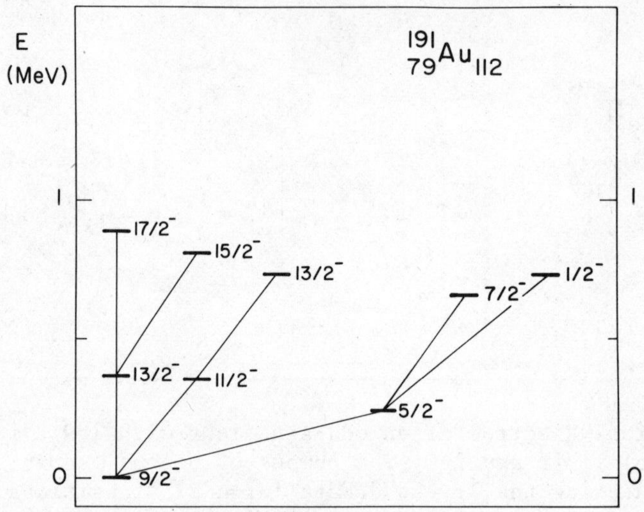

Fig. 9. The experimental spectrum[17] built on the $h_{9/2}$ proton configuration in $^{191}_{79}Au_{112}$. Note the occurrence of the multiplets $13/2^-$, $11/2^-$; $5/2^-$ and $17/2^-$, $15/2^-$, $13/2^-$; $7/2^-$; $1/2^-$. In order to complete the second multiplet a state with $J=9/2^-$ should be found around the energy of the $15/2^-$ state.

5. CONCLUSIONS

In conclusion, I have summarized the present status of the interacting boson fermion model. Although both the schematic (Sect. 4) and realistic (Sect. 2) calculations performed so far seem to indicate that the interacting boson fermion model may provide a framework for a unified description of collective states in odd nuclei, it is clear that much work remains to be done.

Among the many aspects which remain to be investigated in detail, I will mention

(i) a complete study of the super- (Bose-Fermi) symmetries of $H=H_B + H_F + V_{BF}$;

(ii) a study of situations in which the odd particle occupies several non degenerate levels. This study will provide a test of the parametrizations (2.5), (2.6) and (2.7); and

(iii) a study of other properties of odd-A nuclei, especially electromagnetic transition rates and one nucleon transfer intensities.

Finally, an attempt should be made to treat odd-A nuclei in the proton-neutron formalism (IBFA-2), since in this formalism the parameters A_0, Γ_0 and Λ_0 will be directly related to the microscopic structure of the s-d pairs, thus bridging the gap between the macroscopic and microscopic (shell-model) descriptions.

REFERENCES

1. F. Iachello ed., "Interacting Bosons in Nuclear Physics", Plenum Press, New York (1979).
2. S.G. Nilsson, Mat. Fys. Medd. Dan. Vid. Selsk. 29, No. 16 (1955).
3. A. Bohr and B.R. Mottelson, "Nuclear Structure", Vol. II, W.A. Benjamin, Reading, Mass. (1975).
4. A. Bohr, Mat. Fys. Medd. Dan. Vid. Selsk. 26, No. 14 (1952). A. Bohr and B.R. Mottelson, Mat. Fys. Medd. Dan. Vid. Selsk. 27, No. 16 (1953).
5. J. Meyer-ter-Vehn, Nucl. Phys. A249 (1975), 111.
6. F. Iachello and O. Scholten, Phys. Rev. Lett. 43 (1979), 679.
7. A. Arima and F. Iachello, Phys. Rev. Lett. 35 (1975), 1069.
8. A. Arima and F. Iachello, Phys. Rev. C14 (1976), 761.
9. O. Scholten, these Proceedings.
10. I. Talmi, these Proceedings.
11. T. Otsuka and A. Arima, Phys. Lett. 77B (1978), 1; T. Otsuka, Ph.D. Thesis, University of Tokyo, Japan (1979).
12. F. Iachello, these Proceedings.
13. G. Leander, Nucl. Phys. A273 (1976), 286.
14. F. Iachello, Nucl. Phys. A347 (1980), 51.
15. G. LoBianco, private communication.
16. F. Iachello and O. Scholten, Phys. Lett. 91B (1980), 189.
17. J. Wood, private communication.

THE INTERACTING BOSON FERMION MODEL AND SOME APPLICATIONS

O. Scholten

Kernfysisch Versneller Instituut, Groningen
The Netherlands

1. INTRODUCTION

In the IBA model, collective states in even-even nuclei are described in terms of a system of interacting s- and d-bosons. This model can be extended to odd-A nuclei by coupling the degrees of freedom of the odd nucleon to the system of bosons[1]. In general, the Hamiltonian for this coupled system can be written as

$$H = H_B + H_F + V_{BF} , \tag{1.1}$$

where H_B is the usual IBA Hamiltonian which describes the system of s- and d-bosons, and H_F is the Hamiltonian of the odd particle. Since only a single odd particle is coupled to the bosons, it is sufficient to consider only the one-body part of H_F

$$H_F = \sum_{jm} \varepsilon_j a^\dagger_{jm} a_{jm} . \tag{1.2}$$

Here, as well as in the following, the summation index j runs over the shell model orbits in the valence shell, and ε_j represents the single particle energy.

The most general form of the boson-fermion interaction, V_{BF}, has been given in Ref. 1. The main disadvantage of this form is that it gives little insight into the physical origin of the various terms. Furthermore, if all the parameters appearing in it are varied independently, a phenomenological analysis of the observed spectra becomes rather difficult. In the Interacting Boson

Fermion Approximation (IBFA) model[2] only a few, selected terms, having a particular physical significance are taken into account. In this lecture I will describe the main features of the IBFA model and show the results of some realistic calculations performed together with Blasi on the structure of the Europium isotopes.

2. THE MODEL

In the IBFA model, the boson-fermion interaction can be written as[2]

$$V_{BF} = \sum_j A_j \, [(d^\dagger \times \tilde{d})^{(0)} \times (a_j^\dagger \times \tilde{a}_j)^{(0)}]^{(0)} +$$

$$+ \sum_{jj'} \Gamma_{jj'} [\, Q^{(2)} \times (a_j^\dagger \times \tilde{a}_{j'})^{(2)}]^{(0)} + \tag{2.1}$$

$$+ \sum_{jj'j''} \Lambda_{jj'}^{j''} :[\, (d^\dagger \times \tilde{a}_j)^{(j'')} \times (a_{j'}^\dagger \times \tilde{d})^{(j'')}]^{(0)} :$$

where

$$Q^{(2)} = (s^\dagger \times \tilde{d} + d^\dagger \times s)^{(2)} + \chi (d^\dagger \times \tilde{d})^{(2)} \quad . \tag{2.2}$$

On the basis of the microscopic theory[3,4,5], one can estimate the j-dependence of the coefficients A_j, $\Gamma_{jj'}$, $\Lambda_{jj'}^{j''}$, as

$$A_j = \sqrt{5}\,\sqrt{2j+1}\, A_0 ,$$

$$\Gamma_{jj'} = \sqrt{5}(u_j u_{j'} - v_j v_{j'}) \, Q_{jj'} \, \Gamma_0 \, , \tag{2.3}$$

$$\Lambda_{jj'}^{j''} = -2\sqrt{5}\,\frac{1}{\sqrt{2j''+1}} \, (u_j v_{j''} + v_j u_{j''}) \, Q_{jj''} (u_{j'} v_{j''} + v_{j'} u_{j''}) \, Q_{j''j'} \, \Lambda_0 .$$

In (2.3), A_0, Γ_0 and Λ_0 label the strengths of the interaction and u_j and v_j are given by

$$v_j = \alpha_j \sqrt{\frac{N}{\Omega_e}} \quad , \quad u_j = \sqrt{1-v_j^2} \quad , \tag{2.4}$$

where α_j and Ω_e could be calculated using the procedure given by Arima and Otsuka[6]. Moreover, to a good approximation, the single particle matrix elements of the quadrupole operator, $Q_{jj'}$, can be taken proportional to the reduced matrix elements of $Y^{(2)}$

$$Q_{jj'} = \langle \ell\, 1/2\; j \| Y^{(2)} \| \ell'\; 1/2\; j' \rangle \quad , \tag{2.5}$$

if one neglects the radial integrals. It must be noted that, in deriving (2.3), several approximations have been made, such as, for example, the assumption that the D-pair state exhausts the full E2 strength, i.e. $|D\rangle = Q^{(2)}|S\rangle$. A more complete derivation may thus lead to results which differ somewhat from (2.3).

3. LIMITING CASES

In general, spectra calculated in the IBFA model are very complex. They depend both on the structure of the boson-boson and of the boson-fermion interaction. However, some simple features arise whenever the Hamiltonian H_B has a dynamical symmetry. In this section, I will show some typical spectra obtained by coupling a $j = 9/2$ particle to a core with SU(5), SU(3) and SO(6) symmetries. In these numerical calculations, the computer program ODDA[7] has been used.

In the case discussed here of a single shell-model orbit j, the formulas (2.3) can be simplified to

$$\begin{aligned} A \equiv A_j &= \sqrt{5}\,\sqrt{2j+1}\, A_0 \\ \Gamma \equiv \Gamma_{jj} &= \sqrt{5}(u^2-v^2)\; Q\; \Gamma_0 \\ \Lambda \equiv \Lambda^j_{jj} &= -\frac{8\sqrt{5}}{\sqrt{2j+1}}\, u^2v^2Q^2\Lambda_0 \end{aligned} \tag{3.1}$$

where $Q \equiv Q_{jj} = -0.982$ for $j = 9/2$. Since the monopole force does not influence the major properties of the spectra, in this schematic study we have set $A_0 = 0$. For each of the three dynamical symmetries of the core, we have then studied the behavior of the solutions of the IBFA Hamiltonian, as a function of the occupancy v^2 for fixed values of Γ_0 and Λ_0.

3a. The SU(5) Limit

The SU(5) limit of the IBA model corresponds to the vibrational model in the geometric picture. By coupling an odd particle, one expects to recover the main features of the particle-vibration model. In the SU(5) plus particle limit the eigenstates are most conveniently described in the weak coupling scheme

$$|\alpha,L,j,J,M\rangle = [|\alpha,L\rangle_c \otimes |j\rangle]^{(J)}_M \ , \qquad (3.2)$$

where the square brackets denote angular momentum couplings and $|\alpha,L\rangle_c$ represents the wave functions of the even-even core. In the weak coupling limit, the ground state is obtained by coupling the particle to the $|n_d=0,L=0\rangle$ ground state of the core. The lowest excited states are those in which the particle is coupled to the $|n_d=1,L=2\rangle_c$ first excited core state. They form a quintuplet with splitting given by

$$\langle n_d=1,L=2,j,J,M|V_{BF}|n_d=1,L=2,j,J,M\rangle = \qquad (3.3)$$
$$= \sqrt{5}\Gamma\chi \ (-)^{J-j} \begin{Bmatrix} 2 & 2 & 2 \\ j & j & J \end{Bmatrix} + \Lambda\sqrt{2j+1} \begin{Bmatrix} 2 & j & j \\ 2 & j & J \end{Bmatrix} \ .$$

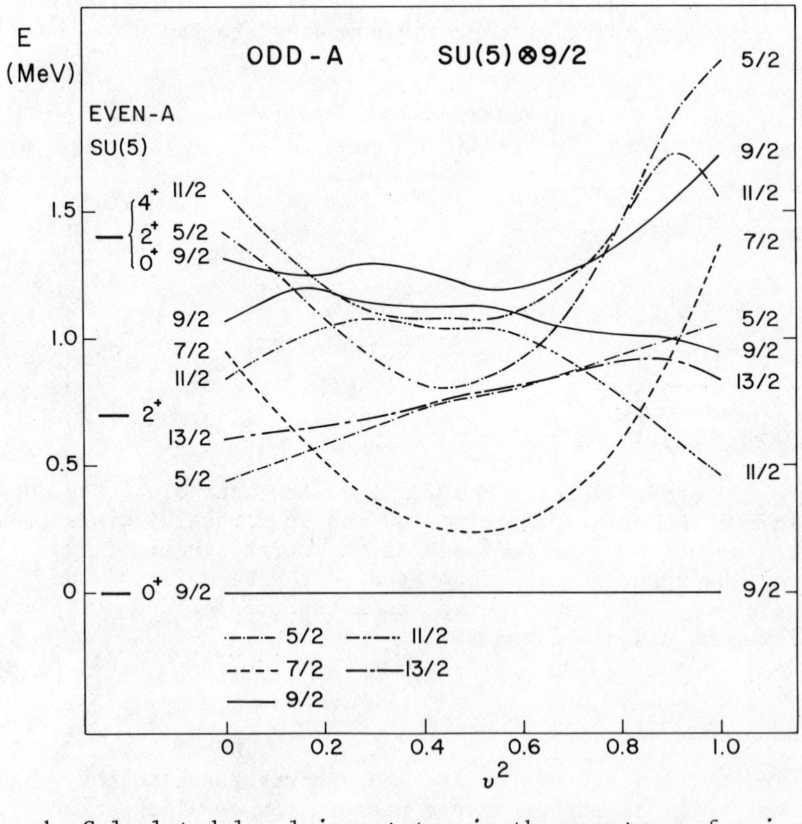

Fig. 1. Calculated low-lying states in the spectrum of a j = 9/2 particle coupled to an SU(5) core (N=5,ε=0.7 MeV) as a function of v^2. Here Γ_0=0.567 MeV,Λ_0=1.83 MeV and $\chi=-\frac{1}{2}\sqrt{7}$.

In order to show the characteristic features of the odd-A spectra in this limit, we plot in Fig. 1 the energies of the low-lying states as a function of v^2.

The values of the parameters used are Γ_0 = 0.567 MeV, Λ_0 = 1.83 MeV and $\chi = -\frac{1}{2}\sqrt{7}$. The core is described by N = 5 and ε = 0.7 MeV in order to simulate a realistic vibrational nucleus. When $v^2 = 0$, the level with J = j-2 is the lowest member of the quintuplet. As the shell gradually fills, the J = j-1 level comes down in energy and it is the lowest member of the quintuplet for $v^2 \simeq 0.5$. An example of this can be found in the positive parity spectrum of ${}^{101}_{45}Rh_{56}$. In the IBFA model this nucleus is described by coupling a $g_{9/2}$ proton to the ${}^{100}_{44}Ru_{56}$ core. For Z = 45, the $g_{9/2}$ level is expected to be just half filled, $v^2 = 0.5$, giving rise to a pure exchange coupling, Fig. 2.

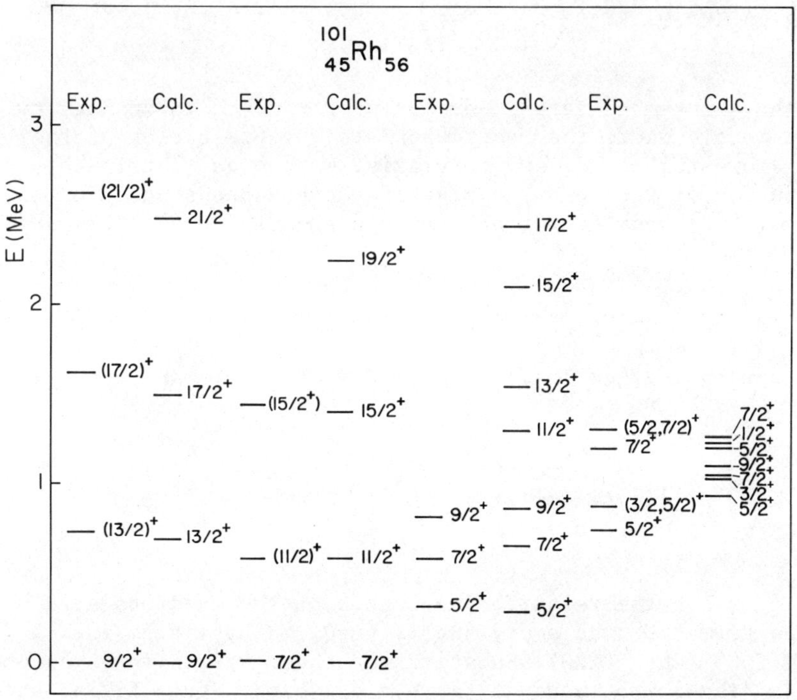

Fig. 2. Calculated and experimental low-lying positive parity levels in ${}^{101}_{45}Rh_{56}$. In the calculation[8] a $g_{9/2}$ proton is coupled to ${}^{100}_{44}Ru_{56}$ using a pure exchange force, $v^2 = 0.5$ and Λ_0 = 2.32 MeV. The core parameters, taken from a best fit to ${}^{100}Ru$, are ε = 0.733 MeV, c_0 = -0.415 MeV, c_2 = -0.283 MeV, c_4 = -0.024 MeV, v_2 = 0.175 MeV and v_0 = -0.31 MeV.

3b. The SU(3) Limit

By coupling a particle to an SU(3) (axially symmetric rotor) core, one can reproduce the basic features of the Nilsson scheme. The spectrum can be ordered into bands, along which the energies increase approximately as J(J+1). In the limit of a large number of bosons, $N \to \infty$, analytic formulas for the energies of the bandheads can be derived. The appropriate coupling scheme for the description of these wave functions is the strong coupling scheme, since in the SU(3) limit the quadrupole force is dominant. In this limit, $N \to \infty$, the eigenstates can be written as

$$|[N],(\lambda,\mu),\ K_c;\ j,\ K_j;\ K\ J\ M> = \\ = \sum_R \sqrt{1+\delta_{K_c,0}}\ \sqrt{2R+1}\begin{pmatrix} j & R & J \\ K_j & K_c & -K \end{pmatrix} |[N],(\lambda,\mu),\ K_c, R,\ j,\ J,\ M>, \tag{3.4}$$

where the quantum numbers K have been introduced in analogy to the Nilsson model, where they correspond to the projection of the angular momentum on the symmetry axis. We now can obtain the contribution of V_{BF} to the energy of the bandheads based on the $(\lambda,\mu) = (2N,0)$ representation of the even-even core[5] ($K_c=0$, $K=K_j$)

$$<[N],(2N,0),0;j,K;\ K\ J\ M|V_{BF}|[N],(2N,0),0;j,K;\ K\ J\ M> = \\ = -2N[\Gamma B_j \sqrt{\tfrac{1}{2}}(3K^2-j(j+1)) + \Lambda\sqrt{2j+1}\ B_j^2\ \tfrac{1}{3}(3K^2-j(j+1))^2] , \tag{3.5}$$

where

$$B_j = [(2j-1)j(2j+1)(j+1)(2j+3)]^{-1/2} . \tag{3.6}$$

Eq. (3.5) shows the relative importance of the quadrupole and exchange terms. For an empty shell, $v^2=0$, we have from (3.1), $\Lambda=0$ and $\Gamma < 0$ (when $\Gamma_0>0$). Substituting these values in (3.5) one sees that the lowest band is the K = 1/2 band. For a filled shell, $v^2=1$, Γ changes sign ($\Gamma>0$), giving rise to a K = j band as the lowest band. In the case of a half filled shell, $v^2=u^2= 1/2$, we have $\Gamma=0$ and, since $\Lambda_0>0$, $\Lambda<0$. From (3.5) it then follows that the lowest state is that for which $K \approx \sqrt{j(j+1)/3}$. For a j = 9/2 orbit, this means that the K = 5/2 and the K = 7/2 bands will be approximately degenerate since $\sqrt{j(j+1)/3} \cong 3$. Also in the case of finite N these features persist, as it is shown in Fig. 3 for N = 9. From this figure one can see some similarities with the Nilsson scheme: (i) the spectrum can be ordered in rotational

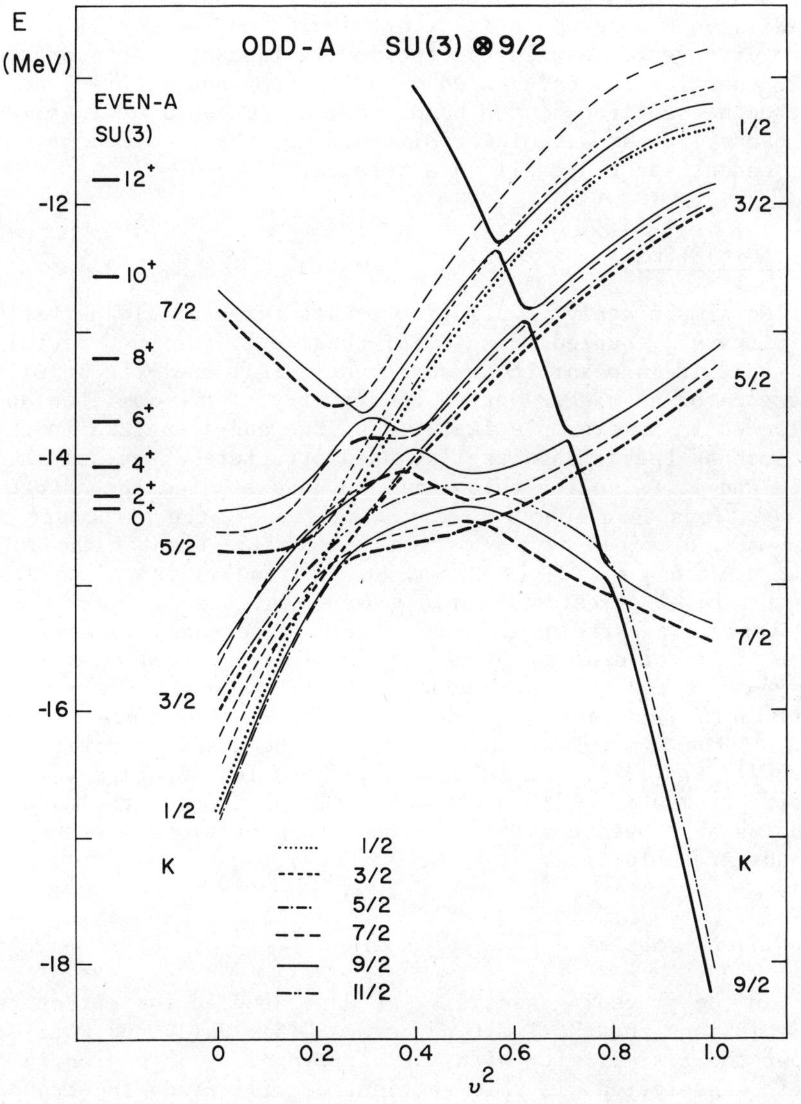

Fig. 3. Calculated levels for the system of a j=9/2 particle coupled to an SU(3) core, N=9, κ=50 KeV, κ'=20.8 KeV, as a function of v^2. The coupling parameters Γ_0, Λ_0 and χ are the same as in Fig. 1.

bands, (ii) as one increases v^2 the K-value of the ground state band increases gradually from K = 1/2 to K = j and (iii) the K = 1/2 band is highly distorted. This distorsion is also present in the Nilsson scheme where it is attributed to the Coriolis force. The main difference between the two models is that in the present case all energies are calculated with the same Hamiltonian, and therefore the position of the bandheads, the moments of inertia of the bands, the amount of Coriolis mixing, the position of the β and γ bands, etc., are all interrelated.

3c. The SO(6) Limit

No simple analytic solution exists for a single particle with arbitrary j coupled to an SO(6) core. Only for the special case j = 3/2 it is possible to construct a full analytic solution[9]. The presence of an higher degree of symmetry of the Hamiltonian[9] is reflected in some simple features of the odd-A spectra in this limit, such as the triangular-like band structure discussed in Refs. 10 and 11. Also for this limit we have studied the nature of the solutions as a function of v^2. In Fig. 4, the structure of the spectrum, obtained by a numerical diagonalization of the IBFA Hamiltonian with j = 9/2, is shown. By changing v^2 from 0 to 0.3 the J = j-1 level comes down rapidly in energy and it forms the ground state in the region $0.3 \lesssim v^2 \lesssim 0.5$. This rapid decrease in energy as a function of the occupancy is also observed in other models, such as the triaxial rotor model of Meyer-ter-Vehn[12] and the γ-unstable plus particle model[13,14,15]. As an example of the lowering of the J = j-1 we show in Fig. 5 the positive parity levels of ${}^{191}_{78}Pt_{113}$. In the IBFA model, these levels are described as arising from a j = 13/2 neutron hole level coupled to ${}^{192}_{78}Pt_{114}$. Here, holes have been considered rather than particles, since the bosons are hole-like.

4. TRANSITIONAL NUCLEI

In the preceding sections, we have studied the structure of the solutions for the limiting cases SU(5), SU(3) and SO(6), and shown their similarity with the corresponding situation in the geometric description. In this section, we will study the transitional region from SU(5) to SU(3). This region is difficult to calculate in conventional models. As an example, we have studied the odd Europium isotopes. The calculations presented here have been done by Blasi and myself and will be published in a forthcoming longer paper[16].

In the IBFA model, the odd-Eu isotopes, Z = 63, are described as a proton coupled to the even Sm isotopes, Z = 62.

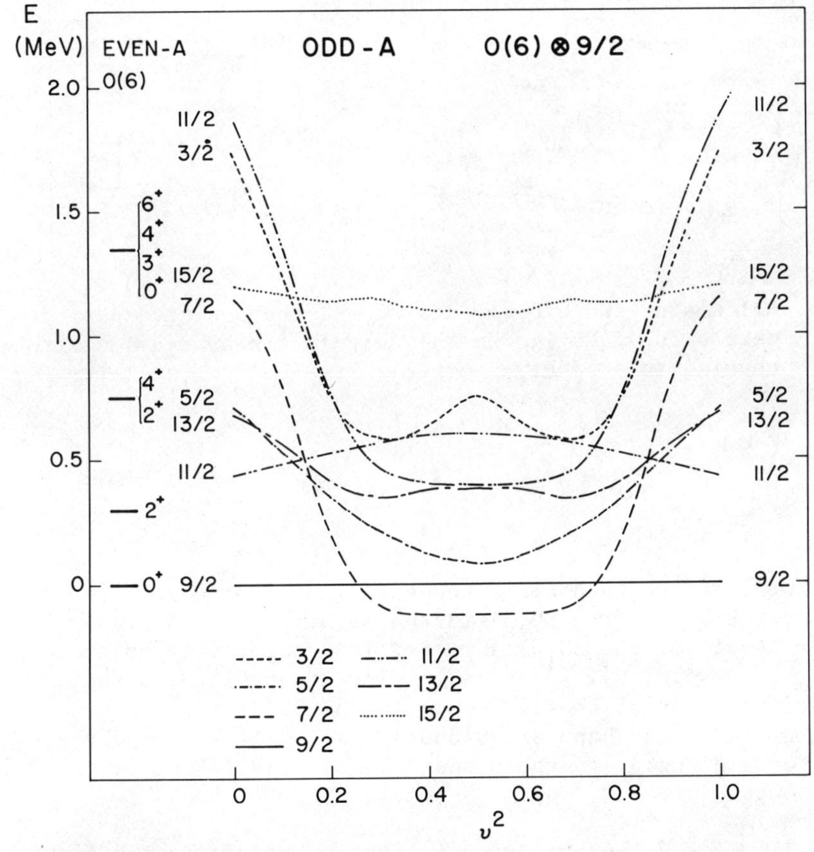

Fig. 4. Calculated low-lying levels of a j = 9/2 particle coupled to an SO(6) core (N=9, A=0.4 MeV, B=0.45 MeV, C=0), as a function of v^2. The coupling parameters are Γ_0=0.567 MeV, Λ_0=1.83 MeV and χ=0.

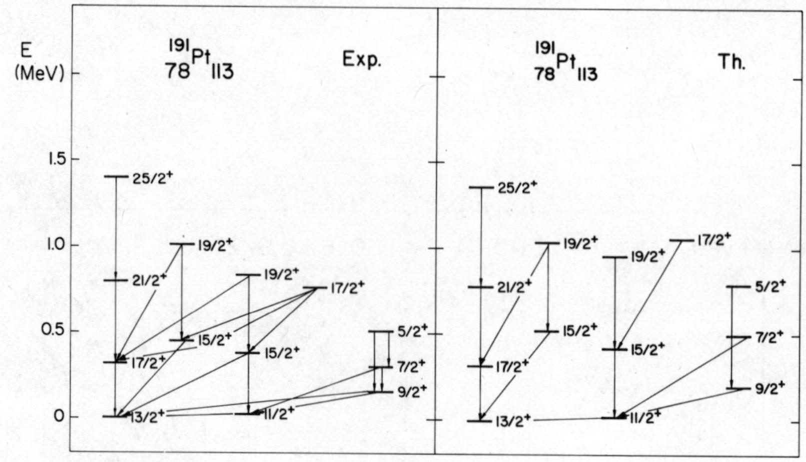

Fig. 5. Calculated and experimental low-lying positive parity levels in $^{191}_{78}Pt_{113}$. In the calculation an $i_{13/2}$ neutron is coupled to a $^{192}_{78}Pt_{114}$ core.

The spectra of the even Sm isotopes show a clear change in structure, Fig. 6, from SU(5) in the lighter to SU(3) in the heavier isotopes. The SU(5) symmetry is recognizable by the triplet of levels, 0^+, 2^+ and 4^+, lying at about twice the excitation energy of the 2_1^+ level. In the SU(3) limit, the energies in the ground state band are proportional to J(J+1), and the 0_2^+ and 2_2^+ states, belonging to the β and γ bands, lie relatively high in the spectrum.

In Fig. 7 the approximate location of the proton single particle levels in the mass region A ≃ 150 is given. The 13 protons outside the Z = 50 closed shell in Eu will primarily occupy the $d_{5/2}$ and $g_{7/2}$ levels to form positive parity states, and the $h_{11/2}$ single particle level to form negative parity states. In the calculation therefore the other levels have been omitted. Calculations including the $d_{3/2}$ and $s_{1/2}$ orbits show that the effect of these levels is not very significant.

In Figs. 8 and 9 the calculated energies are compared with experiment. The parameters used are given in Table I. Since the $g_{7/2}$ and $d_{5/2}$ single particle levels lie very close to each other, v_j^2 has been taken equal for these two levels.

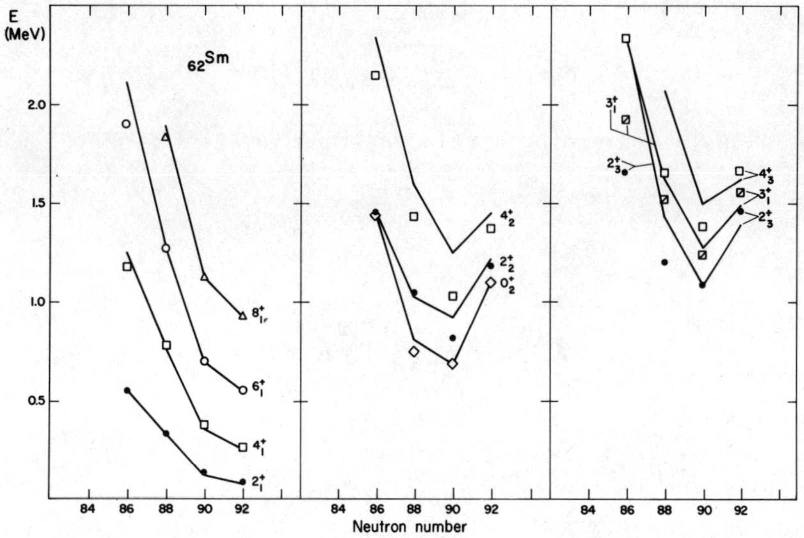

Fig. 6. Calculated and experimental low-lying levels in the even Sm isotopes.

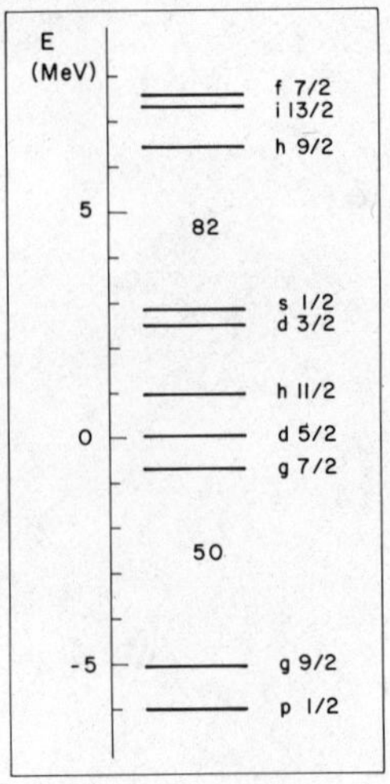

Fig. 7. Relative proton single particle energies for mass number A ≃ 150.

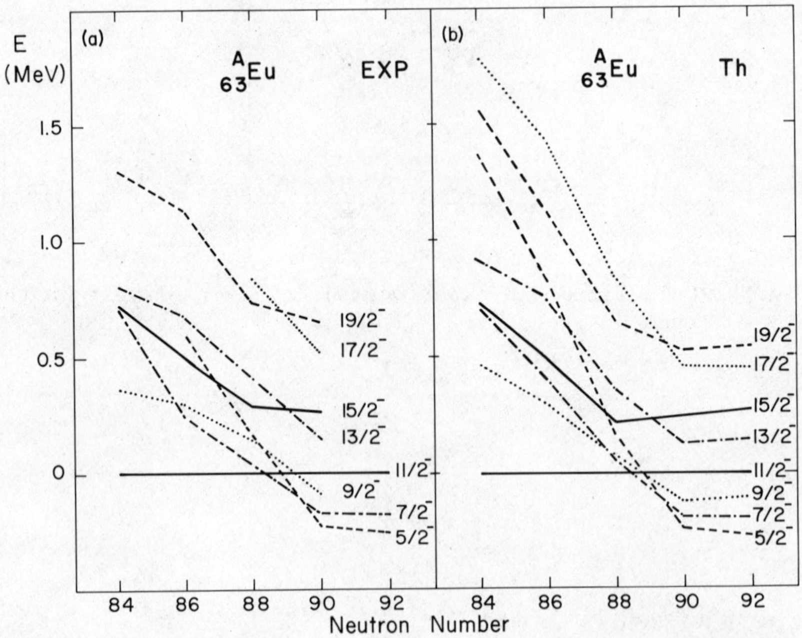

Fig. 8. Experimental and calculated relative energies of the negative parity states in the Eu isotopes.

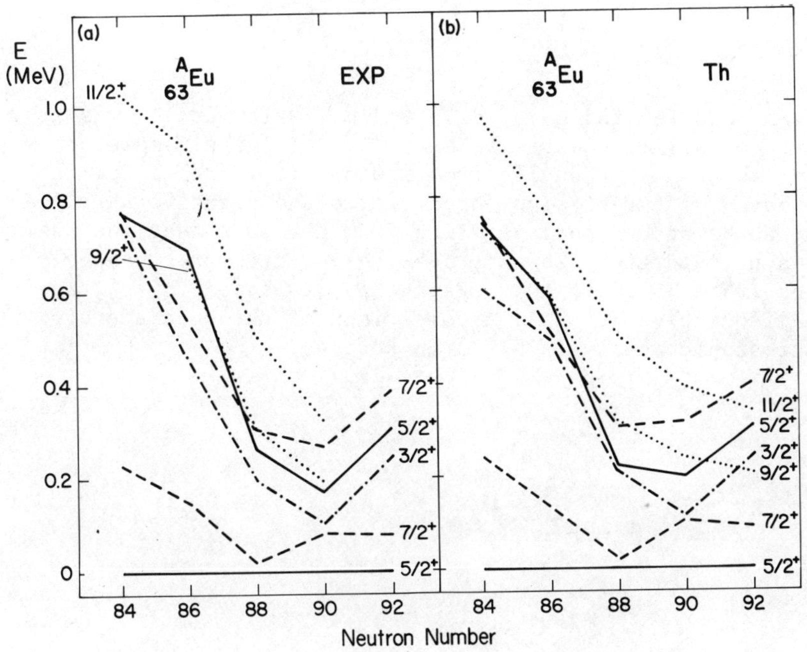

Fig. 9. Experimental and calculated excitation energies of the positive parity states in the Eu-isotopes. For some levels, the experimental spin assignment is uncertain.

As a consequence, the j dependence of A_j, $\Gamma_{jj'}$ and $\Lambda^{j''}_{jj'}$ is given by

$$
\begin{aligned}
A_j &= \sqrt{5}\,\sqrt{2j+1}\,A_0 \;, \\
\Gamma_{jj'} &= \sqrt{5}\,(u^2-v^2)\Gamma_0\, Q_{jj'} \;, \qquad (4.1)\\
\Lambda^{j''}_{jj'} &= -\frac{8\sqrt{5}}{\sqrt{2j''+1}}\, u^2v^2\Lambda_0\, Q_{jj''}Q_{j'j''} \;.
\end{aligned}
$$

For the unique parity level $h_{11/2}$, the expression (3.1) still applies.
In Table I, the values of

$$\gamma = \Gamma_0(u^2-v^2) \qquad (4.2)$$

and

$$\lambda = \Lambda_0 \, u^2 v^2 \tag{4.3}$$

are given both for the $g_{7/2}$, $d_{5/2}$ and $h_{11/2}$ levels. Λ_0 was kept constant for all calculations, $\Lambda_0 = +0.1$ MeV. The parameters γ and λ appear to change gradually from isotope to isotope. The fact that γ has a different sign for positive and negative parity states can be understood as coming from the fact that the positive parity single particle levels, $d_{5/2}$, $g_{7/2}$, are more than half occupied ($v^2 > u^2$), while the negative parity level, $h_{11/2}$, is less than half occupied ($v^2 < u^2$). A comparison of these parameters with the microscopic theory is given in Ref. 4.

Table I. Parameters used in the calculation of the Eu isotopes (in MeV).

A	$d_{5/2} - g_{7/2}$			$h_{11/2}$	
	γ	λ	$\varepsilon_{g_{7/2}} - \varepsilon_{d_{5/2}}$	γ	λ
147	-0.14	0.12	0.25	0.25	0.34
149	-0.17	0.14	0.145	0.33	0.40
151	-0.14	0.20	0.02	0.41	0.57
153	-0.12	0.26	-0.20	0.44	0.77
155	-0.065	0.27	-0.50	0.45	0.80

The transition from SU(5) to SU(3) can be characterized by the behaviour of the $J^\pi = 5/2^-$ level. In the lighter isotopes, this level lies very high in the spectrum, since it belongs to the $n_d = 2$ multiplet. With increasing mass, this level comes down rapidly in energy to become the lowest negative parity state in ^{153}Eu. The fact that in the heavier isotopes the K = 5/2 band is the lowest depends only on the occupancy of the $h_{11/2}$ level. This occupancy also determines the splitting of the $n_d = 1$ quintuplet in the vibrational limit. In the IBFA model these two effects are related. For the positive parity levels, Fig. 9, the changes that occur in the spectra are less spectacular, since for all isotopes the $5/2^+$ level is the ground state. In the light isotopes, the $5/2^+$ and $7/2^+$ levels are rather pure single particle states, in the heavier isotopes the picture is very different. In ^{155}Eu, the $5/2^+_1$ and $7/2^+_1$ levels are members of the $K^\pi = 5/2^+$ ground state band. In order to show the large changes in structure, in Fig. 10 the positive parity part of the spectra of ^{149}Eu and ^{155}Eu are shown in more detail. The levels have been ordered into bands such that the levels within a band are connected by strong E2 transitions.

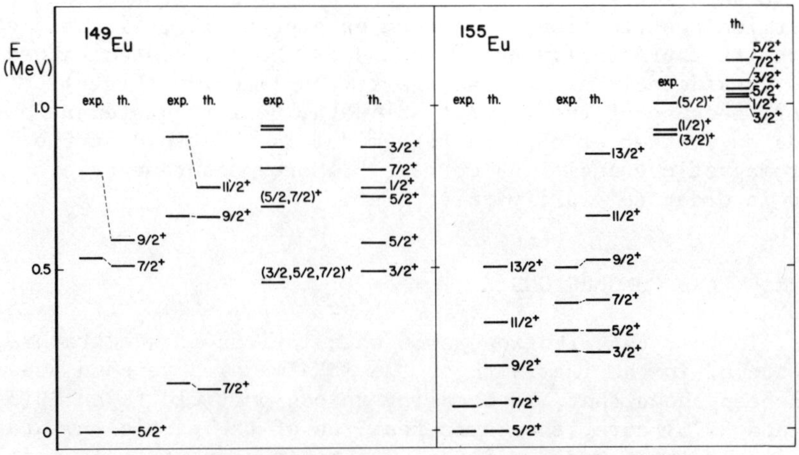

Fig. 10. Experimental and calculated spectra arising from the $d_{5/2}$ and $g_{7/2}$ configurations in $^{149}_{63}Eu_{86}$ and $^{155}_{63}Eu_{92}$. The experimental spin assignment of some levels is tentative.

It is worth noting once more that the spectrum of ^{149}Eu is typical of the particle-vibrator model, while that of ^{155}Eu is typical of the particle-rotor model. In the IBFA, both spectra are calculated using the same Hamiltonian.

5. ELECTROMAGNETIC TRANSITIONS

In the IBFA model, the E2 transition operator can be written as

$$T^{(2)} = e_B\, Q_B^{(2)} + e_F\, Q_F^{(2)}, \tag{5.1}$$

where $Q_B^{(2)} \equiv Q^{(2)}$ is given by (2.2). The same microscopic derivation which gives (2.3), allows one to estimate the j dependence of $Q_F^{(2)}$. If one retains only the one-body part of $Q_F^{(2)}$, one obtains

$$Q_F^{(2)} = \sum_{jj'} Q_{jj'}(u_j u_{j'} - v_j v_{j'})(a_j^\dagger \times \tilde{a}_{j'})^{(2)} . \tag{5.2}$$

Very little is known about E2 matrix elements in these nuclei. We have therefore compared our results only to the measured values of the quadrupole moments, Fig. 11. In the calculation, we have assumed that the boson effective charge is given by the $2_1^+ \rightarrow 0_1^+$ transition in ^{154}Sm, e_B=0.13 eb, and that $\chi = -\frac{1}{2}\sqrt{7}$, as in previous calculations[17]. Since for allowed transitions the matrix elements of $Q_B^{(2)}$ are approximately N times larger than those of $Q_F^{(2)}$, we have set in our calculation $e_F = 0$. The introduction of the fermion contribution, which can be done easily, will not change the results appreciably. As one can see, the calculated values change considerably across the transitional region, Fig. 11. However, because of the scarcity of data, no firm statement can be made on whether or not the IBFA model is able to describe electromagnetic transitions correctly. More measurements are needed in order to clarify this point.

6. SUMMARY AND CONCLUSIONS

In this contribution, some calculations of spectra of odd-A nuclei in the framework of the IBFA model have been presented. It has been shown that, by coupling an odd particle to an SU(5), SU(3) and SO(6) core, the basic features of the particle-vibrator model, the Nilsson model and the γ-unstable plus particle model can be reproduced. The strengths of the quadrupole and exchange terms in the boson-fermion interaction V_{BF} play the role of the Fermi energy, λ_F, and deformation, β, in the geometric model. Especially for the Nilsson model this correspondence is evident.

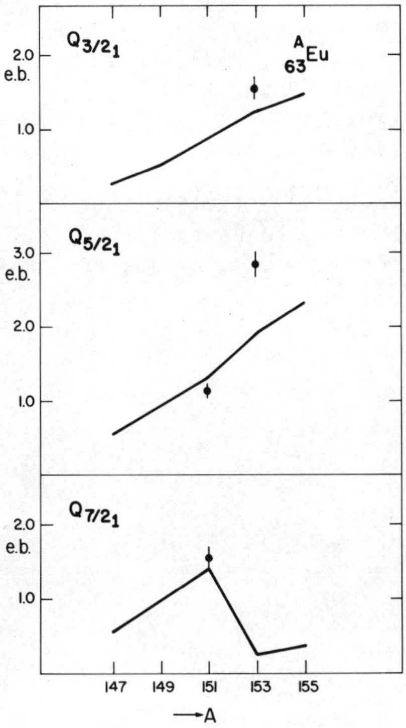

Fig. 11. Experimental and calculated values of the quadrupole moments of the $3/2_1^+$, $5/2_1^+$ and $7/2_1^+$ states.

In order to test the model away from the limits, we have applied it to the study of the Europium isotopes. Good agreement with experiment has been obtained, with parameters changing smoothly from isotope to isotope. This suggests that the IBFA model, may indeed provide the framework for a unified description of collective states in odd-A nuclei.

REFERENCES

1. A. Arima and F. Iachello, Phys. Rev. C14 (1976), 761.
2. F. Iachello and O. Scholten, Phys. Rev. Lett. 43 (1979), 679.
3. I. Talmi, these Proceedings.
4. O. Scholten, these Proceedings.
5. O. Scholten, Ph.D. Thesis, University of Groningen, The Netherlands (1980).
6. T. Otsuka and A. Arima, Phys. Lett. 77B (1978), 1; T. Otsuka, Ph.D. Thesis, University of Tokyo, Japan (1979).
7. O. Scholten, Program ODDA, KVI Internal Report (1980).
8. O. Scholten, KVI Annual Report (1977), 67.

9. F. Iachello, Phys. Rev. Lett. 44 (1980), 772.
10. F. Iachello, these Proceedings.
11. F. Iachello and O. Scholten, Phys. Lett. 91B (1980), 189.
12. J. Meyer-ter-Vehn, Nucl. Phys. A249 (1975), 111.
13. L. Wilets and M. Jean, Phys. Rev. 102 (1956), 788.
14. F. Dönau and S. Frauendorf, J. Phys. Soc. Japan Suppl. 44 (1978), 526.
15. G. Leander, Nucl. Phys. A273 (1976), 286.
16. N. Blasi and O. Scholten, to be published.
17. O. Scholten, F. Iachello and A. Arima, Ann. Phys. (N.Y.) 115 (1978), 325.

DESCRIPTION OF ODD MEDIUM MASS NUCLEI IN THE INTERACTING BOSON-FERMION MODEL

P. von Brentano, A. Gelberg and U.Kaup

Institut für Kernphysik, Universität zu Köln
D-5000 Köln 41, West-Germany

In this paper we will discuss the application of the new Interacting Boson Fermion Model (IBFM)[1] to low lying positive parity levels in Rb, Tc and Ag nuclei. We restrict the discussion on the positive parity levels, because the odd particle is then predominantly in the $g_{9/2}$ shell only, thus providing a rather pure case.

The low lying levels in the odd nucleus were described by the IBFM Hamiltonian of Iachello and Scholten in the following form[2,3].

$$H = H_{core} + H_{coupl.} + \sum_j E_j \, (a_j^+ \tilde{a}_{j'})^{(o)} \tag{1}$$

$$H_{coupl.} = -\sum_j \sqrt{5}\, A_j \, (d^+ \tilde{d})^{(o)} \, (a_j^+ \tilde{a}_j)^{(o)}$$
$$-\sum_{jj'} \sqrt{5}\, \Gamma_{jj'} \, [[(s^+ \tilde{d} + d^+ s)^{(2)} + \chi (d^+ d)^{(2)}](a_j^+ \tilde{a}_{j'})^{(2)}]^{(o)} \tag{2}$$
$$-\sum_{jj'j''} \sqrt{2j''+1}\, \Lambda_{jj'}^{j''} : [(\tilde{d} a_j^+)^{(j'')} (d^+ \tilde{a}_{j'})^{(j'')}]^{(o)} :$$

In calculations with one shell only we will quote the parameters $\Gamma = \sqrt{5}\,\Gamma_{jj}$, $\Lambda = \sqrt{2j+1}\,\Lambda_{jj}^{j}$, $A = \sqrt{5}\,A_j$. The even-even neighbouring nucleus which served as the core was described by the IBA-1 Hamiltonian of Arima and Iachello[4]. The parameters HBAR, C_o, C_2, C_4, F, G given in the figures are the parameters ε, C_o, C_2, C_4, $\tilde{v}_2$, $\tilde{v}_o$ of ref. 4. The even and odd nuclei were fitted using the codes PHINT[5] and ODDA[6]. A special programme oracle[7] was written which allows to get a χ^2-fit for the parameters of the Hamiltonian to the experimental levels and BE2-values. One may try to obtain the IBA-1 parameters from an IBA-2 calculation using a mapping procedure[3]. Although IBA-2 calculations have been carried out by us in most cases discussed here, we did not use this procedure because its validity is restricted to states which are symmetric under proton-neutron boson exchange. This condition seems to be fulfilled well enough for states of heavy nuclei but not for the medium mass nuclei under consideration here.

The Rb isotopes are a good testing case for the IBFM model, because the light neighbouring even Kr[8,9] and Sr[10] isotopes are fair examples of collective nuclei and they are well described in IBA-2 and IBA-1[8]. In addition, after an exploratory IBFM investigation of Rb[11] we had done a rather careful experimental study by in beam gamma spectroscopy[12] which had identified many additional positive parity levels. The results[13] are shown in fig. 1 and 2. About equally good fits were obtained in a one shell calculation with $j_{odd}=g_{9/2}$ and in a 2 shell calculation with $j_{odd}=g_{9/2}$ or $d_{5/2}$. The parameters of the one shell calculation given below differ in an unsystematic way for ^{81}Rb and ^{83}Rb. Also the parameter χ in the IBFM is very different from the χ_ν of the IBA-2 fit of ^{80,82}Kr.

	A/MeV	Γ/MeV	Λ/MeV	χ
^{81}Rb	2.80	2.84	5.79	0.1
^{83}Rb	3.68	2.84	4.90	0.3

These problems have vanished in the two shell calculation. The many free parameters in the multishell IBFM Hamiltonian have been reduced to a few parameters by using the relations given below[18)] which are actually some earlier version of the relations given in ref. 1-3.

$$\Gamma_{jj'} = \Gamma_o \; [(\alpha_j\alpha_{j'})^{1/2} - (1-\alpha_{j'})^{1/2}(1-\alpha_j)^{1/2}] \langle j \| Y_2 \| j' \rangle \tag{3a}$$

$$\Lambda^{j''}_{jj'} = \Lambda_o \; \beta_{jj''} \; \beta_{j'j''} \tag{3b}$$

$$\beta_{jj'} = [\alpha_j \; \alpha_{j'} \; (1-\alpha_j)(1-\alpha_{j'})]^{1/4} \; \langle j \| Y_2 \| j' \rangle \tag{3c}$$

Here α_j is the occupation number of the state j similar to the V_j^2 of the BCS description. For the parameter χ the values of χ_ν of the IBA-2 fit to ^{80,82}Kr were chosen. The values of the parameters used are:*

	A_j/MeV	Γ_o/MeV	Λ_o/MeV	χ	$\alpha_{9/2}$	$\alpha_{5/2}$	Δε/MeV
^{81}Rb	0.5	0.7	3.5	0.7	0.3	0	3.2
^{83}Rb	0.3	0.7	3.5	0.9	0.18	0.005	3.1

Δε is the energy difference between the $g_{9/2}$ and $d_{5/2}$ single particle levels; it was taken from the values given in the Bohr-Mottelson book. 5 parameters are left to be fitted to the odd-A energy levels. As a matter of fact it turns out that only three of them, namely A_j, $\alpha_{9/2}$ and $\alpha_{5/2}$ have to be varied when going from one isotope to the other. This number is not larger than for the one shell calculation, and contrary to the finding in the former case it is systematical and can be physically understood. In addition, the values of the force parameters $\Gamma_{jj'}$, A_j, Λ^j_{jj} with

$j=g_{9/2}$ are smaller by a factor of three in the two shell calculation compared with the one shell calculation. This reduction indicates that the truncation of the model space to one shell can be too severe even for unique parity states. We attribute this effect to the strong $\Delta j=2$ mixing due to the quadropole force. Furthermore, B(E2) values look very different in the two calculations, although energy spectra are similar. We find also that the force parameters are no longer unstable when we include the $g_{7/2}$ state. Thus we conclude, that a two shell IBFM is the appropriate model for these nuclei.

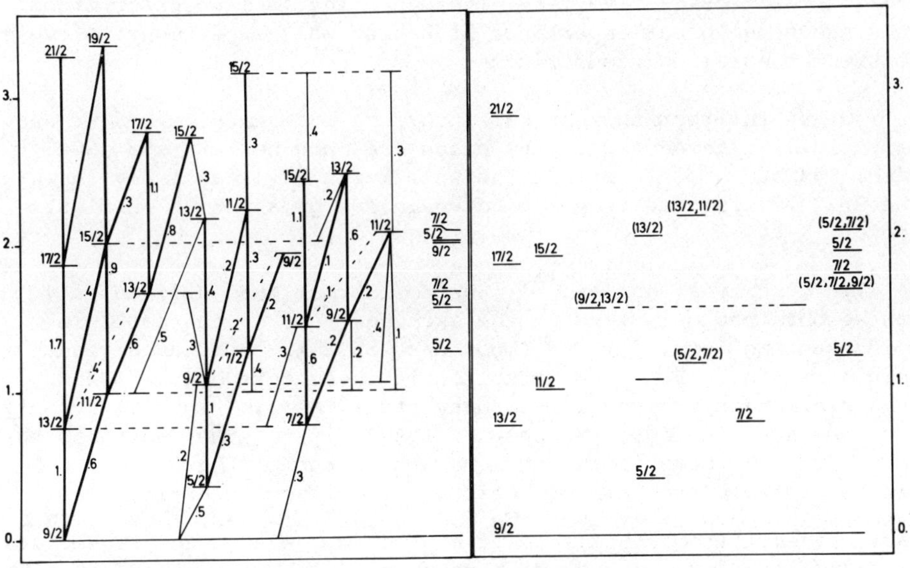

Fig. 1: Right side: Positive parity levels of ^{83}Rb from experiment. Left side: IBFM calcualtions with $j=g_{9/2}$ and $d_{5/2}$ for ^{83}Rb.

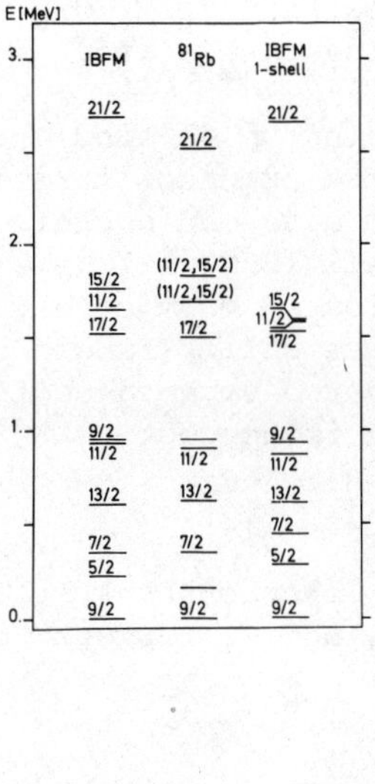

Fig. 2: Comparison of 2 shell IBFM, experimental positive parity levels in ^{81}Rb and one shell IBFM for ^{81}Rb.

The levels of ^{83}Rb in fig. 1 have been arranged in bands according to the limiting SO(6) x $g_{9/2}$ case[14]. However, ^{80,82}Kr are not really pure SO(6) nuclei, and neither is the particle-core coupling SO(6) invariant. Thus we find that the pure SO(6) x $g_{9/2}$ bands are mixing in the actual calculation.

We should mention that a rather satisfactory description of ^{81,83}Rb was obtained also in the frame of the Toki-Faessler model[15]) in which a $g_{9/2}$ particle is coupled to a triaxial core. However, a number of low spin levels observed in beta decay seem to be well described in the IBFM but are missing in the Toki-Faessler model. This shows the great importance of having many experimental levels for such theoretical comparisons.

Other interesting test cases for the IBFM are the odd Tc and Ag nuclei. In these nuclei the cores are much nearer to an SU(5) case. The odd nuclei exhibit the well known (J-1) anormaly in which the $7/2^+$ levels come very much down in energy as shown in figs. 4 and 5.

In our investigation [16,17] we found that the (J-1) anomaly can be obtained only through the exchange term in the particle core coupling (eqn. 2). The importance of this term has already been stressed by Bohr and Mottelson in their textbook. It seems remarkable that one can fit so many states as e.g. in ^{97}Tc by only 4 parameters. In fig. 3 we show an IBA-1 fit to ^{96}Mo, which is the core for ^{97}Tc. Both in the even Mo and Pd nuclei there are many single particle excitations besides the collective excitations, especially above 2 MeV. For the odd-nuclei, however, these single particle excitations are less of a problem, because the levels are either well below 2 MeV or of high spin. We should mention that we have obtained good descriptions also for the nuclei ^{98}Mo, ^{99}Tc, ^{102}Pd, ^{104}Pd, ^{106}Pd and ^{105}Ag. However, in these cases we applied only the one shell IBFM, and there are some anomalies of the $5/2^+$ states, particularly for the Ag isotopes, which point to an influence of the $d_{5/2}$ state.

Summing up we find that the IBFM model with the crucial inclusion of the exchange term gives good fits to the collective levels in medium mass odd nuclei. We believe that the application of a multi shell IBFM will be essential to obtain a really physical description in most cases, even for situations which are dominated by a unique parity single particle state. However, we could investigate only two such cases in detail, thus further work in this direction is needed.

We would like to thank F. Iachello, O. Scholten, and H.W. Schuh, D. Hippe, and R. Vorwerk for discussion.

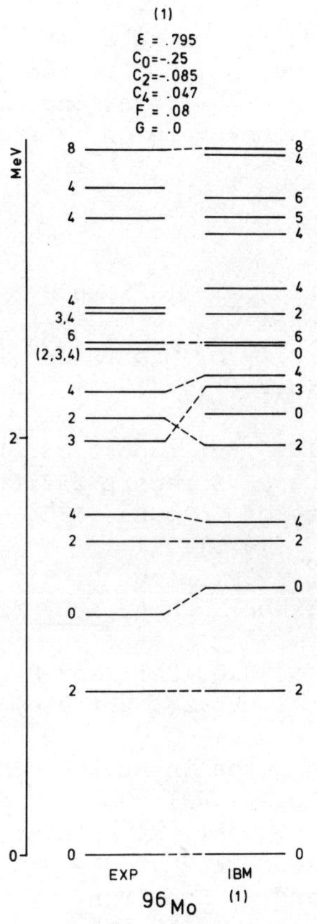

Fig. 3: IBA-1 calculation for ^{96}Mo

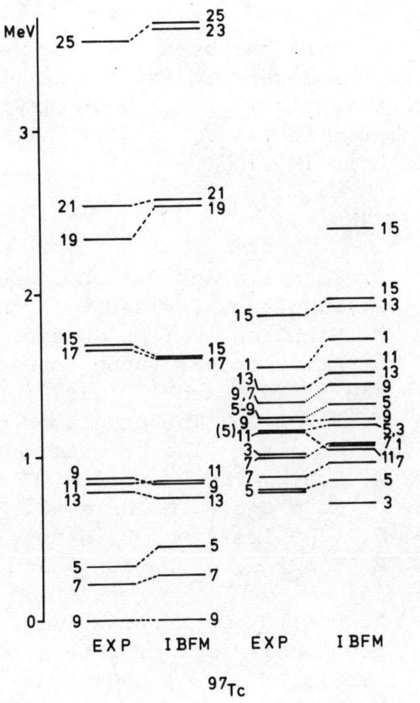

Fig. 4: Comparison of positive parity levels in ^{97}Tc from experiment with IBFM. Parameters: A=2.19 MeV, Γ=1.93 MeV, Λ=5.51 MeV χ=0.3.

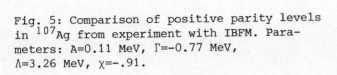

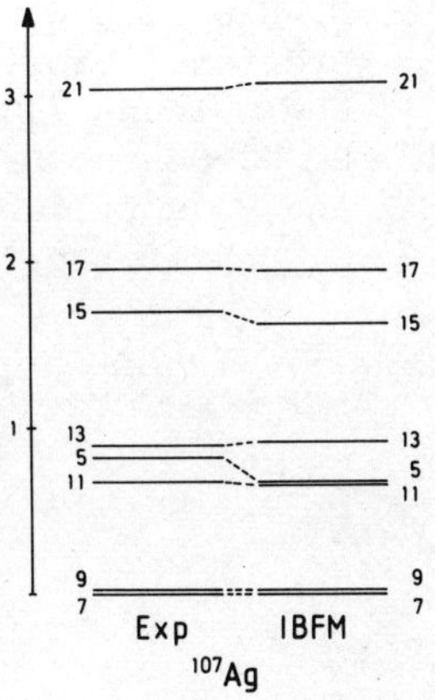

Fig. 5: Comparison of positive parity levels in ^{107}Ag from experiment with IBFM. Parameters: A=0.11 MeV, Γ=-0.77 MeV, Λ=3.26 MeV, χ=-.91.

This work was supported by BMFT.

* As there has been some confusion concerning the relation between the parameters in the formulae and those being input to the different versions of the code ODDA, we give here the relations used by us: $\Gamma_o = (1/5)\sqrt{4\pi}$ VQ, $\Lambda_o = 4\pi/5$ VPDD; the programme version is from October 1979.

REFERENCES

1. F. Iachello and O. Scholten, Phys. Lett., 43 (1979) 679
2. O. Scholten, invited talk this conference
3. O. Scholten, Ph. D. Thesis, The University of Groningen (1980)
4. A. Arima and F. Iachello, Ann. Phys. (N.Y.) 99 (1976) 253, 111 (1978) 201; and 123 (1979) 468
5. O. Scholten "The programme package PHINT" KVI report 63 (1979)
6. O. Scholten "The programme package ODDA" KVI report 252 (1980)
7. W. Schuh, Ph. D. Thesis, The University of Cologne (1980)
8. U. Kaup and A. Gelberg, Z. Phys., A293 (1979) 311
9. H.P. Hellmeister, U. Kaup, J. Keinonen, K.P. Lieb, R. Rascher, R. Ballini, J. Delaunay and H. Dumont, Nucl. Phys. A332 (1979) 241
10. A. Dewald, U. Kaup, W. Gast, A. Gelberg, K.O. Zell and P. von Brentano, Proc. of Intern. Conf. on Nucl. Phys., Berkeley 1980 LBL-11118 (1980) 306
11. A. Gelberg and U. Kaup in "Interacting Boson in Nuclear Physics" ed. by F. Iachello (Plenum Press New York 1979) 59
12. W. Gast, K. Dey, A. Gelberg, U. Kaup, F. Paar, R. Richter, K.O. Zell and P. von Brentano, to be published in Phys. Rev. C
13. U. Kaup, A. Gelberg, P. von Brentano and O. Scholten, to be published in Phys. Rev. C
14. F. Iachello and O. Scholten, Phys. Lett. 91 B (1980) 189
15. H. Toki, A. Faessler, Phys. Lett. 63 B (1976) 121
16. D. Hippe, H.W. Schuh, U. Kaup, K.O. Zell, A. Gelberg and P. von Brentano, Proc. Intern. Conf. on Band Structure and Nucl. Dyn. New Orleans (1980)
17. H. W. Schuh, D. Hippe, R. Vorwerk, U. Kaup, K.O. Zell, A. Gelberg and P. von Brentano to be published.
18. O. Scholten private communication

ODD-A Eu ISOTOPES AND THE INTERACTING BOSON-FERMION MODEL

G. Lo Bianco

Istituto di Scienze Fisiche dell'Università
e Istituto Nazionale di Fisica Nucleare
Milano, Italy

Scholten et al.[1] showed that the interacting boson model gives a satisfactory and unitary description of the even-even Sm isotopes, whose spectra change character with increasing mass number, going from the vibrational to the rotational limit. The neighbouring odd-A Eu isotopes, which also show such a transition, are therefore a suitable test of the Interacting Boson-Fermion Model (IBFM).

I shall only consider the isotopes with A=147, 151 and 153, starting by a breif review of the experimental situation. While the information obtained by radioactive-decay measurments and by particle-transfer reactions has been invaluable, the following discussion will be limited to the recent in-beam γ-ray work.

^{147}Eu has been studied by Fleissner et al.[2] using heavy-ion induced reactions, which are well suited to follow the sequence of the yrast states up to high spin. The complementary information obtained by the use of lighter projectiles, which more easily excite the non yrast states, may however be expecially important for testing the IBFM. ^{147}Eu was studied at the Milan cyclotron by the ^{148}Sm(p,2nγ) reaction (Lo Bianco et al.[3]). While states with spin greater than 19/2 were not populated, the obtained decay scheme includes many levels which were not seen by Fleissner et al. and some of these, at energies above 1 MeV, had not been previously identified.

^{151}Eu received a great deal of interest since results from two particle transfer experiments[4] suggested the possible coexistence of spherical and deformed shapes in this nucleus. Taketani et al.[5],on the basis of a (p,2nγ) experiment, published a partial decay scheme where levels at 415, 600 and 845 keV are interpreted as members of a rotational band based on the $5/2^+$ [413] state at 260 keV. In contrast to that, no evidence of γ rays feeding this state was found in a (d,3nγ) study by Leigh et al.[6], where set of levels connected by strong transitions were interpreted in terms of a band structure built on the $5/2^+$ ground state, on the $7/2^+$ first excited state and on the $11/2^-$ 196 keV level. It should be noted that in the level scheme of Taketani et al. the 845 keV level is supposed to decay through two cascades, one including γ rays having almost the same energies as well known 245 and 340 keV transitions in ^{152}Sm, the other including transitions of 429 and 416 keV, the same as the energies of two coincident γ rays which, according to Leigh et al., connect levels in the negative-parity band. Fleissner et al.[3] studied ^{151}Eu using the (^{6}Li,3n) reaction which selectively populates states belonging to the negative parity band. While Leigh et al. identified a 611 keV $13/2^-$ level fed by 429 and 502 keV transitions, in the decay scheme of Fleissner et al. this level is not included and the 429 keV transition feeds the $15/2^-$ state.

In the hope to shed some light on these discrepances ^{151}Eu was studied again by Lo Bianco et al.[7] using the (p,2nγ) reaction and performing coincidence measurments with better resolution and statistics than in the experiment by Taketani et al.. The existence of the $11/2^+$ 845 keV level proposed by these authors was not supported by the new data, which confirmed the assignment given by Leigh et al. for the 429 and 416 keV transitions and showed the 245 keV γ ray to be in coincidence with the 340 keV $6^+ \rightarrow 4^+$ transition in the ^{152}Sm ground-state band. Several states identified in this experiment were not previously known.

Finally, ^{153}Eu was studied by Dracoulis et al.[8] via the (d,3nγ) reaction. Well developed rotational bands, of both positive and negative parity, were identified in the obtained level scheme. From preliminary measurments at the Milan cyclotron it seems that no essentially new results can be obtained in this case by the (p,2nγ) reaction.

I shall now compare the experimental energies to the predictions of IBFM calculations, starting by considering the negative-parity states, which correspond to an $h_{11/2}$ proton coupled to collective excitations of the core.

In the case of only one active orbit the boson-fermion interaction in the IBFM Hamiltonian[9] may be written as

$$V_{BF} = A\left[(s^\dagger \times s)^{(0)} \times (a_j^\dagger \times \tilde{a}_j)^{(0)}\right]^{(0)} + \Gamma\left\{\left[(d^\dagger \times s + s^\dagger \times \tilde{d})^{(2)} + \chi (d^\dagger \times \tilde{d})^{(2)}\right] \times (a_j^\dagger \times \tilde{a}_j)^{(2)}\right\}^{(0)}$$
$$+ \Lambda : \left[(\tilde{d} \times a_j^\dagger)^{(j)} \times (d^\dagger \times \tilde{a}_j)^{(j)}\right]^{(0)}$$

where $s^\dagger$, $d^\dagger$, $a_j^\dagger$ are the creation operators for s-bosons, for d-bosons and for fermions in the j orbit and $\tilde{d}_\mu = (-1)^\mu d_{-\mu}$, $\tilde{a}_{jm} = (-1)^{j-m} a_{j-m}$. The calculations were performed in Milan and in Groningen using the computer code ODDA written by O. Scholten.

Table I: Best-fit parameters for the negative-parity states.

	Γ	Λ	A
^{147}Eu	0	1386 keV	-500 keV
^{151}Eu	-1100 keV	3464 keV	-300 keV
^{153}Eu	-1050 keV	4677 keV	-400 keV

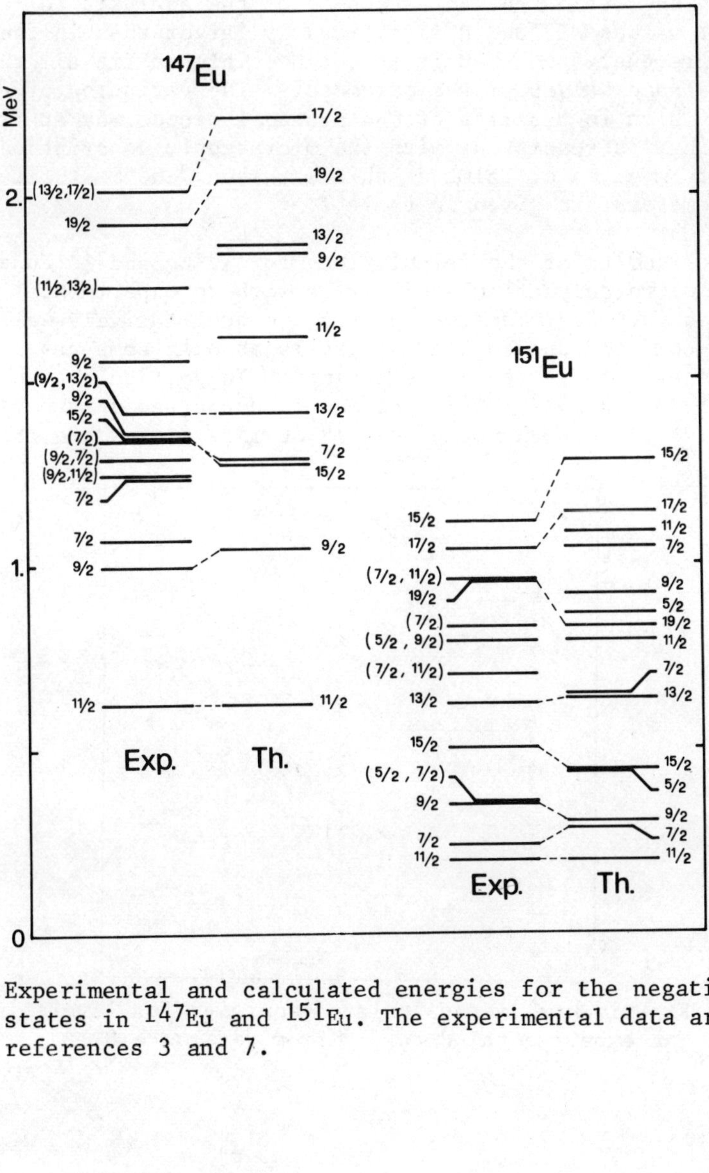

Fig. 1: Experimental and calculated energies for the negative-parity states in ^{147}Eu and ^{151}Eu. The experimental data are from references 3 and 7.

The analysis was at first limited to ^{147}Eu and ^{151}Eu, for which new experimental data had been obtained. The core nuclei were described by a set of parameters which were obtained by Scholten (private communication) on the basis of his study of the Sm isotopes by the model in which two kinds of bosons are considered for neutron and protons pairs. The number of free parameters was reduced by identifying the constant χ , which appears in the quadrupole term of the boson fermion interaction for the Eu isotopes, with the constant in the neutron-boson quadrupole operator for the Sm isotopes. In the case of ^{147}Eu good fits could only be obtained when the strenght of the exchange interaction, Λ , was small, while the results were not very sensitive to Γ, the strength of the quadrupole-quadrupole interaction. In ^{151}Eu instead the order of the levels was found to depend critically on the ratio Γ/Λ. An extension of the analysis to ^{153}Eu gave best-fit values of Γ and Λ significantly larger than in the ^{147}Eu case. It appears therefore that it is not possible to fit all three nuclei with the same values of the parameters. The variation of Λ with mass number, which in a series of odd-proton isotopes may at first appear surprising, is consistent with the microscopic description of the exchange term given by Talmi at this workshop. The best-fit values of the parameters are given in table I.

The results of the calculations for ^{147}Eu and ^{151}Eu are shown in fig. 1, where only calculated levels with an experimental counterpart have been included. The two spectra are qualitatively well reproduced but the computed spacings among levels show in some cases significant differences with respect to experiment. In particular, the spacings 15/2 - 13/2 and 19/2 - 17/2 between favoured and unfavoured states are overestimated in both nuclei. In ^{147}Eu more states are observed below

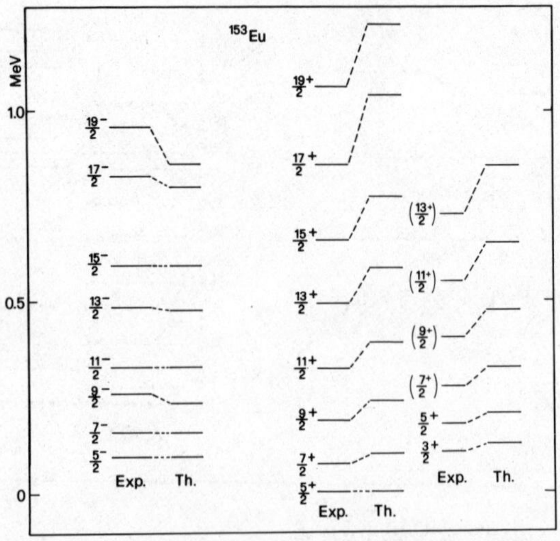

Fig. 2: Experimental and calculated energies for levels in ^{153}Eu. The experimental data are from reference 8.

1.5 MeV than are predicted by the calculations; some states may correspond to octupolar vibrations of the core. The correspondances between experimental and theoretical levels were suggested taking into account the observed decay modes.

The calculations for ^{153}Eu are compared to the experimental spectrum in fig. 2. The negative-parity band, whose irregular spacings are explained in the rotational model by the inclusion of Coriolis coupling, is very well reproduced by the IBFM.

In the case of the positive-parity levels at least two active orbits, $d_{5/2}$ and $g_{7/2}$, have to be considered for the odd proton. The parameters A, Γ and Λ which appear in the IBFM Hamiltonian need then to be replaced by matrices $\{A_j\}$, $\{\Gamma_{jj'}\}$, $\{\Lambda^{j''}_{jj'}\}$, where in our case j, j', j'' = 5/2, 7/2 and j'' is the angular momentum to which the operators $\tilde{d}$ and $a^\dagger_j$ or $d^\dagger$ and $\tilde{a}_j$ are coupled in the exchange term.

The quantities $\Gamma_{jj'}$ and $\Lambda^{j''}_{jj'}$ are assumed to be related to the matrix elements of the spherical harmonic $Y^{(2)}$ by

$$\Gamma_{jj'} = \gamma \langle j \| Y^{(2)} \| j' \rangle$$

$$\Lambda^{j''}_{jj'} = \lambda \, (2j''+1)^{-\frac{1}{2}} \langle j \| Y^{(2)} \| j'' \rangle \langle j'' \| Y^{(2)} \| j' \rangle$$

In the present case the possible dependence of γ and λ on the orbits involved may be neglected since the $d_{5/2}$ and $g_{7/2}$ orbits are almost degenerate. The monopole term was assumed to be orbit independent.

Good fits to the experimental energies of the positive-parity states were obtained with the parameters given in table II.

The results of the calculations for ^{147}Eu and ^{151}Eu are compared to the experimental spectra in fig. 3.

The calculations for ^{147}Eu suggest a weak coupling between the core and the odd particle, giving in good agreement with experiment

Table II: Best-fit parameters for the positive-parity states.

	γ	λ	A	$\varepsilon_{g_{7/2}} - \varepsilon_{d_{5/2}}$
^{147}Eu	-326 keV	2210 keV	-100 keV	250 keV
^{151}Eu	-309 keV	3570 keV	-100 keV	20 keV
^{153}Eu	-259 keV	4621 keV	-100 keV	-200 keV

The last column contains the single-particle energy differences, which have not been adjusted.

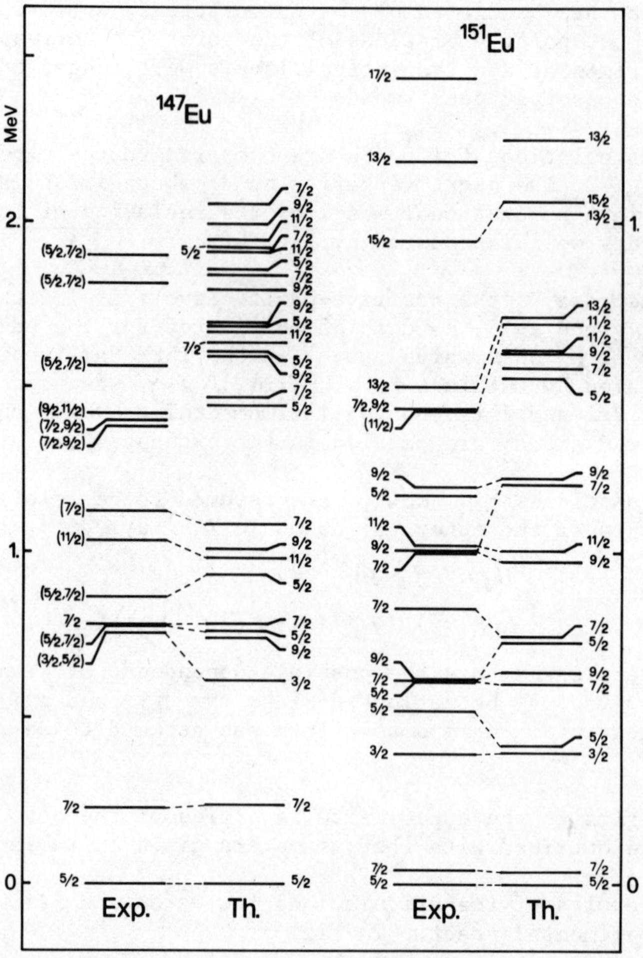

Fig. 3: Experimental and calculated energies for the positive-parity states in ^{147}Eu and ^{151}Eu. The experimental data are from references 3 and 7 and from the Nuclear Data Sheets for A=147.

a group of states at a mean energy of 900 keV which may be interpreted as a multiplet built on the 2^+ level of the core.At higher excitation the comparison is made very difficult by the ambiguities in the experimental spin and parity and by the fact that the predicted energies for the second multiplet are probably overestimated, since the parameters used for the core give the 4^+ state too high by 265 keV.

In the case of ^{151}Eu the correspondances indicated by the broken lines have been estabilished requiring consistency between the computed transition probabilities and the observed decay modes. It may be seen that, even if the theoretical spacings not always agree with the experimental ones, the very complex experimental spectrum is qualitatively well reproduced.

Since also the positive-parity rotational bands in ^{153}Eu are reproduced by the calculations (fig. 2), one may conclude that the IBFM is able to describe, with smoothly varying parameters, both negative and positive-parity states in all three isotopes, from the vibrational-like ^{147}Eu to the well deformed ^{153}Eu.

The experimental work in Milan was done in close collaboration with drs. N. Molho and A. Moroni. The contribution of dr. N. Blasi was essential in performing the calculations. Finally, I wish to thank prof. F. Iachello and dr. O. Scholten for many useful discussions and for the permission to use the computer code ODDA.

REFERENCES

1. O. Scholten, F. Iachello and A. Arima, Ann. of Phys. 115 (1978) 325
2. J.G. Fleissner, E.G. Funke, F.P. Venezia and J.W. Mihelich, Phys. Rev. C16 (1977) 227
3. G. Lo Bianco, N. Molho, A. Moroni, A.Bracco and N. Blasi, to be published in J. Phys. G: Nucl. Phys.
4. D.G. Burke, G. Lovhoiden and J.C. Waddington, Phys. Lett. 43B (1973) 470; H. Taketani, H.L. Sharma and N.M. Hintz, Phys. Rev. C12 (1975) 108
5. H. Taketani, M. Adachi, T. Hattori, T. Matsuzali and H. Nakayama, Phys. Lett. 63B (1976) 154
6. J.R. Leigh, G.D. Dracoulis, M.G. Slocombe and J.O. Newton, J. Phys. G: Nucl. Phys. 3 (1977) 519
7. G. Lo Bianco, N. Molho, A. Moroni, S. Angius, N. Blasi and A.Ferrero, J. Phys. G: Nucl. Phys. 5 (1979) 697
8. G.D. Dracoulis, J.R. Leigh, M.G. Slocombe and J.O. Newton, J. Phys. G: Nucl. Phys. 1 (1975) 853
9. F. Iachello and O. Scholten, Phys. Rev. Lett. 43 (1979) 679

UNIQUE PARITY STATES IN ^{109}Pd AS A TEST OF PARTICLE-ROTOR AND IBFA MODELS

R. F. Casten

Brookhaven National Laboratory
Upton, New York, 11973

Unique parity levels in odd mass heavy nuclei arise, in each shell, from that isolated orbit lying amidst others of opposite parity. Their effective isolation leads to high purity and therefore such states offer an ideal testing ground for nuclear models.

Usually, the known unique parity states are high spin, aligned states disclosed in neutron deficient nuclei through heavy ion, in-beam studies. A simple geometrical picture of these states, valid for moderately deformed nuclei is illustrated in fig. 1. When the Fermi surface is amongst the low K orbits, the coupling scheme approximates a decoupled band picture[1] in which the yrast states of spin I are obtained by a nearly parallel coupling of the particle

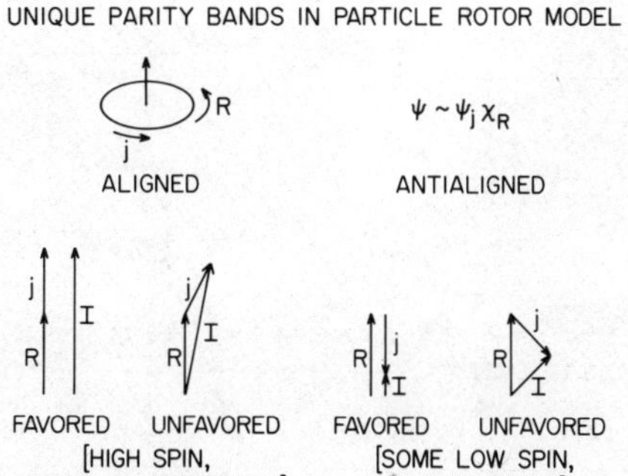

Fig. 1. Decoupled band picture of particle rotor alignment, with indication of typical empirical means of access.

angular momentum $\vec{j}$ and the core rotational angular momentum $\vec{R}$. Successively higher spin states occur for successively larger R values and thus the yrast energies follow those of the ground band in the adjacent even nucleus. These are the favored, aligned levels. At slightly higher energies, unfavored aligned levels are occasionally identified: because of the angle between $\vec{R}$ and $\vec{j}$, a larger R value (hence energy) is required for a given I.

Seldom observed, but a more sensitive means of discriminating various models, are the <u>low</u> spin states arising from the anti-parallel coupling of $\vec{R}$ and $\vec{j}$. As shown in fig. 1, these can be either favored or unfavored. They arise only for $\vec{R} \sim -\vec{j}$ and so are rather restricted in number. Thus, for an $h_{11/2}$ particle, there are only two $3/2^-$ levels compared to six states for each spin $I \geq 11/2$. Due to this and the fact that one expects these levels to occur lower in energy than their high spin counterparts with the same R, they may indeed be expected to be even purer in character.

Unfortunately, they are rarely observed and, until the present study, never has the complete set been established. Exploiting the inherent non-selectivity[2] of the (n,γ) reaction, we have studied the low spin levels in ^{109}Pd and located[3] the complete set of low lying, low spin anti-aligned states of $h_{11/2}$ parentage.

One might not expect, a priori, that the particle-rotor (PR) model would be a suitable framework for interpreting unique parity levels in the A∿100 mass region. However, it is in just such moderately deformed nuclei that the greatest successes[1] of this approach have been achieved since such nuclei have large Coriolis coupling constants, $h^2/2I$, which favor parallel R, j alignment. Furthermore, it has been shown[4] (see fig. 2 (left)) that the model indeed works

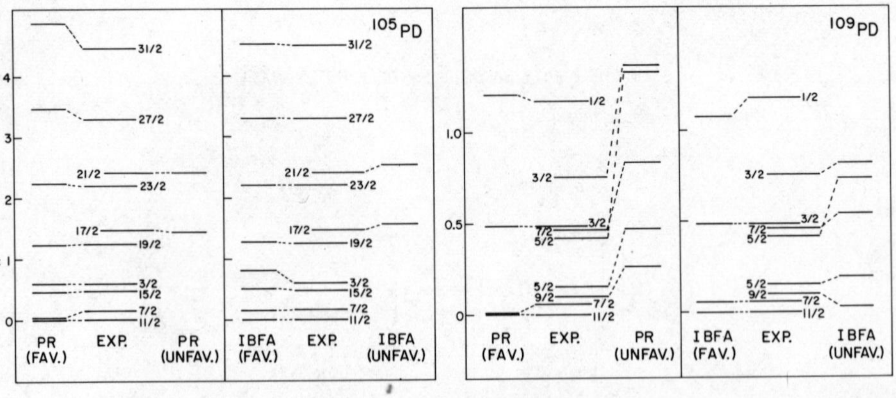

Fig. 2. Unique parity states in ^{105,109}Pd compared with PR and IBFA models. For convenience the favored and unfavored states are compared separately.

exceedingly well for the aligned high spin ($h_{11/2}$ based) states in the even less deformed nucleus ^{105}Pd.

Encouraged by this and by the opportunity presented to perform the first test of the PR model for an extensive set of anti-aligned levels, we performed calculations for ^{109}Pd exactly analogous to those of Rickey and co-workers[4] for the $h_{11/2}$ neutron system in ^{105}Pd. A Nilsson model with pairing, variable moment of inertia and Coriolis coupling was used. The results are shown in fig. 2 (right). As with ^{105}Pd, the favored levels are well reproduced. However, the low spin unfavored states exhibit serious disagreements which cannot be rectified by parameter changes. This is most obvious for the three close lying pairs of states, of spins 3/2, 5/2 and 7/2. The two 3/2 levels, for example, form an isolated pair: Thus, their minimum separation is twice their mixing matrix element. But, in the PR model, the Coriolis interaction is (schematically) given by $h^2/2I\langle f|j-|i\rangle \times \sqrt{(I-K)(I+K+1)}$. The first factor is $\sim$60 keV, the j- matrix elements are $\sim$5 and the square root is 1.5-2. Thus the Coriolis matrix element is approximately 500-600 keV and the closest separation for the $3/2^-$ levels is $\sim$1 MeV. Yet, empirically, they occur <300 keV apart. Strongly attenuating the Coriolis interaction is not an acceptable solution because the other levels could not then be reproduced: in particular, the $11/2^-$ state would not lie lowest.

In view of this, it seemed apt to attempt an IBFA calculation[5] using the code ODDA, written by O. Scholten. For this single j shell ($h_{11/2}$) case, only two parameters, Γ, the strength of the direct or $Q_B \cdot q_F$ interaction, and Λ, the multiplier of the exchange term, were required. The parameters for the even-even Pd core were taken from a systematic survey.[6] The results[7] are shown in fig. 2, where we have also shown the comparison for the high spin states in ^{105}Pd to verify that the agreement for low spin levels in ^{109}Pd is not at the expense of a discrepancy elsewhere. The IBFA calculations are a significant improvement, in particular as regards the splitting of states of common spin.

It is interesting to study the source of this improvement by inspecting the resulting wave functions: their composition, in terms of core states, is summarized in Table I. In the PR model, the odd particle is coupled only with the core states of the quasi ground band. The IBFA, however, allows the full participation of all collective core levels. It is striking (see Table I) that the quasi ground band in fact accounts for only $\sim$40% of the IBFA wave functions. It is therefore not surprising that the PR model could not reproduce the empirical low spin level energies. What is intriguing is why it succeeded for any levels since the core structure of high and low spin states is similar.

One might question whether it is fair to utilize the simple Nilsson PR model instead of, for example, the asymmetric rotor model

Table 1. Probability Distribution Ranges (in %) of Different Categories of Core States in the IBFA and PR Wave Functions (see ref. 7)

^{105,109}Pd States	Core State Category			
	Quasi-ground band		Other	
	PR	IBFA	PR	IBFA
Low spin states in ^{109}Pd	100	28-49	0	72-51
High spin states in ^{105}Pd	100	42-56	0	58-44

so successfully employed in other mass regions. There are two responses appropriate here. First, the same IBFA wave functions show that core states involving the quasi-γ band are rather unimportant (probabilities $\lesssim$10%). Coupling to the quasi β band in fact is far more important ($\sim$40%). Secondly, we performed[3] asymmetric rotor calculations for all γ values and found that, while for $\gamma\sim30°$ it was possible to reduce the splitting between $3/2^-$ states, at the same time the $5/2^-$ splitting increased. For no γ value could a satisfactory fit be obtained.

To conclude, (n,γ) techniques were used to study, for the first time, an essentially complete set of low spin, anti-aligned, unique parity levels in ^{109}Pd. The particle rotor model, notably successful[4] for high spin states of the same $h_{11/2}$ parentage in this mass region, did not, and could not, explain the empirical energy levels, even when core asymmetry was included. IBFA calculations involving only two parameters accounted well for both the high spin levels in ^{105}Pd and the low spin states of ^{109}Pd. Analysis of the resulting IBFA wave functions showed large probabilities for core states other than those of the quasi ground band, in particular those associated with the quasi-β band.

It is a pleasure to acknowledge numerous stimulating discussions with F. Iachello and the collaboration of G. J. Smith in the calculations described here. This work was supported under contract DE-AC02-76CH00016 with the U. S. Department of Energy.

1. F. S. Stephens, Rev. Mod. Phys. 47, 43 (1979).
2. R. F. Casten et al., Phys. Rev. Letters, to be published.
3. G. J. Smith et al., Phys. Letts. 86B, 13 (1979).
4. H. A. Smith, Jr. and F. A. Rickey, Phys. Rev. C14, 1946 (1976).
5. F. Iachello and O. Scholten, Phys. Rev. Letts. 43, 679 (1979).
6. O. Scholten, private communication.
7. R. F. Casten and G. J. Smith, Phys. Rev. Letts. 43, 337 (1979).

INTERACTING BOSON-FERMION MODEL OF COLLECTIVE STATES IN THE ODD-GOLD NUCLEI

J. L. Wood

School of Physics
Georgia Institute of Technology
Atlanta, Georgia 30332

While the limiting cases of the interacting boson-fermion model are now moderately well-known[1,2] and can be related[3] to familiar geometrical situations through the limiting boson symmetries, the many nuclei that lie between these limits have yet to be explored. In the present discussion, the description of collective properties of nuclei intermediate between the O(6) and SU(3) limiting boson symmetries are considered. Specifically, some results are presented for the application[4] of the model to the $h_{9/2}$ collective bands in $^{185-195}$Au using the code[5] ODDA.

The data relevant to this discussion have been taken from a compilation[6] of experimental results on the odd-Au isotopes. The $h_{9/2}$ bands in $^{185-195}$Au are dramatically isolated from the rest of the excitation spectrum, particularly the $h_{11/2}$ bands, by very strongly hindered transitions[7]. This is due to the fact that the $h_{9/2}$ state is a shell model intruder state in the odd-Au isotopes and the hindrance is due to the particle-hole character of the transitions. The particle-hole contrast of the $h_{9/2}$ and $h_{11/2}$ bands has been noted[8,9] also in more subtle contexts. Consequently, in this discussion, the $h_{9/2}$ state is treated as a pure particle state (this is discussed in terms of the interacting boson-fermion model later). In addition to the $h_{9/2}$ intruder state, the $i_{13/2}$ intruder state is known in $^{185-189}$Au (Ref. 6) and ^{195}Au (Ref. 10) This raises the question of where the remaining negative parity intruder states lie, since these will result in j-mixing. The only information available is from the single-proton stripping studies[10] of states in ^{195}Au, which locate the $f_{7/2}$ strength 714 keV above the $h_{9/2}$ state. (This can be compared with the closed-shell case of ^{209}Bi where the $f_{7/2}$ state lies at 896 keV above the $h_{9/2}$ state

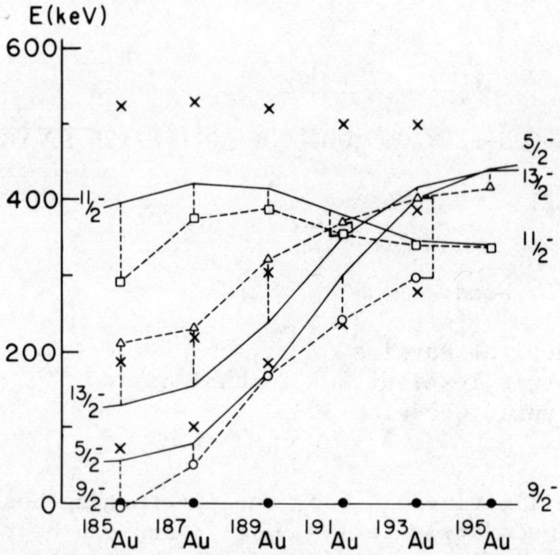

Fig. 1 A comparison between the experimental energies of the lowest $9/2^-$, $5/2^-$, $13/2^-$ and $11/2^-$ members of the $h_{9/2}$ collective bands in the odd-Au isotopes, and the present interacting boson-fermion model (IBFM) and Meyer ter Vehn model calculations. Experimental energies are marked by open points. The energies calculated with the IBFM (using Γ=300 keV and other parameter values defined in the text) are shown as continuous solid lines connected to respective experimental points by vertical dotted lines. The energies calculated with the Meyer ter Vehn model are marked by crosses.

and no other negative parity states are seen below 2.8 MeV (see e. g. Ref. 11).) In this discussion the $h_{9/2}$ state is approximated as a pure-j state.

The $h_{9/2}$ collective bands in $^{185-195}$Au were calculated in the framework of the interacting boson-fermion model using the

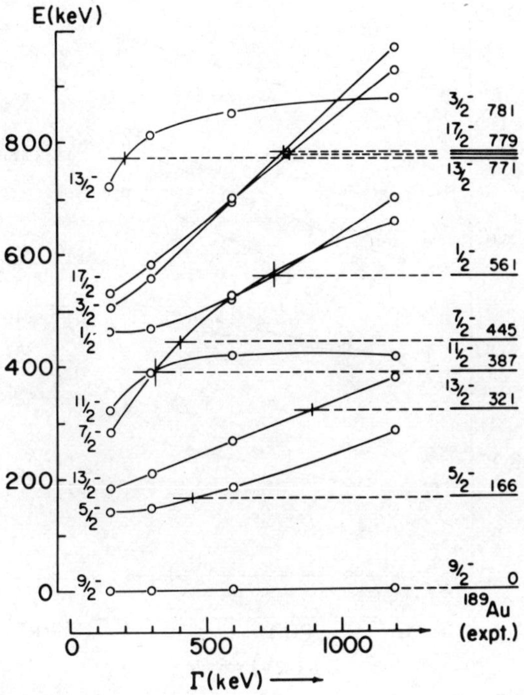

Fig. 2. The dependence of the $h_{9/2}$ collective states on the strength of the quadrupole term in V_{BF}, for A=0 keV and other parameters assigned the values for ^{189}Au. Evidently, there is no obviously preferred value of Γ that gives a "best fit".

restriction that no exchange terms are necessary because of the pure-particle nature of the $h_{9/2}$ state. In addition, the calculations were confined to the case of a pure $j=9/2$ fermion for simplicity (thus, neglecting any possible influence of the $f_{7/2}$ configuration). The boson-fermion interaction was limited to only monopole and quadrupole terms. In Fig. 1 the experimental energies of the $5/2^-$, $11/2^-$ and $13/2^-$ members of the $h_{9/2}$ band are compared with calculations using the Hamiltonian[1]

$$H = H_B + H_F + V_{BF}$$

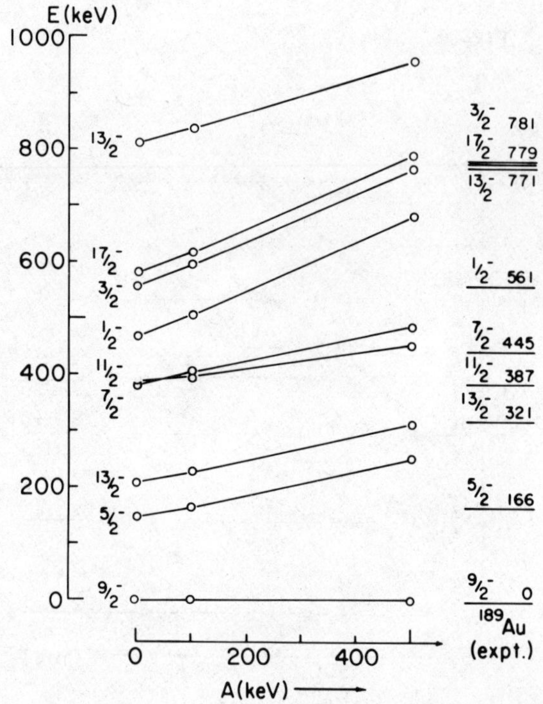

Fig. 3. The dependence of the $h_{9/2}$ collective states on the strength of the monopole term in V_{BF}, for Γ=300 keV and other parameters assigned the values for ^{189}Au.

where H_B was determined by fitting to the neighboring even - Pt core, H_F was taken as arbitrary (energies calculated relative to the band head), and V_{BF} was taken from Ref. 1 with Λ=0 (i.e. no exchange term). Results of calculations using the Meyer ter Vehn single-j rigid triaxial rotor model[12] are also shown for comparison, where the parameters of the model were determined by the neighboring even - Pt core. The fitting of H_B to the Pt cores was done by Bijker, et. al. (Ref. 13) using the interacting proton-neutron boson model (IBA2) and the IBA1 parameterization was extracted using a projection technique[14]. The parameters[1] Γ and A (which give the strength of the quadrupole and monopole boson-fermion interactions, respectively, were taken to be 300 keV and 0 keV for all of these

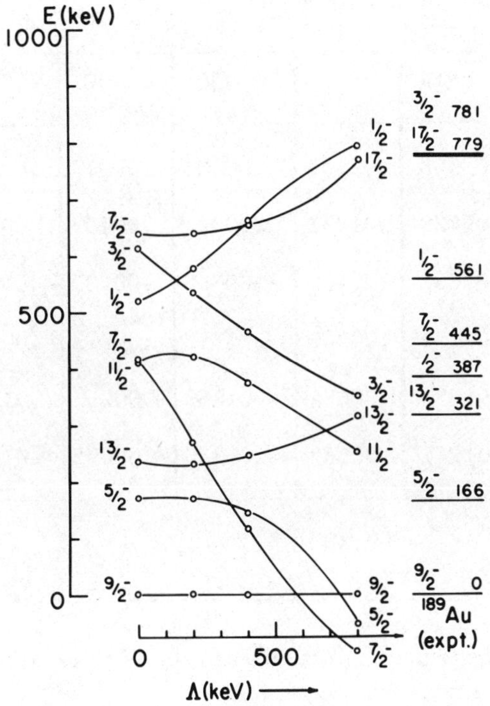

Fig. 4. The dependence of the $h_{9/2}$ collective states on the strength of the exchange term in V_{BF}, for Γ=300 keV, A=0 keV and other parameters assigned the values for ^{189}Au.

odd-Au isotopes. The parameter[1] χ was taken from the fitting[13] of IBA 2 to the Pt cores; both $\chi=\chi_\nu$ and $\chi=\chi_\nu + \frac{1}{3}\chi_\pi$ were investigated and the calculated energies were found to differ by less than 10 keV. A complete listing of the parameters used is given in Table 1. The effect of varying the strength of the quadrupole term, Γ, in V_{BF} is shown in Fig. 2 for the core parameters of ^{189}Au. Although a value of Γ larger than 300 keV would evidently give a better fit in this case, there is clearly no obvious value of Γ that can greatly improve the agreement between theory and experiment. The effect of varying the strength of the monopole term, A, in V_{BF} is shown in Fig. 3 for the core parameters of ^{189}Au. There is no obvious value of A that is

Table 1. The core parameters used in H_B, obtained from an IBA2 fit to the even-Pt isotopes[13)] and projection[14)] of the IBA2 parameters to IBA1 parameters[a].

Core	EPS	PAIR	ELL	QQ	OCT	HEX	χ_ν^b	N
^{194}Pt	.47000	.07548	.02109	.00405	.00317	.00990	0.49	7
^{192}Pt	.51621	.06429	.01832	<.00001	.00152	.00868	0.80	8
^{190}Pt	.61867	.05062	.01449	-.00772	.00283	-.00230	0.45	9
^{188}Pt	.65956	.03225	.00946	-.01398	.00272	-.00918	0.00	10
^{186}Pt	.72575	.01831	.00684	-.01769	-.00261	-.01199	-0.50	11
^{184}Pt	.71370	.01947	.00667	-.01386	-.00095	-.01141	-0.30	12
$\chi_\pi^b = -0.80$								

[a]This parameterization is for a multipole expansion of the IBA1 Hamiltonian.

[b]Taken directly from the IBA2 fit[13)].

preferred over A=0 keV. It is interesting to relax the restriction that no exchange terms are necessary in V_{BF}. The effect of adding an exchange term of strength Λ is shown in Fig. 4 for the core parameters of ^{189}Au. The low-spin band members clearly show a preference for the absence of the exchange term.

For a model with such a large number of possible parameters, it is necessary to indicate clearly where the limitations of the model lie. In Fig. 5 are shown the comparison of theory with experiment for ^{187,189,191}Au (Γ=300 keV, A=0 keV) and their respective even-Pt cores; i.e. the Pt cores were calculated using the values of the IBA 1 parameters given in Table 1. It is evident that the agreement for the odd-A cases is comparable to that for the even-A cases. Thus, to improve the agreement between theory and experiment for the odd-A cases, the first task is to achieve a better core description; and, adjusting the parameters of V_{BF} is pointless. In this particular study, a very serious question arises with regard to the low-lying excited 0^+ state in the lighter

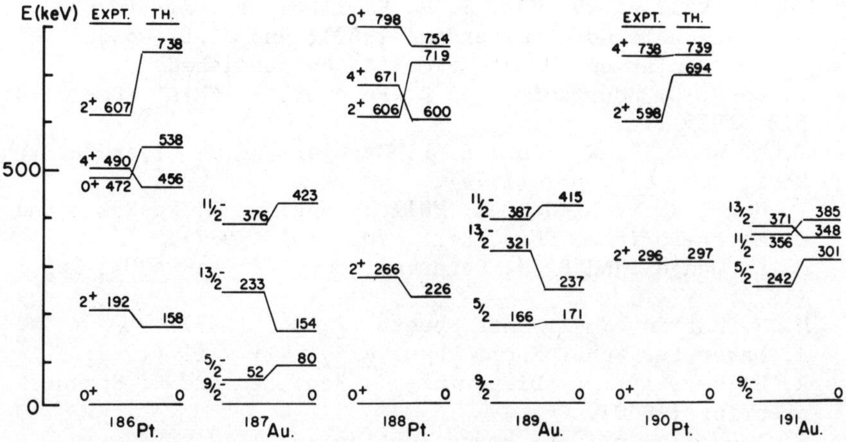

Fig. 5. A comparison of IBA1 and IBFA calculations for the even-Pt and odd-Au isotopes. The parameters used for the even-Pt isotopes and in H_B for the odd-Au isotopes are the same and are given in Table 1.

even-Pt isotopes; namely, is this state in the s,d-boson model space? A partial answer to this is possible since the coupling of the proton to the cores in the odd-Au isotopes can be considered to act as a probe of the core structure. This is currently being investigated[15] for the case of ^{187}Au and ^{186}Pt where the excited 0^+ state lies at 472 keV.

I wish to thank F. Iachello for numerous valuable discussions during this investigation.

This work was supported in part by the U.S. Department of Energy, Contract No. DE-AS05-80ER10599.

REFERENCES

1. F. Iachello and O. Scholten, Phys. Rev. Lett. 43, 679 (1979).
2. F. Iachello and O. Scholten, Phys. Lett. 91B, 189 (1980); and O. Scholten, thesis, Groningen (1980).
3. J. N. Ginocchio and M. W. Kirson, Phys. Rev. Lett. 44, 1744 (1980); A. E. L. Dieperink, O. Scholten and F. Iachello, ibid. 44, 1747 (1980).
4. R. A. Braga, J. L. Wood, M. A. Grimm and E. F. Zganjar, to be published.
5. O. Scholten, unpublished.
6. E. F. Zganjar, Proc. Int. Symposium on Future Directions in Studies of Nuclei Far From Stability, Nashville, Tennessee,

Sept. 1979, p. 49, Eds. J. H. Hamilton, et. al. (North Holland Publ. Co., Amsterdam, 1980); and J. L. Wood, E. F. Zganjar and M. A. Grimm, to be published.

7. V. Berg, C. Bourgeois, and R. Foucher, J. Phys. (Paris) 36, 613 (1975).
8. J. L. Wood, R. W. Fink, E. F. Zganjar and J. Meyer ter Vehn. Phys. Rev. C14, 682 (1976).
9. Y. Gono, R. M. Lieder, M. Müller-Veggian, A. Neskakis and C. Mayer-Böricke, Phys. Lett. 70B, 155 (1977).
10. M. L. Munger and R. J. Peterson, Nucl. Phys. A303, 199 (1978).
11. M. J. Martin, Nucl. Data Sheets 22, 545 (1977).
12. J. Meyer ter Vehn, Nucl. Phys. A249, 111, 141 (1975).
13. R. Bijker, A. E. L. Dieperink, O. Scholten and R. Spanhoff, preprint (1980).
14. F. Iachello, private communication.
15. M. A. Grimm, J. L. Wood and E. F. Zganjar, work in progress.

SHELL MODEL ORIGIN OF THE FERMION-BOSON EXCHANGE INTERACTION

Igal Talmi

The Weizmann Institute of Science
Rehovot, Israel

The interacting boson model has been very successful in the description of collective bands in many nuclei[1). The various limits of collective spectra[2) as well as the transitions between them[3) are very well accounted for. In its detailed version (IBM), collective states of nuclei with valence protons and neutrons outside closed shells are obtained from a boson Hamiltonian with proton s-bosons (with ℓ=0) and d-bosons (ℓ=2) and neutron s- and d-bosons[4). The collective features (in particular enhanced E2 electromagnetic transitions) emerge as a result of coherent admixtures of proton and neutron s- and d-bosons. These admixtures are obtained by diagonalizing the Hamiltonian which contains a strong and attractive quadrupole-quadrupole interaction between proton and neutron bosons. On the other hand, the eigenstates of the proton part of the boson Hamiltonian have definite (generalized) seniorities and are characterized by the numbers of d-bosons. A similar situation holds for the neutrons.

To a certain approximation, it is possible in many cases to consider states in which proton and neutron bosons are coupled in a symmetric way[4). In such cases it is possible to consider states constructed with only one kind of s-bosons and d-bosons as was the case in the earlier version of the model (IBA). In this version the actual calculations are greatly simplified and the various symmetries of the boson Hamiltonian can be clearly displayed.

Recently the boson model has been successfully extended to include calculation of spectra of odd-even nuclei[5). In addition to the system of s- and d-bosons also a single fermion is considered which interacts with the bosons by a quadrupole interaction. As a result systems of collective states emerge also for the odd nucleus.

If the even-even core, represented by the s- and d-bosons, has a rotational spectrum, rotational bands are obtained also for the odd nucleus.

It is clear that in such a description the Pauli principle may be violated. This is not only a matter of principle; the actual results obtained with just a quadrupole interaction cannot reproduce the data. A simple and very useful remedy to these difficulties was found in ref. 5) by introducing an exchange term of the form

$$[(\tilde{d} \times a^{+}_{j})^{(j)} \times (d^{+} \times \tilde{a}_{j})^{(j)}]^{(0)} \tag{1}$$

which corresponds to the exchange diagram in Fig. 1. With appropriately chosen strength of this term very good agreement with experiment was obtained in several regions. In the rotational region, the caulculations reproduce the results obtained with the Nilsson scheme, where the Pauli principle can be easily enforced.

It is thus very interesting to understand the shell model origin of the exchange term (1). In refs. 4) interacting s- and d-bosons were related to states of fermion monopole (J=0) and quadrupole (J=2) pairs of identical fermions in the shell model (S and D pairs respectively). We have thus introduced proton s_p and d_p-bosons. In eq. (1) as well as in Fig. 1 there are unspecified (IBA) s- and d-bosons. Therefore fig. 1 cannot have a direct interpretation

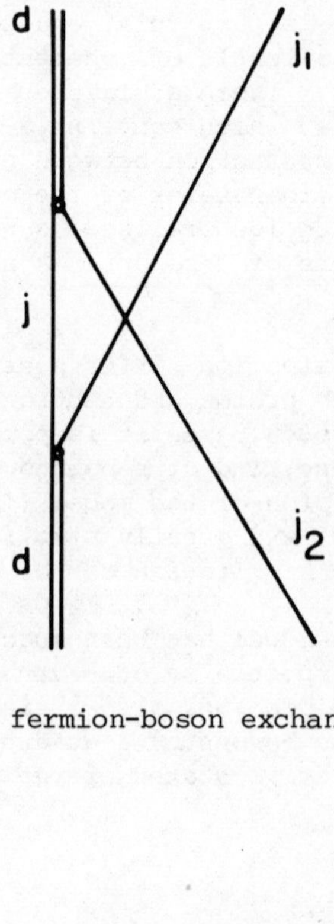

Fig. 1. Diagram of IBA fermion-boson exchange interaction.

in our micriscopic description. If the fermion is a proton and the d-boson a proton boson, there is no quadrupole interaction at the vertices of fig. 1. The interaction between identical nucleons at that vertex is weak and even vanishes for the pairing interaction. If we consider the strong quadrupole interaction between a proton and a neutron d-boson, there are no exchange effects and the diagram in fig. 1 does not appear at all.

These considerations indicate that we should be looking for the origin of the exchange term (1) in the Pauli principle corrections to the quadrupole operator of the odd group of nucleons (protons, say). That quadrupole operator taken between antisymmetric states is not just the sum of a single fermion operator and the operator of the even group of fermions represented by a bosons operator. Due to the antisymmetry, there are also quadrupole exchange terms. We show in the present paper how such terms when scalarly multiplied by the quadrupole operator of the other kind (neutrons in the present example) give rise to terms which can be brought into the form (1).

For the odd group of identical nucleons we take first a group of three protons. We consider matrix elements of the fermion quadrupole operator

$$\sqrt{5}\sum_{jj'}\gamma_{jj'}\sum_{\mu\mu}(-1)^{j-\mu}\begin{pmatrix} j & k & j' \\ -\mu & \kappa & \mu' \end{pmatrix} a^{+}_{j\mu}a_{j'\mu'} \qquad k=2 \tag{2}$$

The reduced matrix elements of the operator (2) between single nucleon-states with j and j' are $\sqrt{5}\gamma_{jj'}$. We take the matrix elements of $a^{+}_{j\mu}a_{j'\mu'}$ between the states

$$\langle 0|B_{J_1M_1}a_{j_1m_1} \qquad \text{and} \qquad a^{+}_{j_2m_2}B^{+}_{J_2M_2}|0\rangle \tag{3}$$

where

$$B^{+}_{JM}=\sum\beta^{(J)}_{jj'}(jmj'm'|jj'JM)a^{+}_{jm}a^{+}_{j'm'} \qquad \beta^{(J)}_{j'j}=(-1)^{j+j'+J}\beta^{(J)}_{jj'} \tag{4}$$

and $B_{JM}=(B^{+}_{JM})^{+}$. The definition (4) avoids the need to restrict the summation to $j\leq j'$. By anticommuting fermion operators these matrix elements can be written as

$$\langle 0|B_{J_1M_1}B^{+}_{J_2M_2}|0\rangle\delta_{j_1j}\delta_{m_1\mu}\delta_{j_2j'}\delta_{m_2\mu'}$$

$$+\langle 0|B_{J_1M_1}a^{+}_{j\mu}a_{j'\mu'}B^{+}_{J_2M_2}|0\rangle\delta_{j_1j_2}\delta_{m_1m_2}$$

$$-\langle 0|B_{J_1M_1}a^{+}_{j_2m_2}a_{j'\mu'}B^{+}_{J_2M_2}|0\rangle\delta_{j_1j}\delta_{m_1\mu}$$

$$-\langle 0|B_{J_1M_1}a^+_{j\mu}a_{j_1m_1}B^+_{J_2M_2}|0\rangle\delta_{j_2j'}\delta_{m_2\mu'}$$

$$-\langle 0|B_{J_1M_1}a^+_{j\mu}a^+_{j_2m_2}a_{j_1m_1}a_{j'\mu'}B^+_{J_2M_2}|0\rangle \qquad (5)$$

The first term in (5) is the matrix element of the operator between single fermion states and will not be further considered. The second term is the matrix element of the operator between two pair states. This term is replaced by an approriate boson operator between boson states. The last three terms are the corrections to the quadrupole matrix elements due to the Pauli principle. They are the subject of the present discussion.

Usint the definition (4) and the anticommutation relations of the fermion operators, these three terms can be evaluated and lead to the following expressions.

$$-4(-1)^{j_1-m_1}\begin{pmatrix} j_1 & k & j' \\ -m_1 & \kappa & \mu' \end{pmatrix}\beta^{(J_1)}_{j_2j''}\beta^{(J_2)}_{j'j''}(j_2m_2j''m''|j_2j''J_1M_1)(j'\mu'j''m''|j'j''J_2M_2) \qquad (6)$$

$$-4(-1)^{j-\mu}\begin{pmatrix} j & k & j_2 \\ -\mu & \kappa & m_2 \end{pmatrix}\beta^{(J_1)}_{jj''}\beta^{(J_2)}_{j'j''}(j\mu j''m''|jj''J_1M_1)(j_1m_1j''m''|j_1j''J_2M_2) \qquad (7)$$

$$-4(-1)^{j-\mu}\begin{pmatrix} j & k & j' \\ -\mu & \kappa & \mu' \end{pmatrix}\beta^{(J_1)}_{jj_2}\beta^{(J_2)}_{j'j_1}(j\mu j_2m_2|jj_2J_1M_1)(j'\mu'j_1m_1|j'j_1J_2M_2) \qquad (8)$$

These three matrix elements correspond to the three diagrams (c), (d) and (e) in fig. 2. The first two terms in (5) correspond to the diagrams (a) and (b) in fig. 2. The wavy line corresponds to the quadrupole operator. Its other end, not shown here is attached to the even group of identical nucleons of the other kind (neutrons).

We now look for an operator constructed of fermion operators and boson operators whose matrix elements between the states

$$\langle 0|b_{J_1M_1}a_{j_1m_1} \qquad \text{and} \qquad a^+_{j_2m_2}b^+_{J_2M_2}|0\rangle \qquad (9)$$

are the sum of (6), (7) and (8). In (9) b^+_{JM}, b_{JM} are creation and annihilation operators of bosons with spin J (which will be 0 or 2). All boson operators commute with all fermion operators. Looking at fig. 2 and introducing in (6), (7) and (8) 3-j symbols, we make use of their symmetry properties to rearrange the coupling of the various spins. We find an operator whose matrix elements are equal to (6) which is given by

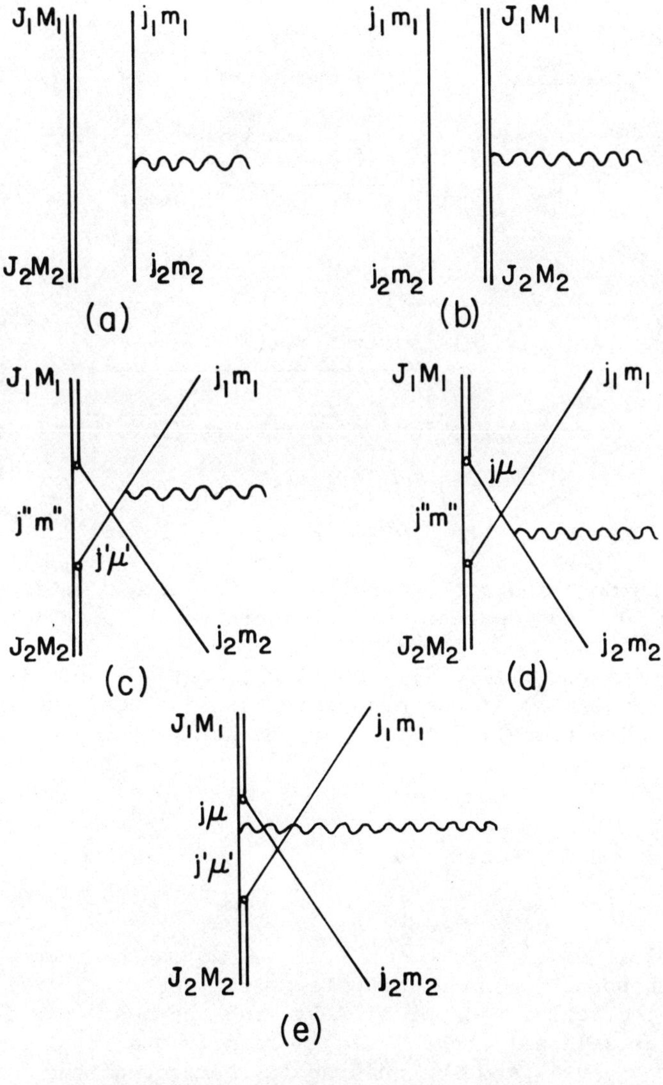

Fig. 2. Diagram of matrix elements of the three fermions quadrupole operator.

$$+ 4\beta^{(J_1)}_{J_2 j''} \beta^{(J_2)}_{j'j''} \sqrt{\frac{(2J_1+1(2J_2+1)}{(2j'+1)(2j''+1)(2k+1)}} \tag{10}$$

$$\left(a^+_{j_1} \times [(b^+_{J_1} \times \tilde{a}_{j_2})^{(j'')} \times \tilde{b}_{J_2}]^{(j')}\right)^{(k)}_{\kappa}$$

Similarly, an operator yielding the matrix elements (7) is

$$-(-1)^{j+j''} 4\beta^{(J_1)}_{jj''} \beta^{(J_2)}_{j_1 j''} \sqrt{\frac{(2J_1+1)(2J_2+1)}{(2j+1)(2j''+1)(2k+1)}} \tag{11}$$

$$([b^+_{J_1} \times (\tilde{b}_{J_2} \times a^+_{j_1})^{(j'')}]^{(j)} \times \tilde{a}_{j_2})^{(k)}_{\kappa}$$

and the one yielding (8) is given by

$$-(-1)^{j_1+j_2} 4\beta^{(J_1)}_{jj_2} \beta^{(J_2)}_{j'j_1} \sqrt{\frac{(2J_1+1)(2J_2+1)}{(2j+1)(2j'+1)(2k+1)}} \tag{12}$$

$$:\left([\tilde{b}_{J_2} \times a^+_{j_1}]^{(j')} \times [b^+_{J_1} \times \tilde{a}_{j_2}]^{(j)}\right)^{(k)}_{\kappa}:$$

In (12) the dots indicate normal order in which creation operators should be let to operate left of annihilation operators.

The operators (10), (11) and (12) should be multiplied by the quadrupole operator of the even group of identical nucleons of the other kind (neutrons). It is given in terms of neutron boson operators by

$$a(s^+_n \tilde{d}_n + s_n d^+_n) + b(d^+_n \times \tilde{d}_n)^{(2)} \tag{13}$$

The coefficients a and b in (13) depend on the number of neutron s- and d-bosons.[4] The quadrupole proton-neutron interaction is given by the scalar product of (13) with the operators (10), (11) and (12) as well as with the single fermion quadrupole operator $\Sigma\gamma_{jj'}(a^+_j \times \tilde{a}_{j'})^{(2)}$ and the quadrupole operator of the even proton group which is similar to (13). The scalar product of (13) with the operators (10), (11) and (12) is a two boson one fermion interaction quite different from the exchange term (1). As long as we use proton bosons and neutron bosons (IBM) it is not possible to simplify the proton-neutron interaction. If, however, we go to the IBA formalism with one kind of boson only, the situation may be greatly simplified.

In the following we will keep only those terms of the proton neutron interaction that will have only one d-boson creation operator and one d-boson annihilation operator. Such terms will yield for states with good F-spin symmetry[4] interactions between one (IBA) d-boson and one fermion. We first notice that the second term in (13) has already two d-boson operators. If we make use of it, both J_1 and J_2 in (10), (11) and (12) must be equal to zero. Such terms, however, turn out to be simply the Pauli corrections to the single fermion quadrupole operator due to the presence of the even fermion group of the same kind. Such corrections to matrix elements between seniority v=1 states are well known for equal α_j values. They will not be further considered here.

The contributions in which we are interested arise from the first term in (13). When that term is scalarly multiplied by (10), (11) and (12), we obtain terms which satisfy the requirement made above as follows. In the expressions (10), (11) and (12) when multiplied by $s_n^+\tilde{d}_n$ we put $J_1=2$, $J_2=0$ and where they are multiplied by $s_n d_n^+$ we put $J_1=0, J_2=2$. Thus, the terms which are selected this way from the scalar product of (13) and (10) are

$$4\beta^{(0)}_{j_2j''}\beta^{(2)}_{j'j''}\frac{1}{\sqrt{(2j'+1)(2j''+1)}}\left(a^+_{j_1}\times[(s^+_p\times\tilde{a}_{j_2})^{(j'')}\times\tilde{d}_p]^{(j')}\right)^{(2)}\cdot s_n d^+_n +$$

$$+\ 4\beta^{(2)}_{j_2j''}\beta^{(0)}_{j'j''}\frac{1}{\sqrt{(2j'+1)(2j''+1)}}\left(a^+_{j_1}\times[(d^+_p\times\tilde{a}_{j_2})^{(j'')}\times s_p]^{(j')}\right)^{(2)}\cdot s^+_n\tilde{d}_n =$$

$$=4\beta^{(0)}_{j_2j_2}\beta^{(2)}_{j'j_2}\frac{1}{\sqrt{(2j'+1)(2j_2+1)}}\ s^+_p s_n[a^+_{j_1}\times(\tilde{a}_{j_2}\times\tilde{d}_p)^{(j')}]^{(2)}\cdot d^+_n$$

$$+\ 4\beta^{(2)}_{j_2j'}\beta^{(0)}_{j'j'}\frac{1}{2j'+1}\ s^+_n s_p[a^+_{j_1}\times(d^+_p\times\tilde{a}_{j_2})^{(j')}]^{(2)}\cdot\tilde{d}_n \qquad (14)$$

The terms obtained from the scalar product of (13) and (11) are

$$-(-1)^{j+j''}4\beta^{(0)}_{jj''}\beta^{(2)}_{j_1j''}\frac{1}{\sqrt{(2j+1)(2j''+1)}}\left([s^+_p\times(\tilde{d}_p\times a^+_{j_1})^{(j'')}]^{(j)}\times\tilde{a}_{j_2}\right)^{(2)}\cdot s_n d^+_n$$

$$-(-1)^{j+j''}4\beta^{(2)}_{jj''}\beta^{(0)}_{j_1j''}\frac{1}{\sqrt{(2j+1)(2j''+1)}}\left([d^+_p\times(s_p\times a^+_{j_1})^{(j'')}]^{(j)}\times\tilde{a}_{j_2}\right)^{(2)}\cdot s^+_n\tilde{d}_n$$

$$=4\beta^{(0)}_{jj}\beta^{(2)}_{j_1j}\frac{1}{2j+1}\ s^+_p s_n[(\tilde{d}_p\times a^+_{j_1})^{(j)}\times\tilde{a}_{j_2}]^{(2)}\cdot d^+_n$$

$$-(-1)^{j+j_1}4\beta^{(2)}_{jj_1}\beta^{(0)}_{j_1j_1}\frac{1}{\sqrt{(2j+1)(2j_1+1)}}\ s^+_n s_p[(d^+_p\times a^+_{j_1})^{(j)}\times\tilde{a}_{j_2}]^{(2)}\cdot\tilde{d}_n \qquad (15)$$

The third group of terms arising from the scalar product of (13) with (12) are

$$-(-1)^{j+j_2}4\beta^{(0)}_{jj_2}\beta^{(2)}_{j'j_1}\frac{1}{\sqrt{(2j+1)(2j'+1)}}[(\tilde{d}_p\times a^+_{j_1})^{(j')}\times(s^+_p\times\tilde{a}_{j_2})^{(j)}]^{(2)}\cdot s_n d^+_n$$

$$-(-1)^{j_1+j_2}4\beta^{(2)}_{jj_2}\beta^{(0)}_{j'j_1}\frac{1}{\sqrt{(2j+1)(2j'+1)}}[(s_p\times a^+_{j_1})^{(j')}\times(d^+_p\times\tilde{a}_{j_2})^{(j)}]^{(2)}\cdot s^+_n\tilde{d}_n$$

$$=-(-1)^{j_1+j_2}4\beta^{(0)}_{j_2j_2}\beta^{(2)}_{j'j_1}\frac{1}{\sqrt{(2j_2+1)(2j'+1)}}s^+_p s_n[(\tilde{d}_p\times a^+_{j_1})^{(j')}\times\tilde{a}_{j_2}]^{(2)}\cdot d^+_n$$

$$-(-1)^{j_1+j_2}4\beta^{(2)}_{jj_2}\beta^{(0)}_{j_1j_1}\frac{1}{\sqrt{(2j+1)(2j_1+1)}}s^+_n s_p[a^+_{j_1}\times(d^+_p\times\tilde{a}_{j_2})^{(j)}]^{(2)}\cdot\tilde{d}_n \qquad (16)$$

The expressions (14), (15) and (16) should be multiplied by the coefficient a in (13).

In order to see the similarity between some of the expressions in (14), (15) and (16) it is worth while to carry out a transformation which will change the order of coupling in the various terms. In all these terms there is a proton d-boson operator coupled to a proton operator and this combination is coupled to another proton operator to form k=2. We can keep that combination but couple the other proton operator to the neutron d operator and then couple both combinations to a total k'=0 (scalar product). Thos change of coupling transformation is given for instance by

$$\langle jj_2(k=2)2k'=0|j,j_2 2(j)k'=0\rangle=(-1)^{j+j_2+2+0}\sqrt{5(2j+1)}\begin{Bmatrix} j & j_2 & 2\\ 2 & 0 & j\end{Bmatrix}=1 \qquad (17)$$

Using this transformation we obtain after slight rearrangements the following expressions, respectively

$$(-1)^{j_2+j'}4\sqrt{5}\frac{\beta^{(0)}_{j_2j_2}\beta^{(2)}_{j'j_2}}{\sqrt{(2j'+1)(2j_2+1)}}s^+_p s_n[(d^+_n\times a^+_{j_1})^{(j')}\times(\tilde{d}_p\times\tilde{a}_{j_2})^{(j')}]^{(0)}$$

$$-4\sqrt{5}\frac{\beta^{(2)}_{j_2j'}\beta^{(0)}_{j'j'}}{2j'+1}s^+_n s_p[(\tilde{d}_n\times a^+_{j_1})^{(j')}\times(d^+_p\times\tilde{a}_{j_2})^{(j')}]^{(0)} \qquad (18)$$

$$(-1)^{j_2+j}\, 4\sqrt{5}\, \frac{\beta^{(0)}_{jj}\, \beta^{(2)}_{j_1 j}}{2j+1}\, s^+_p s_n\, [(\tilde{d}_p \times a^+_{j_1})^{(j)} \times (d^+_n \times \tilde{a}_{j_2})^{(j)}]^{(0)} \;+$$

$$+(-1)^{j_1+j_2}\, 4\sqrt{5}\, \frac{\beta^{(2)}_{jj_1}\, \beta^{(0)}_{j_1 j_1}}{\sqrt{(2j+1)(2j_1+1)}}\, s^+_n s_p\, [(d^+_p \times a^+_{j_1})^{(j)} \times (\tilde{d}_n \times \tilde{a}_{j_2})^{(j)}]^{(0)} \qquad (19)$$

$$(-1)^{j_1+j'}\, 4\sqrt{5}\, \frac{\beta^{(0)}_{j_2 j_2}\, \beta^{(2)}_{j' j_1}}{\sqrt{(2j_2+1)(2j'+1)}}\, s^+_p s_n\, [(\tilde{d}_p \times a^+_{j_1})^{(j')} \times (d^+_n \times \tilde{a}_{j_2})^{(j')}]^{(0)} \;+$$

$$+(-1)^{j_1+j_2}\, 4\sqrt{5}\, \frac{\beta^{(2)}_{jj_2}\, \beta^{(0)}_{j_1 j_1}}{\sqrt{(2j+1)(2j_1+1)}}\, s^+_n s_p\, [(\tilde{d}_n \times a^+_{j_1})^{(j)} \times (d^+_p \times \tilde{a}_{j_2})^{(j)}]^{(0)} \qquad (20)$$

We can now simplify these expressions by going over to the IBA formalism. To do that we should consider matrix elements of the operators (18), (19) and (20) in states where the fermion is coupled to the symmetric state (F=1)

$$\frac{1}{\sqrt{2}}\,(s^+_p d^+_n + s^+_n d^+_p)\,|0> \qquad (21)$$

The same matrix elements are obtained if we couple the fermion to the IBA state $s^+d^+|0>$ (one kind of bosons) and instead of (18),(19) and (20) we use the following operators respectively.

$$(-1)^{j_2+j'}\, 2\sqrt{5}\, \frac{\beta^{(0)}_{j_2 j_2}\, \beta^{(2)}_{j' j_2}}{\sqrt{(2j'+1)(2j_2+1)}}\, s^+ s\, [(d^+ \times a^+_{j_1})^{(j')} \times (\tilde{d} \times \tilde{a}_{j_2})^{(j')}]^{(0)} \;-$$

$$2\sqrt{5}\, \frac{\beta^{(2)}_{j_2 j'}\, \beta^{(0)}_{j' j'}}{2j'+1} : s^+ s\, [(\tilde{d} \times a^+_{j_1})^{(j')} \times (d^+ \times \tilde{a}_{j_2})^{(j')}]^{(0)} : \qquad (22)$$

$$(-1)^{j_2+j}\, 2\sqrt{5}\, \frac{\beta^{(0)}_{jj}\, \beta^{(2)}_{j_1 j}}{2j+1} : s^+ s\, [(\tilde{d} \times a^+_{j_1})^{(j)} \times (d^+ \times \tilde{a}_{j_2})^{(j)}]^{(0)} :$$

$$+(-1)^{j_1+j_2}\, 2\sqrt{5}\, \frac{\beta^{(2)}_{jj_1}\, \beta^{(0)}_{j_1 j_1}}{\sqrt{(2j+1)(2j_1+1)}}\, s^+ s\, [(d^+ \times a^+_{j_1})^{(j)} \times (\tilde{d} \times \tilde{a}_{j_2})^{(j)}]^{(0)} \qquad (23)$$

$$(-1)^{j_1+j'}\, 2\sqrt{5}\, \frac{\beta^{(0)}_{j_2j_2}\beta^{(2)}_{j'j_1}}{\sqrt{(2j_2+1)(2j'+1)}} :s^+s[(\tilde{d}\times a^+_{j_1})^{(j')}\times(d^+\times\tilde{a}_{j_2})^{(j')}]^{(0)}:$$

$$+(-1)^{j_2+j}\, 2\sqrt{5}\, \frac{\beta^{(2)}_{jj_2}\beta^{(0)}_{j_1j_1}}{\sqrt{(2j+1)(2j_1+1)}} : s^+s[(\tilde{d}\times a^+_{j_1})^{(j)}\times(d^+\times\tilde{a}_{j_2})^{(j)}]^{(0)}: \quad (24)$$

Looking at (22), (23) and (24) we see that they contain now two kinds of IBA operators

$$:s^+s[(\tilde{d}\times a^+_{j_1})^{(j)}\times(d^+\times\tilde{a}_{j_2})^{(j)}]^{(0)}: \text{ and } s^+s[(d^+\times a^+_{j_1})^{(j)}\times(\tilde{d}\times\tilde{a}_{j_2})^{(j)}]^{(0)} \quad (25)$$

The first operator in (25) is identical to the Iachello-Scholten exchange operator (1) apart from the factor $n_s=s^+s$. The second operator gives rise to an extra interaction in the state where the single fermion is coupled to a d-boson to form a state with total J=j.

Let us now consider the coefficients of the two operators in (25). Collecting all coefficients we obtain for the coefficient of the exchange term in (25) the expression

$$10\left(\frac{\gamma_{j_1j}\beta^{(2)}_{j_2j}\beta^{(0)}_{jj}}{2j+1} - (-1)^{j+j_2}\frac{\gamma_{jj_2}\beta^{(0)}_{jj}\beta^{(2)}_{j_1j}}{2j+1}\right.$$

$$\left. -(-1)^{j_1+j}\frac{\gamma_{j_2j}\beta^{(0)}_{j_2j_2}\beta^{(2)}_{jj_1}}{\sqrt{(2j+1)(2j_2+1)}} - (-1)^{j_1+j_2}\frac{\gamma_{jj_1}\beta^{(2)}_{jj_2}\beta^{(0)}_{j_1j_1}}{\sqrt{(2j+1)(2j_1+1)}}\right) \quad (26)$$

The coefficient of the second (direct) term in (25) is

$$-10\left\{(-1)^{j_2+j}\frac{\gamma_{j_1j}\beta^{(0)}_{j_2j_2}\beta^{(2)}_{j\,j_2}}{\sqrt{(2j+1)(2j_2+1)}} + (-1)^{j_1+j_2}\frac{\gamma_{jj_2}\beta^{(2)}_{jj_1}\beta^{(0)}_{j_1j_1}}{\sqrt{(2j+1)(2j_1+1)}}\right\} \quad (27)$$

The expressions (26) and (27) can be simplified if we take into account the nature of the quadrupole operator (2). We take this operator to be the one[6] that transforms a fermion pair with J=0 into a pair with J=2.

We take for the operator (2) the quadrupole (k=2) operator which satisfies[6)]

$$[Q_\mu, S^+] = D^+_\mu \tag{28}$$

where S^+ creates a J=0 pair and D^+ a pair with J=2. In refs. 6) the former was defined by

$$S^+ = \Sigma\, \alpha_j S^+_j \tag{29}$$

where we now use the normalization

$$S^+_j = \frac{1}{2}\Sigma\, (-1)^{j-m}\, a^+_{jm} a^+_{j-m} \tag{30}$$

The coefficients $\beta^{(0)}_{jj}$ are given in terms of the α_j by

$$\frac{\beta^{(0)}_{jj}}{\sqrt{2j+1}} = \frac{1}{2}\alpha_j \tag{31}$$

The coefficients $\gamma_{jj'}$ of Q_μ are calculated from (28) to be given by

$$\gamma_{jj'}\alpha_{j'} + (-1)^{j+j'+1}\gamma_{j'j}\alpha_j = \beta^{(2)}_{jj'} + (-1)^{j+j'+1}\beta^{(2)}_{j'j} = 2\beta^{(2)}_{jj'} \tag{32}$$

The last equality in (32) follows from the definition (4). We take the operator Q_μ to be hermitean in a cartesian frame. Its coefficients then satisfy the condition

$$\gamma_{j'j} = \gamma_{jj'}(-1)^{j+j'+1} \tag{33}$$

Using (32) and (33) as well as the definition (31) we obtain for the coefficients of the exchange terms in (25) the expression

$$\frac{20}{\sqrt{2j+1}}\, \gamma_{j_1 j}\gamma_{j_2 j}\, (\alpha^2_j + \alpha_j\alpha_{j_1} + \alpha_j\alpha_{j_2} + \alpha_{j_1}\alpha_{j_2}) \tag{34}$$

The coefficient of the second term in (25) is obtained as

$$-\frac{10}{\sqrt{2j+1}}\, \gamma_{j_1 j}\gamma_{j_2 j}\, (\alpha^2_{j_1} + \alpha^2_{j_2} + \alpha_j\alpha_{j_1} + \alpha_j\alpha_{j_2}) \tag{35}$$

The terms in brackets in (34) and (35) are positive in interesting cases. For equal α_j values they are simply sums of squares. The product $\gamma_{j_1 j}\gamma_{j_2 j}$ are products of appropriate quadrupole interactions at the two vertices of fig. 1. Thus, the reduction of the various terms corresponding to (c), (d) and (e) in fig. 2 into the IBA formalism leads to an exchange term (1) corresponding to fig. 1. The single fermion interacts by a quadrupole interaction with the IBA d-bosons and this interaction is modified by the Pauli principle. This modification includes exchange terms corresponding to fig. 1 as well as direct terms as given in (25).

This completes the discussion of the present case of 3 identical nucleons (protons) and an even number of nucleons of the other kind. The normalization of the states (3) can be carried out in a straightforward manner and will not be presented here. The next question is how to generalize the present results to the case of higher numbers of identical nucleons in the odd group. If all α_j values are equal it is possible to write down the dependence of matrix elements on the number of nucleons. We saw above that the only terms that lead to one boson - one fermion interactions are obtained when $J_1 = 0$ $J_2 = 2$ or $J_2 = 2$ $J_1 = 0$. The corresponding states (3) have thus seniorities v=1 and v=3 (actually we should have projected out from states with $J_1 = 2$, $J = j_1$ or $J_2 = 2$, $J = j_2$ components with v=1; these are small in practical cases and will be ignored here). Hence, matrix elements of the quadrupole operator (k=2) are given by the well known expression for n odd

$$(n\ v{=}1\ J\ \|Q\|\ n\ v{=}3\ J') = \sqrt{\frac{(n-1)(2\Omega-n-1)}{2(2\Omega-4)}}\,(n{=}3\ v{=}1\ J\|Q\|n{=}3\ v{=}3\ J') \qquad (36)$$

where $2\Omega = \Sigma(2j+1)$. The matrix elements on the r.h.s. of (36) have been the subject of the present paper. For values of n higher than 3 they should be multiplied by the coefficient in (36). The situation is more complex when matrix elements are taken between states with higher seniorities. The matrix elements (36) are tho most important ones for the low lying levels. Hence, it is perhaps a good approximation to take in the general case the operators (25) and multiply them by the coefficient in (36). The dependence on the number of nucleons in the even group is introduced through the coefficient a in (13).

The boson model including a single fermion yields good agreement with experiment only when the exchange interaction (1) is introduced. We have now seen how this term is obtained by the effect of the Pauli principle acting between identical nucleons on the quadrupole interaction between protons and neutrons. This lends further support to the shell model description of the interacting boson model.

REFERENCES

1. A. Arima and F. Iachello, Phys. Rev. Lett., 35, 1069 (1975).
2. A. Arima and F. Iachello, Ann. Phys. (N.Y.), 99, 253 (1976); A. Arima and F. Iachello, Ann. Phys. (N.Y.), 111, 201 (1978); A. Arima and F. Iachello, Phys. Rev. Lett. 40, 385 (1978); Ann. Phys. (N.Y.), 123, 468 (1979).
3. O. Scholten, F. Iachello and A. Arima, Ann. Phys. (N.Y.), 115, 325 (1978).

4. A. Arima, T. Otsuka, F. Iachello and I. Talmi, Phys. Lett., 66B, 205 (1977); T. Otsuka, A. Arima, F. Iachello and I. Talmi, Phys. Lett., 76B, 141 (1978).
5. F. Iachello and O. Scholten, Phys. Rev. Lett., 43, 679 (1979).
6. I. Talmi, Nucl. Phys., 172A, 1 (1978); S. Shlomo and I. Talmi, Nucl. Phys., 198A, 81 (1972); I. Talmi, Riv. Nuovo Cimento, 3, 85 (1973).

ON THE BOSON-FERMION INTERACTION

O. Scholten and A.E.L. Dieperink

Kernfysisch Versneller Instituut, University of Groningen, The Netherlands

1. INTRODUCTION

Since it has become increasingly evident that the IBA model and its extension to odd-A nuclei provides at the level of phenomenology an extremely successful description of nuclear collective phenomena, a microscopic understanding of the model is clearly of considerable interest. For the case of even-even nuclei Arima et al.[1] have proposed a method to derive the main properties of the boson-boson interaction from a shell model picture. Application of this method by Otsuka[2] to realistic cases has been shown to give promising results. In the present contribution the boson-fermion[3] interaction, that enters in the description of odd-A nuclei, is derived from the shell model interaction between fermions. The latter has two main characteristics: i) a strong pairing force between the like nucleons, and ii) a quadrupole-quadrupole interaction between neutrons and protons. In terms of boson-boson interactions the former gives rise to attractive binding energies of the s- and d-bosons, whereas the latter leads to the quadrupole-quadrupole force between the neutron and proton bosons. It will be shown in this contribution that this interaction also plays a dominant role in the coupling of the odd-fermion to the bosons in the core[4].

2. DEFINITION OF THE BASIS STATES

First of all it is necessary to define the relation between the shell model fermion space and the IBFA model space. In the case of even-even nuclei this was done by the introduction of the fermion pair creation operators

$$S^\dagger = \sum_j \alpha_j \sqrt{\frac{\Omega_j}{2}} \, (c_j^\dagger c_j^\dagger)^{(0)} , \tag{2.1}$$

and

$$D^\dagger = \sum_{jj'} \beta'_{jj'} \frac{1}{\sqrt{2}} (c_j^\dagger c_{j'}^\dagger)^{(2)} , \tag{2.2}$$

where $c_j^\dagger$ represents a fermion creation operator and $\Omega_j = (2j+1)/2$. With the help of these operators the relation between the shell model and the IBA-model space is defined as[1]

$$(d^\dagger)^{n_d,(L)} (s^\dagger)^{N-n_d} |0\rangle \propto P_{\tilde{v}=2n_d} [(D^\dagger)^{n_d,(L)}] (S^\dagger)^{N-n_d} |0\rangle_c \tag{2.3}$$

where (L) stands for all quantum numbers necessary uniquely to define the states. In Eq. (2.3) the operator P projects onto the component of the state with maximal generalized seniority, $\tilde{v} = 2n_d$. In the shell model space the closed core $|0\rangle_c$ is considered as the vacuum state. The boson vacuum is defined in the usual way as

$$s|0\rangle = d|0\rangle = 0, \tag{2.4}$$

where s and d are the boson annihilation operators.

In the case of an odd number of nucleons a similar definition can be given,

$$[a_j^\dagger (d^\dagger)^{n_d,(L)}]^J (s^\dagger)^{N-n_d} |0\rangle \propto P_{\tilde{v}=2n_d+1} [c_j^\dagger (D^\dagger)^{n_d,(L)}]^J (S^\dagger)^{N-n_d} |0\rangle_c . \tag{2.5}$$

In Eq. (2.5) the odd-nucleon creation operators $a_j^\dagger$ are introduced. They obey by definition the simple commutation rules

$$[d, a_j^\dagger] = [s, a_j^\dagger] = [d^\dagger, a_j^\dagger] = [s^\dagger, a_j^\dagger] = 0 , \tag{2.6}$$

which makes them very convenient to work with in later applications.

3. THE EFFECTIVE SINGLE-NUCLEON OPERATORS

Instead of constructing directly the boson image of a general shell model operator we first construct the IBFA image of the single-nucleon creation and annihilation operators $c^{\dagger}_{j,m}$ and $\tilde{c}_{j,m}$ $(=(-1)^{j-m}\, c_{j,-m})$.

In general it is possible to write the image of $c^{\dagger}_{j}$ in the boson-fermion space as a series of the type

$$c^{\dagger}_{j} \rightarrow A_1\, a^{\dagger}_{j} + A_2(d^{\dagger}\tilde{a}_{j'})^{(j)} + A_3(s^{\dagger}\tilde{a}_{j})^{(j)} + \qquad (3.1)$$
$$+ A_4((s^{\dagger}\tilde{d})a^{\dagger}_{j'})^{(j)} + A_5((d^{\dagger}\tilde{d})^{(L)}a^{\dagger}_{j'})^{(j)} + \ldots .$$

In this series terms of the type $(sd^{\dagger}a^{\dagger}_{j'})^{(j)}$ do not appear since, by definition of our basis states, they correspond to terms changing the seniority in the fermion space by three units, while for a single-particle operator one has $|\Delta\tilde{v}| = 1$.

The coefficients A_i can be determined by equating the matrix elements in the IBFA space[5] to those in the shell model space. As an example we consider the third term at the right-handside of Eq. (3.1). The matrix element of this term between states with lowest number of d-bosons is

$$<s^N\|\, A_3(s^{\dagger}\tilde{a}_{j})^{(j)}\,\|\, s^{N-1}j> = A_3\sqrt{N(2j+1)}. \qquad (3.2)$$

The corresponding matrix elment in the fermion space that must be evaluated is

$$<j^{2N}(\tilde{v}=0)\,\|\, c^{\dagger}_{j}\,\|\, j^{2N-1}(\tilde{v}=1)> = <S^N\|\, c^{\dagger}_{j}\,\|\, S^{N-1}j>$$
$$= \alpha_j \sqrt{\frac{N(2j+1)}{\Omega_e}}, \qquad (3.3)$$

where Ω_e represents the effective shell degeneracy

$$\Omega_e = \sum_j \alpha_j^2(2j+1)/2. \qquad (3.4)$$

In the derivation of Eq. (3.3) the following approximation[2] has been used:

$$<0|S^N(S^{\dagger})^N|0> = N!\ \Gamma(\Omega_e+1)/\Gamma(\Omega_e-N+1) \qquad (3.5)$$

By equating the r.h.s. of Eq. (3.2) and Eq. (3.3) we obtain

$$A_3 = \alpha_j/\sqrt{\Omega_e}. \tag{3.6}$$

Application of this procedure to all terms in Eq. (3.1) for states which correspond to fermion states with $\tilde{v} \leq 2$ leads to the result:

$$c_j^\dagger \to \check{c}_j^\dagger = u_j a_j^\dagger + \frac{v_j}{\sqrt{N}}(s^\dagger \tilde{a}_j)^{(j)} + \sum_{j'} u_j \bar{\beta}_{j'j} \frac{\sqrt{10}}{\hat{j}}(d^\dagger \tilde{a}_{j'})^{(j)} +$$

$$- \sum_{j'} \frac{v_j}{\sqrt{N}} \bar{\beta}_{j'j} \frac{\sqrt{10}}{\hat{j}} s^\dagger (d a_{j'}^\dagger)^{(j)} , \tag{3.7}$$

where

$$v_j = \alpha_j N/\Omega_e$$
$$u_j^2 = 1 - v_j^2$$
$$\bar{\beta}_{jj'} = \beta'_{jj'} u_j u_{j'} \tag{3.8}$$

and

$$\hat{j} = \sqrt{2j+1}.$$

The coefficients $\bar{\beta}jj'$ are normalized

$$\sum_{jj'} (\bar{\beta}_{jj'})^2 = 1. \tag{3.9}$$

The coefficients v_j^2 can be regarded as the occupation probability of the shell model state $|j\rangle$ and play the same role as the corresponding quantities that occur in the BCS theory.

The operator $\check{c}_j^\dagger$, introduced in Eq. (3.7) has to be distinguished from the true single-particle creation operator. It corresponds to the truncated image of $c_j^\dagger$ in the boson-fermion space and will hereafter be referred to as the effective single-particle creation operator.

In arriving at Eq. (3.7) we have restricted ourselves to states with $\tilde{v} \leq 2$ since only in that case the lowest order contributions to the four different terms in the image of $c_j^\dagger$ can be determined. These four terms correspond to the fermion seniority increasing and decreasing parts in the transfer operator leading from even to odd and from odd to even seniority states, respectively. The operator is of lowest order in the d-boson creation and

and annihilation operators. The result (3.7) can therefore be considered as an expansion in terms of the quantity n_d/Ω_e.

Within the IBFA model space $\check{c}_j^\dagger$ is the operator to be used for the calculation of single-particle transfer amplitudes. This operator is however only a truncated image of the real single-particle creation operator $c_j^\dagger$.

The effective single-particle operators can also be looked upon in a different way. As was noted above the odd-nucleon operators $a_j^\dagger$ commute with the boson operators (Eq. (2.6)), while on the other hand the shell model single-particle creation operators obey rather complicated commutation relations with the pair operators, Eq. (2.12), and in general

$$[c_j^\dagger, D] \neq 0 \quad \text{and} \quad [c_j^\dagger, S] \neq 0 \; . \tag{3.10}$$

The operator $\check{c}_j^\dagger$ has been constructed in such a way that for the lowest generalized seniority states the matrix elements of the commutators in the boson and in the fermion space are equal:

$$\begin{aligned} <|\,[c_j^\dagger, S]\,|>_F &= <|\,[\check{c}_j^\dagger, s]\,|>_B \quad , \\ <|\,[c_j^\dagger, D]\,|>_F &= <|\,[\check{c}_j^\dagger, s]\,|>_B \quad . \end{aligned} \tag{3.11}$$

The reason that the operator $\check{c}_j^\dagger$ has a much more complicated structure than $c_j^\dagger$ is due to the fact that in the shell model space the Pauli principle is taken into account in the basis states, being fully antisymmetric. In the boson space the effects of the Pauli principle have to be put in explicitly into the operators, giving rise to coefficients depending on the number of nucleons occupying a given shell (v_j^2) and moreover to additional terms in the boson image.

4. CONSTRUCTION OF THE BOSON QUADRUPOLE OPERATOR

The boson image of general fermion operators such as the quadrupole operator can readily be constructed by replacing the operators $c_j^\dagger$ and $\tilde{c}_j$ directly by $\check{c}_j^\dagger$ and $\check{c}_j$, respectively. Since in our approach we always equate matrix elements for the lowest seniority states, it is consistent to make a further simplification by replacing the expectation value of $(s^\dagger s)$ by N. The boson image of the fermion quadrupole operator,

$$q^{(2)} = \sum_{jj'} Q_{jj'} (c_j^\dagger \tilde{c}_{j'})^{(2)} \; ,$$

can thus be written as

$$Q^{(2)} = Q^{(2)}_{B} + Q^{(2)}_{BF} , \tag{4.1}$$

where the boson image is separated into a pure boson operator, $Q^{(2)}_{B}$, and a term $Q^{(2)}_{BF}$, acting on the odd-particle:

$$Q^{(2)}_{B} = - \mathcal{N}_{\beta} \frac{1}{\sqrt{2N}} \left[(s^{\dagger}\tilde{d}+d^{\dagger}s)^{(2)} + \chi(d^{\dagger}\tilde{d})^{(2)} \right] , \tag{4.2}$$

where

$$\mathcal{N}_{\beta} = \sum_{jj'} Q_{jj'} \, \overline{\beta}_{jj'} (u_j v_{j'} + v_j u_{j'}) , \tag{4.3}$$

and

$$\chi = - \frac{10\sqrt{2N}}{\mathcal{N}_{\beta}} \sum_{jj'j''} \overline{\beta}_{j''j} \, \overline{\beta}_{j'j''} \, Q_{jj'} \begin{Bmatrix} 2 & 2 & 2 \\ j & j' & j'' \end{Bmatrix} (v_j v_{j'} - u_j u_{j'}) , \tag{4.4}$$

and

$$Q^{(2)}_{BF} = \sum_{jj'} Q_{jj'} \, (u_j u_{j'} - v_j v_{j'}) (a^{\dagger}_j \tilde{a}_{j'})^{(2)} +$$
$$- \sqrt{\frac{10}{N}} \sum_{jj'j''} Q_{jj'} (u_j v_{j'} + v_j u_{j'}) \; \overline{\beta}_{j''j} \frac{1}{\hat{j}} \left[(d^{\dagger} \tilde{a}_{j''})^{(j)} (s a^{\dagger}_{j'})^{(j')} \right]^{(2)}$$
$$+ \text{h.c.} \tag{4.5}$$

We note that the operator $Q^{(2)}_{BF}$ consists of two different terms. The first term is equivalent to the usual quadrupole operator that is considered in particle-phonon coupling models[6]. The second term in Eq. (4.5) is new and is entirely due to the fact that the nucleons which form the bosons are in distinguishable from the odd-nucleon. This second term appears to be crucial in the description of odd-A nuclei in the framework of the IBFA-model.

Under the assumption that the $|d\rangle$ state exhausts the full quadrupole strength the coefficients $\overline{\beta}_{jj'}$ can be estimated by requiring that the matrix element $\langle s^N \| Q^{(2)} \| ds^{N-1} \rangle$ is maximal.

From this one obtains (using eq. 4.3))

$$\overline{\beta}_{jj'} = Q_{jj'}(u_j v_{j'}+v_j u_{j'})/\mathcal{N}_\beta \ , \tag{4.6}$$

where $\overline{\beta}_{jj'}$ is normalized according to Eq. (3.9).

5. THE BOSON-FERMION INTERACTION

As has been noted above the neutron-proton quadrupole force plays a dominant role in the spectra of collective nuclei. In the following we derive a boson-fermion interaction using the quadrupole force between neutrons and protons

$$V^{\pi\nu} \simeq \kappa_{\pi\nu}\, q^{(2)}_\pi . q^{(2)}_\nu \ . \tag{5.1}$$

For concreteness we assume that the number of neutrons outside the closed shell is even, $n_\nu = 2N_\nu$, and the number of protons is odd, $n_\pi = 2N_\pi+1$.

The boson image of the interaction (5.1) can easily be constructed by using Eqs. (4.1-5). For the pure boson-boson interaction we obtain

$$V^{\pi\nu}_B = \kappa\, Q^{(2)}_\pi . Q^{(2)}_\nu \ , \tag{5.2}$$

where

$$Q^{(2)}_\rho = (s^\dagger\tilde{d}+d^\dagger s)^{(2)}_\rho + \chi_\rho (d^\dagger\tilde{d})^{(2)}_\rho \quad ; \ \rho = \pi,\nu \ , \tag{5.3}$$

$$\kappa = \kappa_{\pi\nu}\, \mathcal{N}_{\beta_\nu} \mathcal{N}_{\beta_\pi}/(2\sqrt{N_\nu N_\pi}) \ , \tag{5.4}$$

and χ_ρ is given by Eq. (4.4). The part of the quadrupole interaction involving the odd-proton degree of freedom can now be written as

$$V^{\pi\nu}_{BF} = \overline{\kappa} \sum_{jj'} Q_{jj'}(u_j u_{j'}-v_j v_{j'})(a^\dagger_j \tilde{a}_{j'})^{(2)}_\pi . Q^{(2)}_\nu \ +$$

$$- \overline{\kappa}\sqrt{\frac{10}{N_\pi}}\,\mathcal{N}_{\beta_\pi} \sum_{jj'j''} \overline{\beta}_{jj'}\overline{\beta}_{j''j}\,\frac{1}{\hat{j}}\,[\,(d^\dagger\tilde{a}_{j''})^{(j)}(s a^\dagger_{j'})^{(j')}]^{(2)}_\pi . Q^{(2)}_\nu \ +$$

$$+ \text{h.c.} \ , \tag{5.5}$$

where

$$\overline{\kappa} = -\kappa_{\pi\nu}\, \mathcal{N}_{\beta_\nu}/\sqrt{2N_\nu} \ . \tag{5.6}$$

The interaction is written explicitly in terms of neutron and proton degrees of freedom. However, for numerical reasons we wish to perform the actual calculations in the IBA-1 formalism. To calculate the image V_{BF} of $V_{BF}^{\pi\nu}$ in the IBA-1 model space we require that the matrix elements of $V_{BF}^{\pi\nu}$ between symmetric[3] states in the ν-π boson model are equal to the corresponding ones in IBA-1. By considering furthermore only matrix elements between states with $n_d \leq 1$ we retain only the terms with at most one d-boson annihilation and/or creation operator resulting in a further simplification of the interaction. (The approximations made in this last step are however of the same order as those made in arriving at Eq. (5.5)). The final result can be expressed as

$$V_{BF} = \sum_{jj'} \Gamma_{jj'} (a_j^\dagger \tilde{a}_{j'})^{(2)} . Q^{(2)} + \tag{5.7}$$

$$+ \sum_{jj'j''} \Lambda_{jj'}^{j''} \frac{1}{\hat{j}} : \left[(d^\dagger \tilde{a}_j)^{(j'')} (a_{j'}^\dagger \tilde{d})^{(j'')} \right]_0^{(0)} : ,$$

where

$$\begin{aligned} Q^{(2)} &= (s^\dagger \tilde{d} + d^\dagger s)^{(2)} + \chi_\nu (d^\dagger \tilde{d})^{(2)} , \\ \Gamma_{jj'} &= \Gamma_0 \, Q_{jj'} (u_j u_{j'} - v_j v_{j'}) , \\ \Lambda_{jj'}^{j''} &= -2\sqrt{5}\, \Lambda_0 \, \beta_{jj''} \, \beta_{j'j''} , \\ \beta_{jj'} &= \mathcal{N}_\beta \, \overline{\beta}_{jj'} = Q_{jj'} (u_j v_{j'} + v_j u_{j'}) , \end{aligned} \tag{5.8}$$

and

$$\begin{aligned} \Gamma_0 &= \overline{\kappa} \, N_\nu / (N_\nu + N_\pi) , \\ \Lambda_0 &= \overline{\kappa} \sqrt{2N_\pi} \, \mathcal{N}_\beta^{-1} \, N_\nu / (N_\nu + N_\pi) . \end{aligned} \tag{5.9}$$

A simple estimate for the quantities $Q_{jj'}$, related to the single-particle matrix elements of the quadrupole operator, can be obtained by taking it equal to the reduced matrix elements of the rank two spherical harmonic,

$$Q_{jj'} = \langle \ell \tfrac{1}{2} j \| Y^{(2)} \| \ell' \tfrac{1}{2} j \rangle . \tag{5.10}$$

The first term in Eq. (5.7) can be regarded as the conventional particle-phonon quadrupole force. Since in our approach V_{BF} is related to the neutron-proton quadrupole force the constant χ in the boson quadrupole operator Q is equal to χ_ν for the coupling of an odd-proton. The second term in V_{BF} which is not present in conventional models for odd-A nuclei, originates from the second term in the quadrupole operator (4.5) and solely expresses the fact that the building blocks of the boson are indistinguishable from the odd-nucleon. Because this term can also be thought to arise from a process in which the odd nucleon is exchanged with one of the nucleons that make up the d-boson[2], this interaction will be referred to as the exchange force. The normal ordering operator (: :) indicates that in calculating matrix elements of the exchange force, contributions arising from commuting $a_j^\dagger$ and $\tilde{a}_j$ must be neglected. It should be noted that the coefficients $\beta_{jj'}$ as appearing in Eq. (5.8) are not normalized since this is more convenient from a phenomenologic point of view.

6. APPLICATION TO THE Eu-ISOTOPES

In order to test the validity of our approach we have applied it to the Europium (Eu) isotopes. In the IBFA-model the Eu-isotopes (Z=63) are described as a proton coupled to the Sm(Z=62) isotopes. A comparison of calculated and experimental excitation energies is given in the preceding contribution to this conference[7].

The parameters for the core hamiltonian were taken from a previous calculation for the even-even Sm isotopes. To determine the values of the parameters Γ_o, Λ_o and v_j^2 (which are not linearly independent) appearing in the boson-fermion interaction V_{BF}(Eqs. 5.7-10) from a fit to the excitation energies it has been assumed that the ratio $R = \Gamma_o/\Lambda_o$ is approximately constant for all isotopes, and that $\tilde{v}_j$ varies smoothly. The reason for taking R (which is a proton parameter in the present case) constant for a series of isotopes is that this is consistent with the assumption that χ_π is constant.

In fig. 1 the resulting parameters Λ_o and Γ_o are shown along with the occupation numbers v_j^2. It is seen that whereas the N_ν dependence of Γ_o and Γ_o/Λ_o agrees reasonably well with the trend predicted by Eq. (5.9) there is a systematic difference of about a factor two between the values for the even-parity states and those for the odd-parity states (which have been fitted independently). We note, however, that the values of both Γ_o and

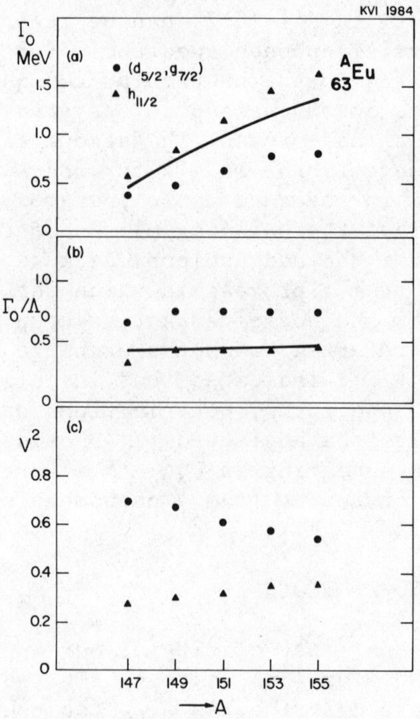

Fig. 1. Parameters determined from a phenomenological calculation for the Eu-isotopes[8]. The full lines are calculated using Eq. (5.9) with $\bar{\kappa}$ = 3 MeV.

and Γ_0/Λ_0 for these two cases can be brought in mutual agreement by increasing the value of Q_{jj} for the odd-parity states (j=11/2) by a factor 1.7 over the estimate (5.10). The need for such an enhancement can partially be explained by the fact that in Eq. (5.10) all radial matrix elements (which should be larger for the j = 11/2 orbit) are taken equal.

Finally we note that with the use of Eqs. (5.4), (5.6) and (5.9) one can relate Γ_0 to κ, the strength of the neutron-proton boson-boson interaction. From the present fit we find $\bar{\kappa}$ = 2 MeV, from which we obtain κ = −1.1 MeV (since $N_\beta \simeq 2$). This value should be compared with the values of κ obtained from calculations for even-even nuclei, $\kappa \sim 0.15$ MeV. The origine of this large difference is not clear.

8. SUMMARY AND CONCLUSIONS

In this paper we proposed a simple and yet a rather detailed microscopic derivation of the boson-fermion interaction. It is based on the strong neutron-proton quadrupole interaction. The resulting boson-fermion interaction contains two terms, a quadrupole and an exchange term. With these two terms in V_{BF} it is possible to reproduce the wide variety of features, observed in the spectra of collective odd-even nuclei[5]. As a result of our derivation we obtained a specific dependence of the strength parameters on the occupation probabilities v_j^2 of the different shell model orbits and on the boson number N_ν, N_π. The predicted parameters qualitatively agree with those determined from phenomenological calculations of the Europium isotopes. The relative strength of the quadrupole and exchange force agrees with the prediction. The relation between the strength of the particle-boson and the boson-boson interaction however requires further investigation.

REFERENCES

1. A. Arima, T. Otsuka, F. Iachello and I. Talmi, Phys. Lett. 66B (1977), 205;
   T. Otsuka, A. Arima and F. Iachello, Nucl. Phys. A309 (1978), 1.
2. T. Otsuka, Ph.D. Thesis, University of Tokyo, 1979 and to be published;
   T. Otsuka in "Interacting Bosons in Nuclear Physics", F. Iachello, ed., Plenum Press, New York (1978), p. 93 and this conference.
3. F. Iachello and O. Scholten, Phys. Rev. Lett. 43 (1979), 679.
4. I. Talmi, private communication and preceding contribution to the conference.
5. O. Scholten, Ph.D. Thesis, University of Groningen, 1980.
6. L.S. Kisslinger and R.A. Sorensen, Rev. Mod. Phys. 35 (1963), 853;
   L.S. Kisslinger and K. Kumar, Phys. Rev. Lett. 19 (1967), 1239.
7. O. Scholten, contribution to this conference.
8. O. Scholten, KVI Ann. Rep. 1979, p. 105.

APPROXIMATE DIAGONALIZATION OF THE PAIRING PLUS QUADRUPOLE HAMILTONIAN IN THE FRAMEWORK OF THE NUCLEAR FIELD THEORY. ODD-A NUCLEI

Andrea Vitturi

Istituto di Fisica dell'Università and INFN
Padova, Italy*, and
Oak Ridge National Laboratory
Oak Ridge, Tennessee 37830, USA

The Nilsson model[1] provides with a good description of the single-particle spectrum in deformed nuclei. An essential feature of this model is that pairs of fermions couple to all possible angular momenta, giving rise to the aligned states (cf. e.g. Refs. 2) and 3); also Ref. 4) and 5)). One way to obtain the correct occupancy of the Nilsson levels is to allow the particles around the Fermi surface to interact through a monopole pairing force (cf. e.g. Ref. 6)).

It has recently been suggested[7-10] that a convenient description of the nuclear deformation can be obtained only in terms of pairs of fermions coupled to angular momentum zero and two (S-D fermion subspace). This is the ansatz which is at the basis of the IBM, and which seems to be directly contradictory to the fermion aligned model (cf. e.g. 4) and 5)).

In an attempt to assess the differences between the two models in a quantitative way, the IBM was translated[11] into the language of the nuclear field theory[12]. A microscopic version of the IBM is thus obtained based on the particle-vibration coupling scheme. Interpreting the s- and d-bosons as pairing vibrations, and utilizing for the strength of the quadrupole force the self-consistent value, renormalized by the exchange of collective surface modes, the model has no free parameter. Another microscopic parameter free version of the IBM can be obtained[8-10] by mapping the S-D fermion subspace into a boson s-d space and the corresponding fermion operators into boson operators.

* Permanent address.

The basic feature of the NFT is that it deals with both boson and fermion degrees of freedom on par. Thus the Pauli principle and overcompleteness of the basis are properly taken into account at each order of perturbation. The extension to odd-systems of the translation of the IBM carried out for the even system, is uniquely determined by the NFT rules. This program is carried out below.

We start from a fermion Hamiltonian of the form

$$H = H_0 + \sum_{\lambda=0,2} H_{\text{pairing}}(\lambda) + H_{\text{quadrupole}} \tag{1}$$

where

$$H_{\text{pairing}}(\lambda) = -G_\lambda (2\lambda+1) \sum_\mu P^\dagger_{\lambda\mu} P_{\lambda\mu} \tag{2}$$

$$H_{\text{quadrupole}} = -\kappa \sum_\mu (-1)^\mu Q_{2\mu} Q_{2-\mu} \tag{3}$$

For a single j-shell

$$P^\dagger_{\lambda\mu} = \left(\frac{\pi}{2\lambda+1}\right)^{1/2} \langle j \| T_\lambda \| j \rangle \left[c^\dagger_j c^\dagger_j\right]^\lambda_\mu \tag{4}$$

$$Q_{2\mu} = \frac{1}{\sqrt{5}} \langle j \| Q_2 \| j \rangle \left[c^\dagger_j c_j\right]^2_\mu \tag{5}$$

The monopole and quadrupole collective modes are obtained by separately diagonalizing in RPA the monopole and quadrupole pairing interactions. They are completely determined by their collective energies ω_λ and particle-vibration coupling vertices Λ_λ. For a single j-shell we have

$$\omega_\lambda = \varepsilon - 2\pi G_\lambda |\langle j \| T_\lambda \| j \rangle|^2 = \varepsilon - Z_\lambda \tag{6}$$

and

$$\Lambda_\lambda \langle j \| T_\lambda \| j \rangle = -\sqrt{2}(\varepsilon - \omega_\lambda) = -\sqrt{2} Z_\lambda \tag{7}$$

where ε is the unperturbed energy of the two-particle system. Thus a unique quantity Z_λ determined both the vertex strength Λ_λ and the energy denominators $\varepsilon-\omega_\lambda$ of all the graphs involving pairing vertices.

The basis states are chosen to be combinations of s and d bosons coupled to the odd particle, i.e. of the type

$$|n_s n_d \alpha R; IK\rangle = \sum_{Mm} \langle RMjm|IK\rangle a^{+}_{jm} |n_s n_d \alpha R\rangle \tag{8}$$

where $2(n_s+n_d)+1$ is the number of valence particles, R the angular momentum of the even-A part, I the total angular momentum after the coupling to the odd particle, and α the usual additional quantum numbers characterizing the coupling of the d-bosons to the angular momentum R .

The NFT provides with the graphical perturbative rules to determine the matrix elements of the Hamiltonian between these basic collective states. In the first approximation we consider all the graphs in which the phonons and the odd particle interact up to the order $1/\Omega$, Ω being the pair degeneracy $(2j+1)/2$ of the shell. This corresponds to taking into account processes in which only two bosons interact at a time, or those in which one boson interacts with the odd particle. The graphs in which the odd particle propagates unperturbed, are those we have dealt with in the even case, and clearly correspond to the boson-boson part of the effective boson-fermion Hamiltonian. The values of these matrix elements have thus the same structure as those listed in Ref. 13; the effective degeneracy $\Omega' = \Omega-2nd$ has to be replaced by $\Omega' = \Omega-2nd-1$ because of the blocking effect of the odd particle. The other graphs arise directly from the coupling to the odd particle, and correspond to the boson-fermion part of the effective Hamiltonian (cf. e.g. Ref. 14). They are displayed in Fig. 1.

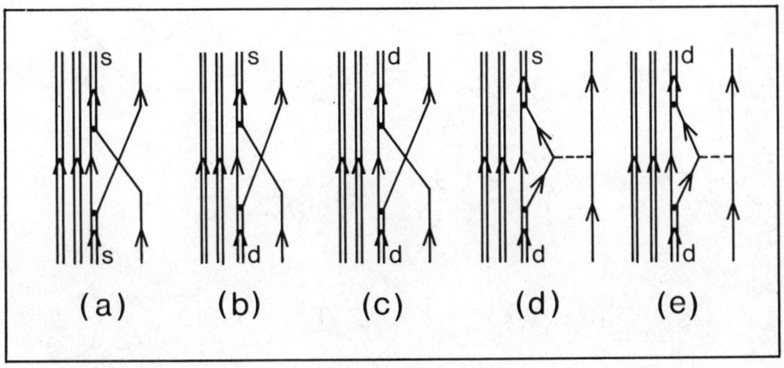

Fig. 1

Graphs (1a), (1b) and (1c) are exchange-type diagrams, which take into account the effect of the Pauli principle present in the fermion structure underlying the bosons, and correspond to an effective repulsive interaction. These graphs obviously act only between like particles. The other graphs (1d) and (1e) are due to the quadrupole particle-hole interaction. This interaction acts in all the channels. As discussed in Ref. 13 the $1/\Omega$ contributions have to be supplemented by the contributions arising from the application of the following empirical rules: a. in case that the d-bosons or the odd particle do not interact in lowest order with the s-bosons, the next order graphical contribution has to be included, b. the contribution of each graph has to be "symmetrized" by replacing $n_s \to n_s\,(1 - \frac{n_s}{\Omega'})$, where $\Omega' = \Omega - 2n_d$ (or $\Omega' = \Omega - 2n_d - 1$ in the odd-case) is the effective degeneracy left for the correlation of the s-bosons. The values of the contributions to the matrix elements of the Hamiltonian are displayed in Table 1.

Table 1. Contributions to the Matrix Elements of the Hamiltonian Coming from Coupling to the Odd-Particle

Monopole Pairing

$$\langle n_s n_d \alpha R; I | H_{pairing}(\lambda=0) | n_s n_d \alpha R; I \rangle = Z_s \frac{n_s}{\Omega}$$

Quadrupole Pairing

$$\langle n_s n_d \alpha R; I | H_{pairing}(\lambda=2) | n_s n_d \alpha' R'; I \rangle = -Z_d n_d \left\{ \frac{(\Omega' - n_s)^2}{\Omega'^2} + \frac{n_s^2}{\Omega'^2} \right\}$$

$$\cdot \sum_{\alpha'' R''} (n_d - 1, \alpha'' R'' | \} n_d \alpha R)(n_d - 1, \alpha'' R'' | \} n_d \alpha' R') \; (2R+1)^{1/2} (2R'+1)^{1/2}$$

$$\cdot \sum_{\rho} 10\,(2\rho+1) \begin{Bmatrix} j & j & 2 \\ j & \rho & 2 \end{Bmatrix} \begin{Bmatrix} j & 2 & \rho \\ R'' & I & R' \end{Bmatrix} \begin{Bmatrix} j & 2 & \rho \\ R'' & I & R \end{Bmatrix}$$

$$\langle n_s n_d \alpha R; I | H_{pairing}(\lambda=2) | n_s+1, n_d-1, \alpha' R'; I \rangle = 2\sqrt{2}\; \frac{Z_s Z_d}{Z_s + Z_d}$$

$$\cdot \sqrt{n_d} \left[\frac{(n - n_d + 1)(\Omega - n - n_d + 1)}{\Omega(\Omega - 2n_d - 1)} \right]^{1/2} (n_d - 1, \alpha' R' | \} n_d \alpha R)$$

$$\begin{Bmatrix} R' & 2 & R \\ j & I & j \end{Bmatrix} \sqrt{5}\; (2R+1)^{1/2} \; (-)^{R'+j+I}$$

Quadrupole-Quadrupole

$$\langle n_s n_d \alpha R; I | H_{\text{quadrupole}} | n_s n_d \alpha' R'; I \rangle$$

$$= -4Kq^2 \, \frac{n_s(\Omega' - n_s)}{\Omega'(\Omega'-1)} \, n_d \, (2R+1)^{1/2} (2R'+1)^{1/2}$$

$$\cdot \sum_{\alpha'' R''} (n_d - 1, \alpha'' R'' | \} n_d \alpha R)(n_d - 1, \alpha'' R'' | \} n_d \alpha' R')$$

$$\cdot \sum_{\rho} 10\,(2\rho+1) \begin{Bmatrix} j & j & 2 \\ j & \rho & 2 \end{Bmatrix} \begin{Bmatrix} j & 2 & \rho \\ R'' & I & R' \end{Bmatrix} \begin{Bmatrix} j & 2 & \rho \\ R'' & I & R \end{Bmatrix}$$

$$\langle n_s n_d \alpha R; I | H_{\text{quadrupole}} | n_s + 1, n_d - 1, \alpha' R'; I \rangle$$

$$= -2\sqrt{2}\, K q^2 \sqrt{n_d}$$

$$\cdot \left[\frac{(n - n_d + 1)(\Omega - n - n_d + 1)}{\Omega(\Omega - 2n_d + 1)} \right]^{1/2} (n_d - 1, \alpha' R' | \} n_d \alpha R)$$

$$\cdot \begin{Bmatrix} R' & 2 & R \\ j & I & j \end{Bmatrix} \sqrt{5}\, (2R+1)^{1/2} (-)^{R' + j + I}$$

Starting from the matrix elements given in Table 1 and the contributions coming from the even part (see Ref. 13) it is possible to write an "equivalent" Hamiltonian, acting in the space defined by the states (8), and leading to the same values of the matrix elements. We obtain

$$H = H_{\text{bos}} + H_{\text{bos-ferm}} \tag{9}$$

where

$$\mathcal{H}_{bos} = w_s s^+s + w_d d^+d + b_1 (s^+s^+)_{00}(ss)_{00}$$
$$+ \left[b_2 (d^+d^+)_{00}(ss)_{00} + h.c.\right] + \left[b_3 \sum_\mu (d^+d^+)_{2\mu}(ds)_{2\mu} + h.c.\right] \quad (10)$$
$$+ b_4 \sum_\mu (d^+s^+)_{2\mu}(ds)_{2\mu} + \sum_{\lambda\mu} b_5(\lambda)\, (d^+d^+)_{\lambda\mu}(dd)_{\lambda\mu}$$

and

$$\mathcal{H}_{bos-ferm} = \epsilon_j\, a_j^+ a_j + f_1 \sum_m (s^+a_j^+)_{jm}(sa_j^+)_{jm}$$
$$+ \left[f_2 \sum_m (d^+a_j^+)_{jm}(sa_j)_{jm} + h.c.\right] \quad (11)$$
$$+ \sum_{\lambda m} f_3(\lambda)\, (d^+a_j^+)_{\lambda m}(da_j)_{\lambda m}$$

All the constant b_i and f_i of the effective Hamiltonian (9) which are directly related to the phenomenological IBA Hamiltonian, are only dependent on the strengths of the fermion interactions. We have

$$b_1 = \frac{Z_s}{\Omega}$$

$$b_2 = 2\sqrt{5}\left(\frac{Z_s Z_d}{Z_s + Z_d} - Kq^2\right)\left[\frac{(\Omega - n - \hat{n}_d)(\Omega - n - \hat{n}_d + 1)}{\Omega^2(\Omega - 2\hat{n}_d - 1)}\right]^{1/2}$$

$$b_3 = 20\sqrt{2}\begin{Bmatrix}2 & 2 & 2\\ j & j & j\end{Bmatrix}\left[\frac{16 Z_d^4 Z_s}{(9Z_d^2 - Z_s^2)(Z_s + Z_d)^2} - Kq^2\right]\left[\frac{\Omega - n - \hat{n}_d}{\Omega(\Omega - 2\hat{n}_d - 1)}\right]^{1/2}$$

$$b_4 = \frac{2Z_s}{\Omega} + \frac{2}{\Omega}(Z_d - 2Kq^2)\,\frac{\Omega - n - \hat{n}_d}{\Omega - 2\hat{n}_d - 1}$$

$$b_5(\lambda) = C_\lambda\left\{ Z_d\left[\frac{(\Omega' - \hat{n}_s)^2}{\Omega'^2} + \frac{\hat{n}_s^2}{\Omega'^2}\right] + 4Kq^2\,\frac{\hat{n}_s(\Omega' - \hat{n}_s)}{\Omega'(\Omega' + 1)}\right\} \quad (12)$$

$$c_\lambda = 50 \begin{Bmatrix} j & j & 2 \\ j & j & 2 \\ 2 & 2 & \lambda \end{Bmatrix} - (1-\delta_{\lambda,0}) \frac{100}{\Omega-2} \begin{Bmatrix} \lambda & 2 & 2 \\ j & j & j \end{Bmatrix}^2 + \delta_{\lambda,0} \frac{5}{\Omega(\Omega-1)}$$

$$f_1 = \frac{Z_s}{\Omega}$$

$$f_2 = \frac{2\sqrt{5}}{\Omega} \left(\frac{Z_s Z_d}{Z_s+Z_d} - \kappa q^2 \right) \left(\frac{\Omega-n-\hat{n}_d-1}{\Omega-2\hat{n}_d-1} \right)^{1/2}$$

$$f_3 = -10 \begin{Bmatrix} 2 & j & j \\ 2 & \lambda & j \end{Bmatrix}$$

$$\cdot \left\{ Z_d \left[\frac{(\Omega'-\hat{n}_s)^2}{\Omega'^2} + \frac{\hat{n}_s^2}{\Omega'^2} \right] + 4\kappa q^2 \frac{\hat{n}_s(\Omega'-\hat{n}_s)}{\Omega'(\Omega'-1)} \right\}$$

As an example of the procedure we consider now the matrix element of the quadrupole operator. In addition to the term coming from the boson component, as in the even case, we have an additional term coming from the odd particle. The zeroth order contribution corresponds to graph (2a), in which the quadrupole field acts on the odd particle without any interaction with the boson part. According to rule (a) we have to add the contributions of higher order graphs taking into account the coupling to the s-bosons during the time the quadrupole field is acting (the effect of the coupling before and after the electromagnetic field has acted is taken into account through the diagonalization). The total contribution is given by $\langle n_s n_d\ \alpha R;\ I \| Q_2 \| n_s n_d\ \alpha' R';\ I' \rangle = S(1 - \frac{2n_s}{\Omega'})$ where $S = \langle j \| Q_2 \| j \rangle \sqrt{2I+1}\ \sqrt{2I'+1}\ (-1)^{j+R+I} \begin{Bmatrix} 2 & j & j \\ R & I & I' \end{Bmatrix} \delta_{\alpha\alpha'}\ \delta_{RR'}$. Note that the same matrix element can be obtained from an "equivalent" quadrupole operator of the form

$$Q_{2\mu} = \langle j \| Q \| j \rangle \Big\{ \Big(\frac{2}{\Omega} \frac{\Omega-n-\hat{n}_d}{\Omega-2\hat{n}_d-1} \left[(d^\dagger s)_{2\mu} + (s^\dagger d)_{2\mu} \right]$$
$$-10 \begin{Bmatrix} 2 & 2 & 2 \\ j & j & j \end{Bmatrix} \left(1 - \frac{2\hat{n}_s}{\Omega-2\hat{n}_d-1}\right) (d^\dagger d)_{2\mu} \qquad (13)$$
$$+ \left(1 - \frac{2\hat{n}_s}{\Omega-2\hat{n}_d-1}\right) (a_j^\dagger a_j)_{2\mu} \Big\}$$

$$\langle n_s n_d \alpha R; I \| Q_2 \| n_s n_d \alpha' R'; I' \rangle$$

(a)	(b)	(c)
S	$-\dfrac{n_s}{\Omega} S$	$-\dfrac{n_s}{\Omega} S$

$$= S\left(1 - 2\frac{n_s}{\Omega}\right) \cong S\left(1 - 2\frac{n_s}{\Omega'}\right)$$

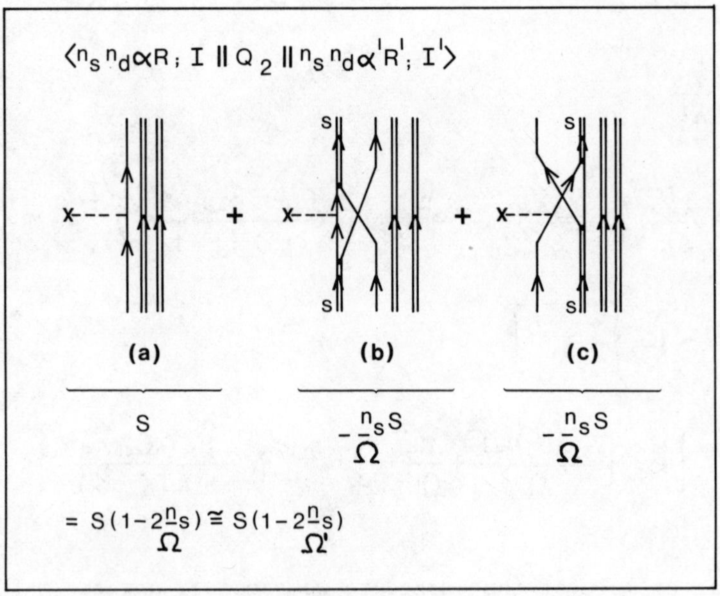

Fig. 2

The same method can be used for the evaluation of the one-nucleon transfer amplitudes. The zeroth order contribution (even nuclei → odd nuclei) is given by graph (3a): the external field (e.g. the one-particle transfer field generated in a (d,p) reaction) creates a particle which propagates without interacting with the even part. Graph (3b) gives the next order contribution, allowing the created particle to interact with the s-bosons during the time it is created. The final expression for the one-nucleon transfer amplitude connecting basis states with the same number of s and d bosons is therefore given by

$$\langle n_s n_d \, \alpha R; I \| c_j^+ \| n_s n_d \alpha R \rangle = x_o \, (2I+1)^{1/2} \left(1 - \frac{n_s}{\Omega}\right) \tag{14}$$

where x_0 is the normalization factor coming from the different normalizations of the initial and final states. This factor is given by[15]

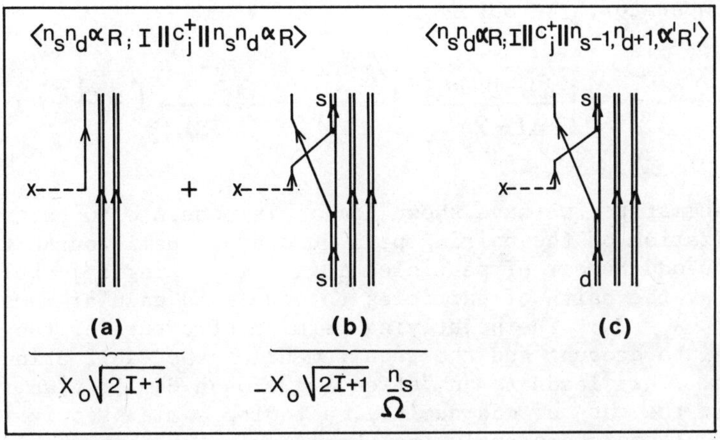

Fig. 3

$$x_o = \left(1 + \frac{dM}{dE}\right)^{1/2} = \left(1 + \frac{n_s}{\Omega}\right)^{1/2} \tag{15}$$

which leads to the final value

$$\begin{aligned} \langle n_s n_d \alpha R; I \| c_j^+ \| n_s n_d \alpha R \rangle &= (2I+1)^{1/2}\left(1 - \frac{n_s}{\Omega}\right)\left(1 + \frac{n_s}{\Omega}\right)^{1/2} \\ &\simeq (2I+1)^{1/2}\left(1 - \frac{n_s}{\Omega}\right)^{1/2} \simeq (2I+1)\left(\frac{\Omega - n - n_d}{\Omega - 2n_d}\right)^{1/2} \end{aligned} \tag{16}$$

At the same order of perturbation we have also the possibility of changing a d-boson into an s-boson during the creation of the odd-particle. This process is shown in Fig. (3c) and the corresponding value is

$$\begin{aligned} \langle n_s n_d \alpha R; I \| c_j^+ \| n_s - 1, n_d + 1, \alpha' R' \rangle &= \left[\frac{\Omega (n - n_d)}{\Omega - 2n_d}\right]^{1/2} (n_d + 1)^{1/2} \\ &\cdot \sqrt{10}\,(2I+1)^{1/2}(2R'+1)^{1/2} \begin{Bmatrix} j & j & 2 \\ R & R' & I \end{Bmatrix} (n_d \alpha R |\} n_d + 1, \alpha' R') \end{aligned} \tag{17}$$

As in the case of the Hamiltonian and of the quadrupole operator, we can write an expression for the "equivalent" one-nucleon transfer operator. We obtain

$$c_j^+ \rightarrow \mathfrak{c}_j^+ = \left[\frac{\Omega - n - \hat{n}_d}{\Omega - 2\hat{n}_d}\right]^{1/2} c_j^+ - \frac{\sqrt{5}}{\Omega - 2\hat{n}_d}(s^+ d c_j^+) \qquad (18)$$

To summarize, we have shown how it is possible to perform the diagonalization of the pairing plus quadrupole Hamiltonian for the case of an odd number of particles moving in a single j-shell and restricting the pairs of particles to couple to angular momentum $\lambda = 0$ or $\lambda = 2$. The underlying fermion structure of the bosons is taken into account and the requirement of the Pauli principle fulfilled. This leads to an "effective" boson Hamiltonian (boson-fermion in the case of odd-nuclei), allowing a microscopic interpretation of the phenomenological Hamiltonian used in the IBA.

References

1) S. G. Nilsson, Mat. Fys. Medd. Dan. Vid. Selsk. 29, no 16 (1955)
2) A. Bohr and B. R. Mottelson, Nuclear Structure, Vol. II, Addison-Wesley, Reading, Mass. (1975)
3) B. R. Mottelson, Proceedings of the International School of Physics "E. Fermi" on Nuclear Spectroscopy, Course XV, Academic Press, N.Y. (1962)
4) A. Bohr and B. R. Mottelson, Preprint NORDITA-80/19 (1980)
5) R. A. Broglia, Contribution to this Proceeding
6) D. R. Bes and R. A. Sorensen, The Pairing-Plus-Quadrupole Model, Advances in Nuclear Physics (Plenum Press), Vol. 2 (1969) 129, and references therein
7) F. Iachello, in Interacting Bosons in Nuclear Physics, Ed. F. Iachello, Plenum Press, New York (1979) 1, and references therein
8) T. Otsuka, A. Arima, F. Iachello and I. Talmi, Phys. Lett. 66B (1977) 205; 76B (1978) 139
9) T. Otsuka, A. Arima and F. Iachello, Nucl. Phys. A309 (1978) 1
10) T. Suzuki, M. Fuyuki and K. Matsuyanagi, Prog. Theor. Phys. 61 (1979) 1682
11) D. R. Bes and R. A. Broglia, in Interacting bosons in Nuclear Physics, Ed. F. Iachello, Plenum Press, New York (1979) 143
12) P. F. Bortignon, R. A. Broglia, D. R. Bes and R. Liotta, Phys. Rep. 30C (1977) 305
13) R. A. Broglia, K. Matsuyanagi, H. M. Sofia and A. Vitturi, Nucl. Phys., to be published
14) O. Civitarese, R. A. Broglia and D. R. Bes, Phys. Lett. 72B (1977) 45
15) D. R. Bes, G. G. Dussel and H. M. Sofia, Am. Journal of Physics 45 (1977) 191

BOSE-FERMI SYMMETRIES IN NUCLEI

F. Iachello

Physics Department, Yale University, New Haven, Ct.06520
and
Kernfysisch Versneller Instituut, Groningen
The Netherlands

1. INTRODUCTION

In the last few years, several examples of dynamical symmetries have been found in the spectra of even-even nuclei. These symmetries can be called of "normal" type, since they are related to properties of a system of identical particles (bosons). Within the framework of an algebraic description, these symmetries arise whenever the following set of conditions is met:

(i) the Hamiltonian, H, which describes the system, has group structure G.
(ii) H can be written in terms only of Casimir invariants of a group chain $G \supset G' \supset G'' \supset \dots$.

For the interacting boson model, $G \equiv U(6)$, and three dynamical symmetries are possible[1-3], related to the group chains

$$U(6) \begin{cases} \nearrow U(5) \supset O(5) \supset O(3) \supset O(2) \quad , & \text{I} \\ \rightarrow U(3) \supset O(3) \supset O(2) \quad , & \text{II} \\ \searrow O(6) \supset O(5) \supset O(3) \supset O(2) \quad . & \text{III} \end{cases} \qquad (1.1)$$

Correspondingly, the energy levels and other properties of the system, are given by simple, analytic relations. For example, the energy levels for the three group chains I, II and III are given by[1-3]

$$E([N],n_d,v,n_\Delta,L,M) = \varepsilon n_d+\alpha\frac{1}{2}n_d(n_d-1)+\beta(n_d-v)(n_d+v+3)+\gamma[L(L+1)-6n_d], \quad \text{I}$$

$$E([N],(\lambda,\mu),K,L,M) = (\tfrac{3}{4}\kappa-\kappa')L(L+1)-\kappa[\lambda^2+\mu^2+\lambda\mu+3(\lambda+\mu)], \qquad \text{II}$$

$$E([N],\sigma,\tau,\nu_\Delta,L,M) = A\tfrac{1}{4}(N-\sigma)(N+\sigma+4) + \tfrac{B}{6}\tau(\tau+3) + C\,L(L+1). \qquad \text{III}$$

When treating odd-A nuclei within the framework of the interacting boson-fermion model[4] (IBFA), one is faced with a system containing both bosons and fermions. One may then inquire whether or not some sort of symmetry is possible for this coupled system. These symmetries are not of the "normal" type, since they involve particles of different character (bosons and fermions), and have become of considerable interest in physics[5]. In the course of this lecture, I will investigate the possible occurrence of these, more complex symmetries, in nuclei.

2. BOSE-FERMI SYMMETRIES

A possible "new" type of symmetry which can occur in a mixed system of bosons and fermions is the following. Consider a set of boson operators $b^\dagger_\alpha(\alpha=1,\ldots,n)$. The bilinear products $b^\dagger_\alpha b_{\alpha'} \equiv G^{(B)}_{\alpha\alpha'}$ generate the algebra of the group $U^{(B)}(n)$. I have added a superscript B to U(n) in order to indicate that this group is generated by boson operators. Similarly, consider a set of fermion operators $a^\dagger_i(i=1,\ldots.,m)$. The bilinear products $a^\dagger_i a_{i'} \equiv G^{(F)}_{ii'}$ also generate a closed algebra, that of the group $U^{(F)}(m)$, where the superscript F stands for fermion. A mixed system of bosons with n degrees of freedom and fermions with m degrees of freedom is thus described by the group $U^{(B)}(n) \otimes U^{(F)}(m)$. In the interacting boson-fermion model, n=6, and $m=\sum_j(2j+1)$, where j goes over all the possible values of the angular momentum of the odd particle(s). In general, no further symmetry, in addition to $U^{(B)}(n) \otimes U^{(F)}(m)$, will be present. However, suppose that the group structure in the fermion space is similar to that in the boson space. Then, a peculiar type of symmetry may arise. As an example, consider the case in which

(i) the bosons have O(6) symmetry, chain III in (1.1);
(ii) the fermions can occupy only one single-particle level with j=3/2.

In this case, the appropriate boson group structure is

$$U^{(B)}(6) \supset O^{(B)}(6) \supset O^{(B)}(5) \supset O^{(B)}(3) \supset O^{(B)}(2) \quad , \qquad (2.1)$$

while the appropriate fermion group structure is

$$U^{(F)}(4) \supset Sp^{(F)}(4) \supset SU^{(F)}(2) \supset O^{(F)}(2) \ . \qquad (2.2)$$

Because of the isomorphisms SU(4) ≈ SO(6) and Sp(4) ≈ SO(5), and of the homomorphism SU(2) ≈ SO(3), the group structure in the fermion space is identical to the group structure in the boson space! Thus, the following group chain is possible for the mixed system[6]

$$U^{(B)}(6) \otimes U^{(F)}(4) \supset SO^{(B)}(6) \otimes SU^{(F)}(4) \supset$$
$$\supset Spin(6) \supset Spin(5) \supset Spin(3) \supset Spin(2) \, . \quad (III.BF) \qquad (2.3)$$

Here I have introduced the spinor groups Spin(6), Spin(5), Spin(3) and Spin(2)[7]. Within the framework of a creation (annihilation) operator formalism, the group Spin(6) can be thought of as being generated by

$$\begin{aligned} G^{(1)}_\mu &= B^{(1)}_\mu - \frac{1}{\sqrt{2}} A^{(1)}_\mu \, , \\ G^{(2)}_\mu &= B^{(2)}_\mu + A^{(2)}_\mu \, , \\ G^{(3)}_\mu &= B^{(3)}_\mu + \frac{1}{\sqrt{2}} A^{(3)}_\mu \, , \end{aligned} \qquad (2.4)$$

where

$$\begin{aligned} B^{(J)}_\mu &= (d^\dagger \times \tilde{d})^{(J)}_\mu \, , \quad J = 1,3 \, , \\ B^{(2)}_\mu &= (s^\dagger \times \tilde{d} + d^\dagger \times \tilde{s})^{(2)}_\mu \, , \\ A^{(J)}_\mu &= (a^\dagger \times \tilde{a})^{(J)}_\mu \, , \quad J = 1,2,3. \end{aligned} \qquad (2.5)$$

In (2.5), $s^\dagger$, $d^\dagger_\mu$(μ= 0, ±1, ±2) are creation operators for bosons, $a^\dagger_m$(m=±1/2,±3/2) are creation operators for fermions, $\tilde{d}_\mu=(-)^\mu d_{-\mu}$, $\tilde{s} = s$, $\tilde{a}_m = (-)^{3/2-m} a_{-m}$ and the brackets denote tensor products. Similarly, the group Spin(5) is generated by the 10 operators $G^{(1)}_\mu$, $G^{(3)}_\mu$ and Spin(3) by the three operators $G^{(1)}_\mu$.
As I have indicated above[7]

$$\begin{aligned} Spin(6) &\approx SU(4) \, , \\ Spin(5) &\approx USp(4) \, , \\ Spin(3) &\approx SU(2) \, . \end{aligned} \qquad (2.6)$$

The introduction of the spinor groups is the simplest extension of symmetry ideas to mixed systems of bosons and fermions. This extension can also be seen as a generalization of the concept of total angular momentum, $\vec{J} = \vec{L} + \vec{S}$, to larger groups. The corresponding symmetry will be denoted in the following Bose-Fermi symmetry (BF).

3. CLASSIFICATION OF STATES

The group chain (III.BF) provides the framework for a group theoretical classification of the states of the combined system. This classification scheme can be constructed using standard group theoretical techniques. The quantum numbers needed to classify the representations of the various groups appearing in the chain are

$$
\begin{array}{ll}
U^{(B)}(6) & [N] \\
U^{(F)}(4) & \{M\} \\
O^{(B)}(6) & \Sigma \\
Spin(6) & (\sigma_1,\sigma_2,\sigma_3) \\
Spin(5) & (\tau_1,\tau_2) \\
Spin(3) & J \\
Spin(2) & M_J
\end{array}
\qquad (3.1)
$$

Here [N] denotes a totally symmetric representation of U(6), while {M} denotes a totally antisymmetric representation of U(4). For the spinor groups Spin(6) and Spin(5) I have used the notation of Murnagham[8] in order to stress the difference between even (integer $\sigma_1,\sigma_2,\sigma_3$) and odd (half integer $\sigma_1,\sigma_2,\sigma_3$) representations. The representations of $O^{(B)}(6)$ are characterized only by one number $\Sigma\equiv\Sigma_1$ since the other two numbers Σ_2 and Σ_3 are zero for totally symmetric representations. As in the case of even-even nuclei previously discussed, it turns out that the step from Spin(5) to Spin(3) is not fully reducible and therefore an extra label is needed. This label will be called ν_Δ in the following. The complete set of quantum numbers needed to classify uniquely the states is thus $|[N],\{M\},\Sigma,(\sigma_1,\sigma_2,\sigma_3),(\tau_1,\tau_2),\nu_\Delta,J,M_J\rangle$. In addition to specifying the labels which characterize the states, one also needs to construct the representations of each subgroup in the chain contained in their respectively larger group. For $U^{(B)}(6)\supset O^{(B)}(6)$ the reduction is the same as that given in Refs. 3 and 9. A simple algorithm can be constructed for the subsequent reduction into Spin(6) ⊃ Spin(5) ⊃ Spin(3). For M=0 (no fermions) and M=1 (one fermion), which are the cases of interest for the applications presented in this article, the appropriate algorithm has been given in Refs. 3 and 6. Accordingly, even and odd nuclei are characterized by the quantum numbers $\sigma_2=\sigma_3=0$, $\tau_2=0$ and $\sigma_2=\sigma_3=1/2$, $\tau_2=1/2$, respectively. In other words, even nuclei belong to tensor representations of Spin(6) ⊃ Spin(5), while odd nuclei belong to spinor representations.

Like in the previous case, in which only bosonic degrees of freedom were present, one can now construct analytic solutions to the eigenvalue problem for the Hamiltonian[4]

$$H = H^{(B)} + H^{(F)} + V^{(BF)} . \tag{3.2}$$

This is done by writing H in terms of Casimir operators of the groups appearing in (III.BF). The corresponding expression can be written as

$$E([N],\{M\},\Sigma,(\sigma_1,\sigma_2,\sigma_3),(\tau_1,\tau_2),\nu_\Delta,J,M_J)=E_3N+E_4N^2+E_5M+E_6M^2 +$$
$$+E_7\Sigma(\Sigma+4)-\frac{A}{4}[\sigma_1(\sigma_1+4)+\sigma_2(\sigma_2+2)+\sigma_3^2]+\frac{B}{6}[\tau_1(\tau_1+3)+\tau_2(\tau_2+1)] +CJ(J+1), \tag{3.3}$$

where E_3, E_4, E_5, E_6, E_7, A, B and C are arbitrary constants which label the interactions. Here, I have used the fact that the eigenvalues of the linear and quadratic Casimir operators of the groups U(n) in the representations [N] and {M} contain only linear a quadratic term in N and M, and the standard forms of the eigenvalues of the quadratic Casimir operators of Spin(6), Spin(5) and Spin(3) in the representations $(\sigma_1,\sigma_2,\sigma_3)$, (τ_1,τ_2) and J, given, apart from a constant factor, by[10]

$$\begin{aligned}
\langle C_6\rangle &= \sigma_1(\sigma_1+4) + \sigma_2(\sigma_2+2) + \sigma_3^2 , \\
\langle C_5\rangle &= \tau_1(\tau_1+3) + \tau_2(\tau_2+1) , \\
\langle C_3\rangle &= J(J+1) .
\end{aligned} \tag{3.4}$$

For a given nucleus, N (the number of bosons) and M (the number of fermions) are fixed and thus the first four terms do not contribute to the excitation energies. The fifth term, E_7, can be, for most purposes, absorbed into the A term. Thus, one can limit oneself to a consideration of[6]

$$E'([N],\{M\},\Sigma,(\sigma_1,\sigma_2,\sigma_3),(\tau_1,\tau_2),\nu_\Delta,J,M) = -\frac{A}{4}[\sigma_1(\sigma_1+4)+\sigma_2(\sigma_2+2)+\sigma_3^2] +$$
$$+ \frac{B}{6}[\tau_1(\tau_1+3)+\tau_2(\tau_2+1)] + CJ(J+1). \tag{3.5}$$

It should be noted that this expression reduces to that given in Ref. 3 when M=0. In fact, in this case, $\sigma_1=\sigma$, $\sigma_2=\sigma_3=0$, $\tau_1=\tau$, $\tau_2=0$ and J=L and (3.5) reduces to the expression (3.7) of Ref. 3, apart from the constant term $(A/4)\,N(N+4)$. The spectra corresponding to (3.5) for N=3, M=0 and N=2, M=1 are shown in Fig. 1. A remarkable feature of some observed spectra is that they display the Bose-Fermi symmetry III.BF. Consider, for example, the Ir nuclei, Z=77. The odd proton can occupy the single-particle levels $d_{3/2}$, $s_{1/2}$,

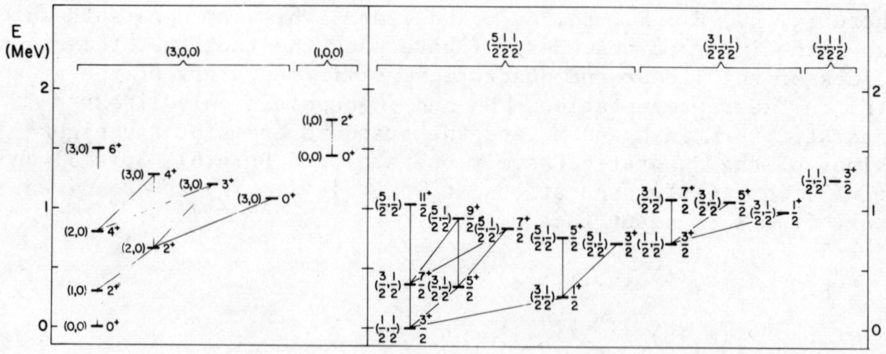

Fig. 1. Typical even and odd spectra (N=3,M=0 and N=2,M=1) corresponding to the Bose-Fermi symmetry III.BF. The energy levels are given by Eq.(3.5) with A/4=90 keV, B/6=60 keV and C=10 keV. The ground state is taken as zero of the energy. The numbers in parentheses next to each level denote the Spin(5) quantum numbers (τ_1,τ_2). The numbers on the top of the figure denote the Spin(6) quantum numbers $(\sigma_1,\sigma_2,\sigma_3)$. The lines connecting the levels denote large electromagnetic (E2) transitions.

$h_{11/2}$, $d_{5/2}$ and $g_{7/2}$. In the ground state, the most important configuration is the $d_{3/2}$ configuration. In addition, the even-even nuclei in this region, Pt and Os, are reasonably well described by a boson O(6) symmetry[11].

Thus, one may expect to encounter the situation described by symmetry III.BF. This is shown in Fig. 2 where the experimental spectra of the even (^{192}Pt) and odd (^{191}Ir) nuclei are shown. It is interesting to note that not only the two nuclei have spectra describable by (3.5) but also the same values of the parameters B and C describe both even and odd spectra, as shown in Fig. 3.

4. ELECTROMAGNETIC TRANSITION RATES

In a similar way one can calculate analytic expressions for electromagnetic transition rates. The transition operator contains now both a bosonic and a fermionic part. For example, the E2 operator can be written as

$$T^{(E2)} = T^{(B)\,(2)} + T^{(F)\,(2)} \quad . \tag{4.1}$$

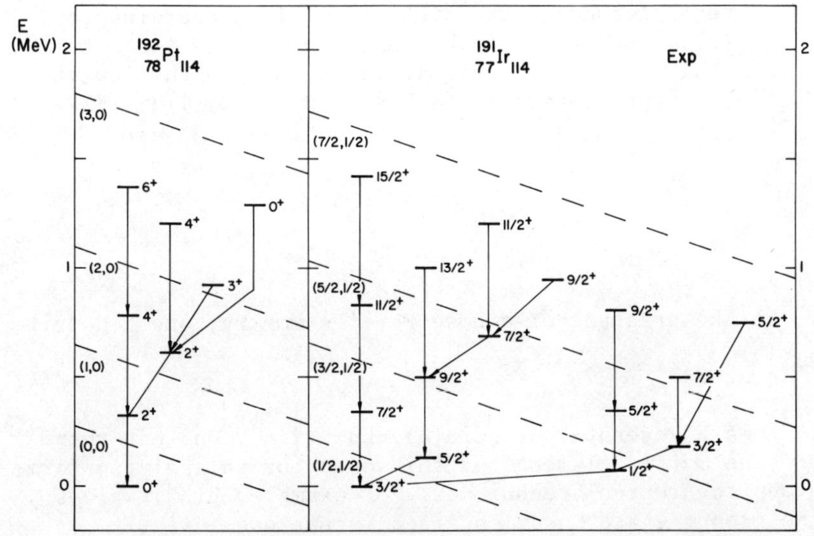

Fig. 2. The experimental spectra of $^{192}_{78}Pt_{114}$ (even nucleus, Ref.12) and $^{191}_{77}Ir_{114}$ (odd nucleus, Ref. 13). The lines connecting the levels denote observed electromagnetic transitions (E2 and M1).

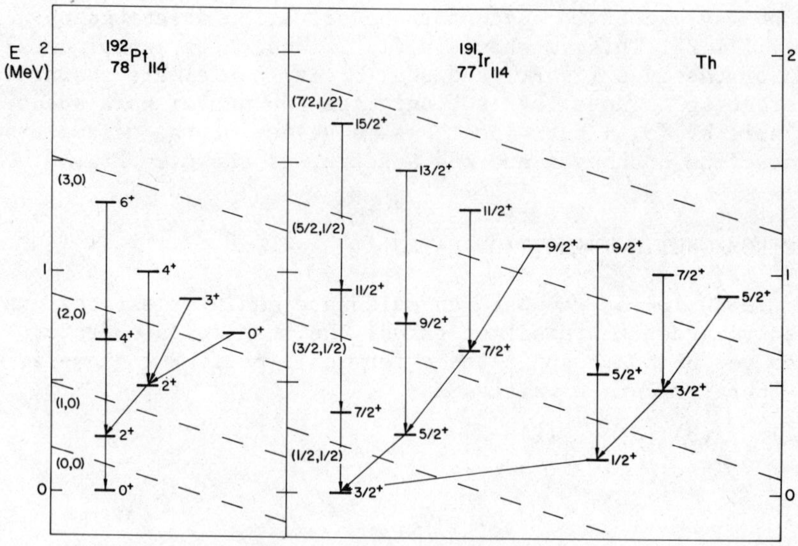

Fig. 3. Theoretical spectra of ^{192}Pt and ^{191}Ir, according to Eq. (3.5) with B/6 = 40 keV and C = 14 keV. The A term does not appear since all states belong to the lowest Spin(6) representations (8,0,0) and (17/2, 1/2, 1/2).

However, in the presence of a Bose-Fermi symmetry, one can write

$$T^{(E2)} = \alpha_2 \, G^{(2)} \tag{4.2}$$

where $G^{(2)}$ is a generator of Spin(6), Eq. (2.4). One can then calculate the matrix elements of this operator in analytic form using group theoretical techniques. For example, B(E2) values along the ground state band, obtained in the usual way,

$$B(E2;\, J_i \to J_f) = \frac{1}{2J_i+1}\, |\langle J_i || T^{(E2)} || J_f \rangle|^2 \;, \tag{4.3}$$

are given by

$$B(E2;\ \tau+1 \rightarrow \tau) = \alpha_2^2\ \frac{(\tau+1)}{(2\tau+5)}(N-\tau)\ (N+\tau+4)\ ,\ \text{even-nuclei,} \qquad (4.4)$$

and

$$B(E2;\ \tau+1 \rightarrow \tau) = \alpha_2^2\ \frac{(\tau+1/2)}{(2\tau+4)}\ (N-\tau+1/2)(N+\tau+9/2),\ \text{odd nuclei} \qquad (4.5)$$

where $\tau \equiv \tau_1$, $\tau = 0,1,2...$, in even nuclei, and $\tau = 1/2, 3/2, 5/2,...$, in odd nuclei.
Eq. (4.4) is the same as Eq. (5.12) of Ref. 3. The formula for odd nuclei, (4.5), has been obtained by Kuyucak[14]. Moreover, the transition operator (4.2) satisfies the following selection rules

$$\Delta\sigma = 0, \quad \Delta\tau = 0,\ \pm 1 \quad , \qquad (4.6)$$

the first being a consequence of the fact that $G^{(2)}$ is a generator of Spin(6) and thus cannot connect different Spin(6) representations. Thus, for example, the transitions $7/2_2^+ \rightarrow 3/2_1^+$ and $3/2_2^+ \rightarrow 3/2_1^+$ are forbidden, Fig. 4. A preliminary test of the analytic expressions for E2 transitions has been done and the results are shown in Table I. Note that, once more, the same coefficient α_2^2 seems to describe both even and odd nuclei.

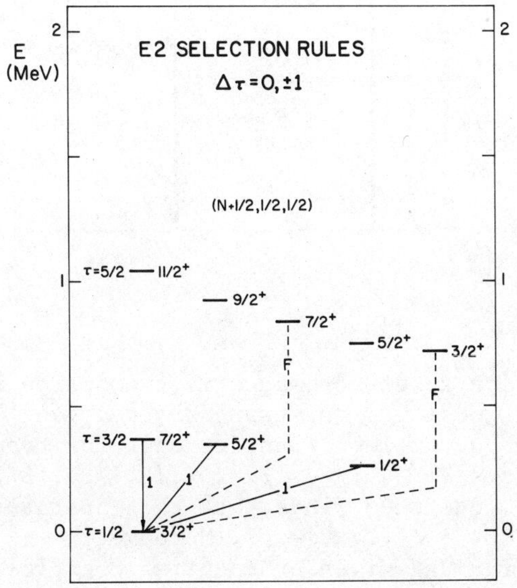

Fig. 4. A schematic illustration of the E2 selection rules.

Table I. Experimental test of Spin(6) symmetry in electromagnetic transition rates[15,16]. The numbers indicate $B(E2;g.s \to J^{\pi})$ values normalized to $B(E2;0_1^+ \to 2_1^+)$ in ^{194}Pt.

Nucleus	State	Spin(6)	(p,p')	Coulex
$^{194}Pt(\sigma=7)$	2_1^+	1.00	1.00	1.00
	2_2^+	0.00	<0.01	0.005
$^{193}Ir(\sigma=15/2)$	$1/2_1^+$	0.11	≈0.08	0.068
	$5/2_1^+$	0.33	0.48	0.44
	$3/2_2^+$	0.00	0.07	0.052
	$7/2_1^+$	0.44	0.40	0.30
	$3/2_2^+$	0.00	0.019	0.017
	$7/2_2^+$	0.00	0.10	0.056
$^{191}Ir(\sigma=17/2)$	$1/2_1^+$	0.14	0.04	0.051
	$5/2_1^+$	0.41	0.64	0.55
	$3/2_2^+$	0.00	0.08	0.068
	$7/2_1^+$	0.54	0.48	0.31
	$7/2_2^+$	0.00	0.07	<0.01

In addition to electromagnetic transition rates one can also calculate intensities for transfer reactions[14]. If one assumes that the one nucleon transfer operator transforms like the representations (1/2, 1/2, 1/2), (1/2,1/2), 3/2 of Spin(6), Spin(5), Spin(3), one then finds that this operator has selection rules $\Delta\sigma = \pm 1/2$, $\Delta\tau = \pm 1/2$, Fig. 5. Moreover, ratios of intensities for transfer are given by coupling coefficients for the appropriate groups. Some of these predictions are compared with experiment in Table II. The large asymmetry in the population of the $3/2_3^+$ state should be noted.

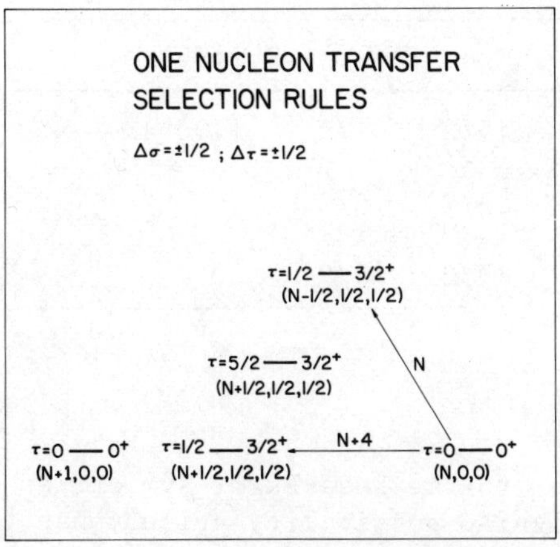

Fig. 5. Schematic illustration of the selection rules of the one nucleon transfer operator.

Table II. Experimental test of Spin(6) symmetry in transfer reactions[17,18]. The numbers indicate transfer intensities normalized to the g.s. → g.s. transfer.

(a) $^{194}_{78}Pt_{116} \rightarrow ^{193}_{77}Ir_{116}$						
J^{π}	E(keV)	σ	τ	Spin(6)	(d, ^{3}He)	(t,α)
$3/2^+$	0	15/2	1/2	1.00	1.00	1.00
	180	15/2	5/2	0	0.08	0.05
	460	13/2	1/2	0.64	0.74	0.72

Table II cont'd

(b) $^{192}_{76}Os_{116} \rightarrow ^{193}_{77}Ir_{116}$					
J^{π}	E(keV)	σ	τ	Spin(6)	(α,t)
$3/2^+$	0	15/2	1/2	1.00	1.00
	180	15/2	5/2	0	0.07
	460	13/2	1/2	0	<0.01

5. SUPERSYMMETRIES

In addition to the Bose-Fermi symmetries described above, an even more intriguing possibility arises when treating mixed systems of bosons and fermions. This possibility is suggested by the fact that the empirical coefficients A, B and C in the eigenvalue expression (3.5) appear to be approximately the same both for even and odd nuclei, Figs. 2 and 3. Thus, it may be that both even and odd nuclei belong to the same multiplet of the group larger than $U^{(B)}(6) \otimes U^{(F)}(4)$. The intriguing possibility is that this larger group, called supergroup in the following, may be identified with a graded Lie group. Graded Lie groups $U(n|m)$ can be generated by means of the operators[19]

$$\begin{aligned} G^{(B)}_{\alpha\alpha'} &= b^{\dagger}_{\alpha} b_{\alpha'} , \quad \alpha,\alpha' = 1,\ldots.n , \\ G^{(F)}_{ii'} &= a^{\dagger}_{i} a_{i'} , \quad i,i' = 1,\ldots.m , \\ F_{\alpha i} &= b^{\dagger}_{\alpha} a_{i} , \\ F^{\dagger}_{i\alpha} &= a^{\dagger}_{i} b_{\alpha} . \end{aligned} \tag{5.1}$$

It is customary to place these operators in a matrix

$$\left(\begin{array}{c|c} b^{\dagger}b & b^{\dagger}a \\ \hline a^{\dagger}b & a^{\dagger}a \end{array} \right) . \tag{5.2}$$

The operators $b^{\dagger}b$ and $a^{\dagger}a$ are said to form the Bose sector of the superalgebra, while $a^\dagger b$ and $b^\dagger a$ form the Fermi sector. Thus, the Bose sector of $U(n|m)$ is $U^{(B)}(n) \otimes U^{(F)}(m)$. In the particular case of Sect. 2, the obvious candidate for the graded group is $U(6|4)$, with Bose sector $U^{(B)}(6) \otimes U^{(F)}(4)$. If $U(6|4)$ is the appropriate supergroup, the complete group chain is now

$$U(6|4) \supset U^{(B)}(6) \otimes U^{(F)}(4) \supset SO^{(B)}(6) \otimes SU^{(F)}(4) \supset$$
$$\supset Spin(6) \supset Spin(5) \supset Spin(3) \supset Spin(2). \tag{5.3}$$

A few months ago, Bars and Balantekin[20] have succeeded in constructing the representations of all unitary supergroups $U(n|m)$. In the example discussed here, in which the representations of $U^{(B)}(6)$ are fully symmetric and those of $U^{(F)}(4)$ are fully antisymmetric, the representations of $U(6|4)$ are labelled by one integer $\mathcal{N} = N+M$, the total number of bosons plus fermions. For $U(6|4)$ the possible values of M are restricted to be 0, 1, 2, 3, 4. Thus, for given , $N=\mathcal{N}$, $\mathcal{N}-1$, $\mathcal{N}-2$, $\mathcal{N}-3$, $\mathcal{N}-4$. If now there is a supersymmetry $U(6|4)$, then one immediate consequence is that the coefficients A, B and C must be the same for all members of the supermultiplet which include both even (M = 0, 2, 4) and odd (M = 1, 3) nuclei. Although the observed spectra shown in Fig. 2 seem to indicate that this is the case, further experimental work must be done in order to identify which representation of $U(6|4)$ is realized in practice, i.e. how many nuclei belong to the supermultiplet and what are their quantum numbers. Work in this direction is being presently performed both experimentally[21,22] and theoretically[23]. A full account both of the results based on the Bose-Fermi, Spin(6), algebra and on the graded, $U(6|4)$, superalgebra will be presented in forthcoming longer publications.

6. CONCLUSIONS

I have here presented some evidence that dynamical symmetries of a "new" type are present in the spectra of complex nuclei. The experimental evidence available at the moment indicates that Bose-Fermi symmetries of the type described in Sect. 2 are certainly present. On the other side, it is not yet clear to what extent supersymmetries of the type described in Sect. 5 are present. Preliminary indications are encouraging, but further work must be done in order to elucidate this point.

It is also clear that the example shown here is only one in a vast number of other possible cases. All these cases are being investigated. Among the other cases, it is worthwhile mentioning here the Bose-Fermi symmetry associated with the group chain II in (1.1). A non trivial realization of it occurs when

(i) the bosons have SU(3) symmetry, II;
(ii)the fermions can occupy two levels with j = 1/2, 3/2.
Then, the boson group structure is

$$U^{(B)}(6) \supset SU^{(B)}(3) \supset O^{(B)}(3) \supset O^{(B)}(2) \ , \qquad (6.1)$$

while the fermion group structure can be written as

$$U^{(F)}(6) \supset SU^{(F)}(3) \otimes SU^{(F)}(2) \supset SU^{(F)}(2) \supset O^{(F)}(2). \qquad (6.2)$$

Here, the fermion space with j = 1/2, 3/2 has been split into an orbital, L = 1 and spin, S = 1/2, contribution. The orbital contribution then forms the (1,0) representation of SU(3). One can now combine the boson $SU^{(B)}(3)$ and the fermion $SU^{(F)}(3)$ into a single SU(3) to give the group chain

$$U^{(B)}(6) \otimes U^{(F)}(6) \supset SU^{(B)}(3) \otimes SU^{(F)}(3) \otimes SU^{(F)}(2) \supset SU(3) \otimes SU^{(F)}(2) \supset$$

$$\supset Spin(3) \supset Spin(2) \ . \qquad II.FB \qquad (6.3)$$

Similar techniques can be used to construct Bose-Fermi chains, I.BF, based on the chain I of (1.1).

An interesting aspect of the Bose-Fermi symmetries discussed here is that they appear to be directly related to the composite nature of the s-d boson pairs. In fact, the energy formula (3.3) can also be obtained in purely fermionic models, such as the Ginocchio model[24], without any reference to bosons. The Ginocchio model assumes a set of degenerate nucleon single-particle levels with specific values of the angular momenta (j = 7/2, 5/2, 3/2, 1/2, for example). The results presented here indicate that Bose-Fermi symmetries persist even in the case in which the angular momenta of the single-particle levels are not precisely as in the Ginocchio model and, moreover, they are not degenerate.

In conclusion, I would like to emphasize that, in addition to providing the basis for establishing the existence of dynamical symmetries in physics, the various analytic solutions discussed here give considerable insight into the structure of odd-A nuclei. This is particularly important since spectra of odd-A nuclei are very complex.

REFERENCES

1. A. Arima and F. Iachello, Ann. Phys. (N.Y.) 99 (1976), 253.
2. A. Arima and F. Iachello, Ann. Phys. (N.Y.) 111 (1978), 201.
3. A. Arima and F. Iachello, Ann. Phys. (N.Y.) 123 (1979), 468.

4. F. Iachello and O. Scholten, Phys. Rev. Lett. 43 (1979), 679.
5. L. Corwin, Y. Ne'eman and S. Sternberg, Rev. Mod. Phys. 47 (1975), 573 and references therein.
6. F. Iachello, Phys. Rev. Lett. 44 (1980), 772.
7. R. Gilmore, "Lie Groups, Lie Algebras and some of their Applications", J. Wiley, New York, (1974), p. 111.
8. F.D. Murnagham, "The Theory of Group Representations", John Hopkins, Baltimore, (1938).
9. A. Arima and F. Iachello, Phys. Rev. Lett. 40 (1978), 385.
10. V.S. Popov and A.M. Perelomov, Sov. J. Nucl. Phys. 5 (1967), 489. C.O. Nwachuku and M.A. Rashid, J. Math. Phys. 18 (1977), 1387.
11. J.A. Cizewski, R.F. Casten, G.J. Smith, M.L. Stelts, W.R. Kane, H.G. Börner and W.F. Davidson, Phys. Rev. Lett. 40 (1978), 167.
12. M. Finger, R. Foucher, J.P. Husson, J. Jastrzebski, A. Johnson, G. Astzer, B.R. Erdal, A. Kjelberg, P. Patzelt, Å. Hoglund, S.G. Malmskog and R. Henck, Nucl. Phys. A188 (1972), 369.
13. J. Lukasiak, R. Kaczarowski, J. Jastrzebski, S. André and J. Treherne, Nucl. Phys. A313 (1979), 191.
14. S. Kuyucak, private communication.
15. M.N. Harakeh, P. Goldhoorn, Y. Iwasaki, J. Lukasiak, L.W. Put, S.Y. van der Werf and F. Zwarts, KVI preprint 240 (1980).
16. C. Baktash, J.X. Saladin, J.J. O'Brien and J.G. Alessi, Phys. Rev. C18 (1978), 131.
17. Y. Iwasaki, E.H.L. Aarts, M.N. Harakeh, R.H. Siemssen and S.Y. van der Werf, KVI preprint (1980).
18. Y. Yamazaki, R.K. Sheline and D.G. Burke, Z.Physik A285 (1978), 191.
19. P.G.D. Freund, and I. Kaplanski, J. Math. Phys. 17 (1976), 228.
20. I. Bars and A. Baha Balantekin, Yale Preprint YTT 80-06 (1980).
21. J. Wood, these Proceedings.
22. J.A. Cizewski, these Proceedings.
23. A.B. Balantekin, I. Bars and F. Iachello, to be published.
24. J.N. Ginocchio, Phys. Lett. 79B (1978), 173; ibid. 85B (1979), 9; Ann. Phys. (N.Y.) 126 (1980), 234.

DYNAMICAL SUPERSYMMETRIES AND THE POSITIVE PARITY STATES IN THE ODD-GOLD ISOTOPES

J. L. Wood

School of Physics
Georgia Institute of Technology
Atlanta, Georgia 30332

The recent suggestion[1)] that dynamical supersymmetries may be present in the spectra of heavy odd-mass nuclei provides an exciting new avenue for pursuing the classification of the low-energy structure of complex nuclei; and, for the first time, offers the possibility of a concrete realization of dynamical (Bose-Fermi) supersymmetries. In the investigation discussed here, the applicability of the supersymmetric classification scheme to positive parity states in ^{193,195,197}Au is considered; and, some E2 transitions that are predicted to be forbidden (in terms of selection rules for the quantum numbers of the supersymmetry) are compared with experimental results.

The data relevant to this discussion have been taken from Nuclear Data Sheets[2,3,4)], together with more recent experimental results on ^{193}Au (Refs. 5,6,7,8,9), ^{195}Au (Ref. 10) and ^{197}Au (Ref. 10,11). The portion of the supersymmetric spectrum[1)] together with the forbidden E2 transitions, of interest to this discussion, are shown in Fig. 1. All the levels in ^{193}Au up to 1.3 MeV that are known are shown in Fig. 2. From systematics[5)] and the quality of the decay data on high-spin[7)] and low-spin[5,6)] ^{193}Hg, all states up to 1 MeV are believed to be located. Since the situation considered in Ref. 1 is for a j=3/2 fermion coupled to L=0,2 bosons with 0(6) symmetry, in order to classify the positive parity states in ^{193}Au in terms of this supersymmetry, it is necessary to identify the location of the single-proton configurations $s_{1/2}$, $d_{3/2}$, $d_{5/2}$, $g_{7/2}$. In particular, for the supersymmetric classification scheme to be valid, we require that the ground state of ^{193}Au is $d_{3/2}$, and the other single-proton configurations are not strongly mixed with the low-lying spectrum

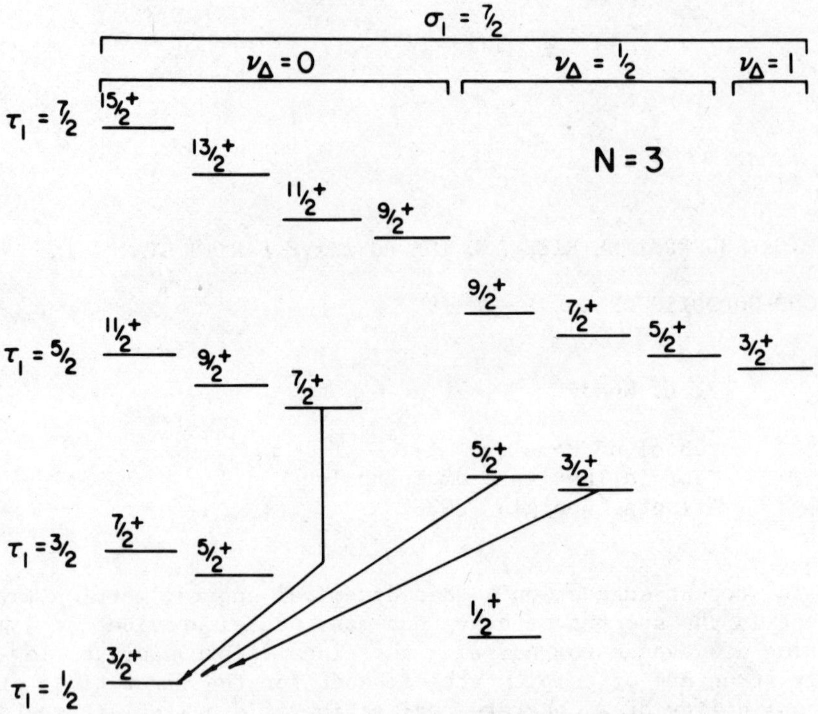

Fig. 1. Part of the spectrum of a system possessing a supersymmetry generated by L=0,2 bosons with O(6) symmetry and a j=3/2 fermion. The labelling quantum numbers are defined in Ref. 1. Only the states with maximum value of σ_1 are shown. The forbidden E2 decays of the τ_1=5/2 states are indicated by arrows.

built on the $d_{3/2}$ parent state. (It should be noted that the boson symmetry in this region is well known[12)] to be O(6) and in fact ^{193}Au is isotonic to ^{191}Ir and ^{192}Pt which were discussed in Ref. 1.) Unfortunately, information on the location of the single-proton strength in ^{193}Au is not available and the only possibility is to consider data for ^{195}Au from single-proton stripping studies[10)]. The known positive parity states in ^{195}Au up to 1.5 MeV are shown in Fig. 3. On the right-hand side are presented data on the location of the single-proton transfer strength with ℓ=0 and ℓ=2 angular distributions ($s_{1/2}$, $d_{3/2}$, $d_{5/2}$). No ℓ=4 states were identified in the proton-stripping studies[10)], and it is assumed $g_{7/2}$ lies too high in energy to be relevant to this discussion. Evidently, the $d_{3/2}$ and $d_{5/2}$ configurations dominate the ground state and 439 keV state and the main part (∿80%) of the $s_{1/2}$ configuration is concentrated in the energy range 850-1500 keV contrary to labelling used in Ref. 5). Although these data support

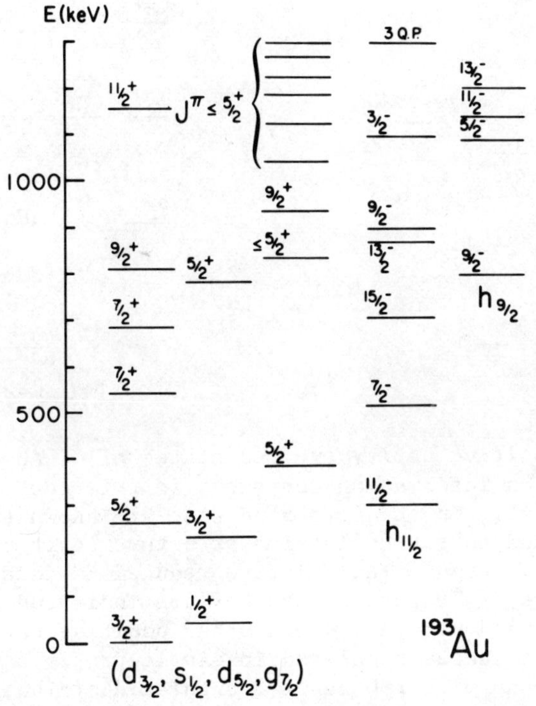

Fig. 2. The excited states of ^{193}Au up to 1.3 MeV based on information contained in references given in the text. The two left-hand columns contain states assigned to the collective structure built on the $d_{3/2}$ configuration. The central column contains other positive parity states, including the $d_{5/2}$ state at 382 keV. The states at 828, 929 and 1040 keV are discussed in the text. The negative parity states are only included to give a complete picture.

the validity of a supersymmetric classification of the states shown on the left-hand side of Fig. 3, there is a lack of information on E2 transitions between these states. A similar limitation exists for ^{197}Au. Thus, the present discussion is mainly confined to E2 transitions in ^{193}Au. Some further discussion of ^{195,197}Au is made later.

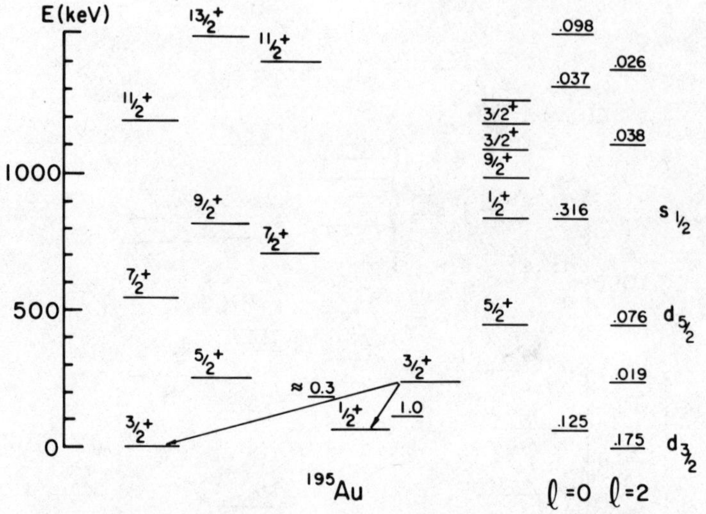

Fig. 3. The positive parity excited states of ^{195}Au up to 1.5 MeV, based on information contained in references given in the text. On the left-hand side are shown the states assigned to the collective structure built on the $d_{3/2}$ configuration. The relative reduced E2 decay probabilities for the $3/2_2^+$ state at 242 keV are indicated. On the right-hand side are shown other positive parity states; and the states populated in single-proton stripping reactions with ℓ=0 and ℓ=2 angular distributions. The C^2S factors given in Ref. 10 are included. The main components of the $d_{3/2}$, $d_{5/2}$ and $s_{1/2}$ configurations are marked.

It is possible to discuss the location of the single-proton configurations in ^{193}Au using the proton-stripping data for ^{195}Au by employing systematics. The positive parity states in ^{193}Au and ^{195}Au are remarkably similar[5,13)], the energies of many corresponding states only differing by 20 keV or less. Thus, it is asserted that (see Fig. 2) the ground state of ^{193}Au is $d_{3/2}$ and the state at 382 keV is predominantly the $d_{5/2}$ configuration. The lowest candidate for a $1/2^+$ state in ^{193}Au lies at 1040 keV and it is proposed that this is the analog of the 841 keV $1/2^+$ state in ^{195}Au, which is the most strongly populated state with ℓ=0 seen in the proton-stripping study. The positive parity states in ^{193}Au are shown classified as a supersymmetric scheme on the left-hand side of Fig. 4. The states are labelled by the quantum numbers

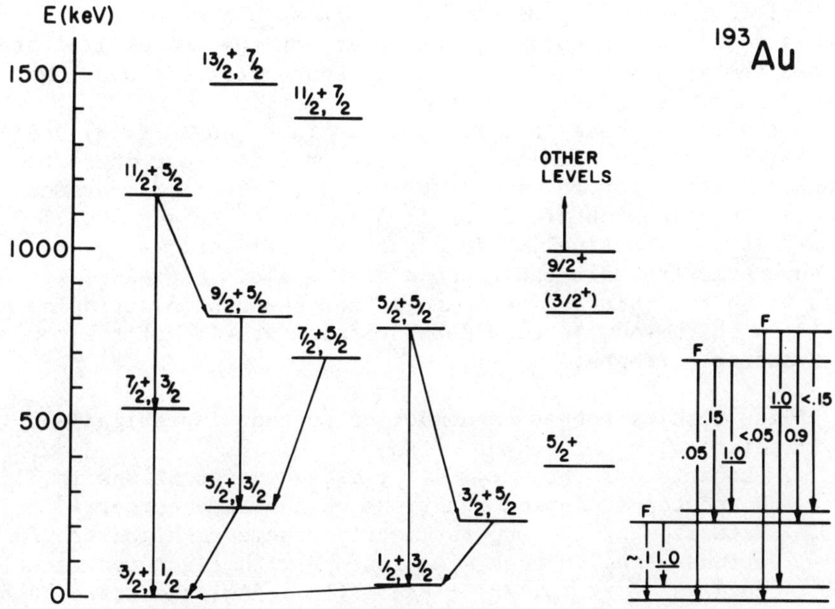

Fig. 4. The positive parity excited states of ^{193}Au. The collective structure built on the $d_{3/2}$ configuration is shown on the left-hand side with the observed major E2 decay branches indicated by arrows. On the right-hand side are shown other positive parity states (see text for discussion) and the relative reduced E2 decay probabilities for the states with $\tau_1=5/2$, $J^\pi=3/2^+$, $5/2^+$, $7/2^+$. The forbidden transitions are marked with an F.

J^π, τ_1, where τ_1 is defined in Ref. 1. The strong E2 transitions between the low-lying levels are shown as arrows. To the right of center are shown levels that are not assigned to the supersymmetric scheme. The $5/2^+$ state at 382 keV and the $9/2^+$ state at 929 keV form the beginning of a band which is interpreted as a $d_{5/2}$ collective structure. The state at 828 keV, with a possible spin of $3/2$, is discussed later. On the right-hand side of Fig. 4 are shown the reduced E2 decay branches of the low-spin states with $\tau_1=5/2$. These states have forbidden E2 transitions to the ground state according to the supersymmetric E2 selection rule $\Delta\tau_1=0,\pm1$ The transitions that are forbidden have an F above them. Clearly, the classification scheme of Ref. 1 is, to a good approximation, applicable to the positive parity states in ^{193}Au. The branching ratio for the decay of the $J^\pi,\tau_1=3/2^+$, $5/2$ state at 225 keV can be compared with the predictions of particle-asymmetric rotor and particle-vibrator model calculations which predict 1.1:<u>1.0</u> (Ref. 14) and 0.7:1.0 (Ref. 15), respectively. It should be noted that these calculations[14,15)] adopt $s_{1/2}$ character for the lowest $1/2^+$ state, whereas the $s_{1/2}$ strength is probably concentrated above 1 MeV.

Although the data on γ transitions in ^{195}Au (Ref. 3) and ^{197}Au (Ref. 4,11) are very limited, it is interesting to compare the branching ratios for the decays of the $J^\pi, \tau_1 = 3/2^+, 5/2$ states with ^{193}Au. The values obtained for ^{195}Au and ^{197}Au are ∿0.3:1.0, and 0.7:1.0, respectively. This quantity evidently suggests that the supersymmetry is breaking down at N ∿ 116. A possible explanation for this can be found in the one-proton stripping data for ^{197}Au which show[10)] considerable fragmentation of the ℓ=0 and ℓ=2 transfer strength.

These results suggest a number of further investigations in the immediate neighborhood of ^{193}Au.

a) Location of the $s_{1/2}$, $d_{3/2}$, $d_{5/2}$ configurations in ^{193}Au by proton-stripping reactions on the rare target ^{192}Pt.
b) Extension of the supersymmetric scheme to higher representations using in-beam spectroscopy with light-ion reactions such as ^{194}Pt (p, 2n) ^{193}Au. (In ^{193}Au, the state at 828 keV is likely to belong to a higher representation.)
c) Application of the classification scheme to the positive parity states in ^{191}Au. (This is currently being investigated by the author.)

Ultimately, it will be necessary to incorporate nuclei in this region into supermultiplets, such as ^{190}Os, ^{191}Ir, ^{192}Pt, ^{193}Au, ^{194}Hg which have approximate $d_{3/2}$ subshell occupancies of 0,1,2,3,4. Finally, it should be noted that the supersymmetry used in the present discussion can be related to a microscopic model proposed by Ginocchio[16)].

I wish to thank F. Iachello for the many discussions during this study, and J. Ginocchio for pointing out the relation between the above supersymmetry and his microscopic model. I am particularly indebted to the Georgia Tech Research Institute for a travel grant.

This work was supported in part by the U.S. Department of Energy, Contract No. DE-AS05-80ER10599.

REFERENCES

1. F. Iachello, Phys. Rev. Lett. 44, 772 (1980).
2. M. B. Lewis, Nucl. Data Sheets B8, 389 (1972).
3. B. Harmatz, Nucl. Data Sheets 23, 607 (1978).
4. B. Harmatz, Nucl. Data Sheets 20, 73 (1977).
5. E. F. Zganjar, J. L. Wood, R. W. Fink, L. L. Riedinger, C. R. Bingham, B. D. Kern, J. L. Weil, J. H. Hamilton, A. V. Ramayya, E. H. Spejewski, R. L. Mlekodaj, H. K.

Carter and W. D. Schmidt - Ott, Phys. Lett. 58B, 159 (1975).
6. J. L. Wood, unpublished data.
7. C. Vieu, thesis, Orsay (1974).
8. P. O. Tjøm, M. R. Maier, D. Benson, F. S. Stephens and R. M. Diamond, Nucl. Phys. A231, 397 (1974).
9. H. Strusny, F. Dubbers, L. Funke, P. Kemnitz, E. Will and G. Winter, Zfk-295, p. 38 (1975) Rossendorf Rept.
10. M. L. Munger and R. J. Peterson, Nucl. Phys. A303, 199 (1978).
11. H. H. Bolotin, D. L. Kennedy, B. J. Linard, A. E. Stuchbery, S. H. Sie, I. Katayama and H. Sakai, Nucl. Phys. A321, 231 (1979).
12. R. F. Casten and J. A. Cizewski, Nucl. Phys. A309, 477 (1978).
13. C. Vieu and J. S. Dionisio, J. Phys. (Paris) Colloq. 36-C5, 87 (1975).
14. C. Vieu, S. E. Larsson, G. Leander, I. Ragnarsson, W. De Wieclawik and J. S. Dionisio, J. Phys. G4, 531 (1978).
15. V. Paar, C. Vieu and J. S. Dionisio, Nucl. Phys. A284, 199 (1977).
16. J. N. Ginocchio, Ann. Phys. (NY) 126, 234 (1980).

TESTS OF DYNAMICAL SUPERSYMMETRIES VIA CHARGED PARTICLE TRANSFER REACTIONS*

J. A. Cizewski†

Los Alamos Scientific Laboratory
Los Alamos, New Mexico 87545

ABSTRACT

We have investigated the (t, p) and $(\vec{t}, \alpha)$ reactions on enriched ^{191,193}Ir targets. The resultant spectroscopic strengths are compared and contrasted with the expectations of the dynamical supersymmetry scheme proposed for the Pt-Ir region.

INTRODUCTION

One of the most exciting developments in nuclear structure models has been the suggestion of Iachello that dynamical supersymmetries[1] may exist in nuclei. The first solution of the supersymmetric structure was obtained for coupling a $j = 3/2$ fermion to an O(6) boson core and was predicted[1] to occur in the Pt-Ir region where the even-Pt nuclei have been well described[2] as exhibiting the O(6) symmetry[3] of the IBA and Ir ground states have $J^{\pi} = 3/2^{+}$. The spinor representation for describing the structure of a particular nucleus in this region is Spin (6).

A schematic level diagram for the equivalent even- and odd-mass nuclei is shown in Fig. 1. The eigenvalue equation for the Spin (6) spectrum is given by[1]

$$E[N, (\sigma_1\sigma_2\sigma_3), (\tau_1\tau_2), \nu_\Delta, J, M] = -\tfrac{1}{4}A[\sigma_1(\sigma_1+4)+\sigma_2(\sigma_2+2)+\sigma_3^{\,2}]$$
$$+B[\tau_1(\tau_1+3)+\tau_2(\tau_2+1)] + C\,J\,(J+1) \qquad (1)$$

The quantum numbers are very similar to those obtained for O(6) boson spectra. N is the total number of bosons; σ_1 is similar to σ of O(6) with $\sigma_2 = \sigma_3 = 0(1/2)$ for even (odd) mass nuclei; τ_1 is analogous to τ of the O(6) limit, with $\tau_2 = 0(1/2)$ for even (odd) nuclei; and J is the total angular momentum with projection M. As in the case of O(6) nuclei, ν_Δ is necessary to uniquely identify states and is not important in determining transition probabilities, for example. Given the simple forms of σ_2, σ_3 and τ_2 (the additional Spin (6) quantum numbers that do not occur in the O(6) boson symmetry) eq. 1 reduces to the well-known O(6) eigenvalue equation for even nuclei. Since the character of the states is determined by σ_1 and τ_1, in Fig. 1, for simplicity, we have labeled states with σ, τ, and ν_Δ only.

The Spin (6) level scheme in an odd-mass nucleus is not a simple weak-coupling picture. An example is given by the first $\tau = 3/2$ multiplet which would seem to be analogous to coupling the $3/2^+$ ground state to the $2^+ \tau = 1$ state in the even core. However, weak coupling would give a 7/2, 5/2, 3/2, 1/2 multiplet; in the present case the 3/2 state is "missing", with the lowest $3/2^+$ state being of $\tau = 5/2$ character. In analogy to the 0^+ state of the traditional two-phonon triplet in the even nucleus being pushed up in energy to become the "band head" of the $\sigma < \sigma_{max}$ sequence, the "weak coupling" $3/2^+$ state becomes the "band head" of the first $\sigma < \sigma_{max}$ sequence in the odd nucleus.

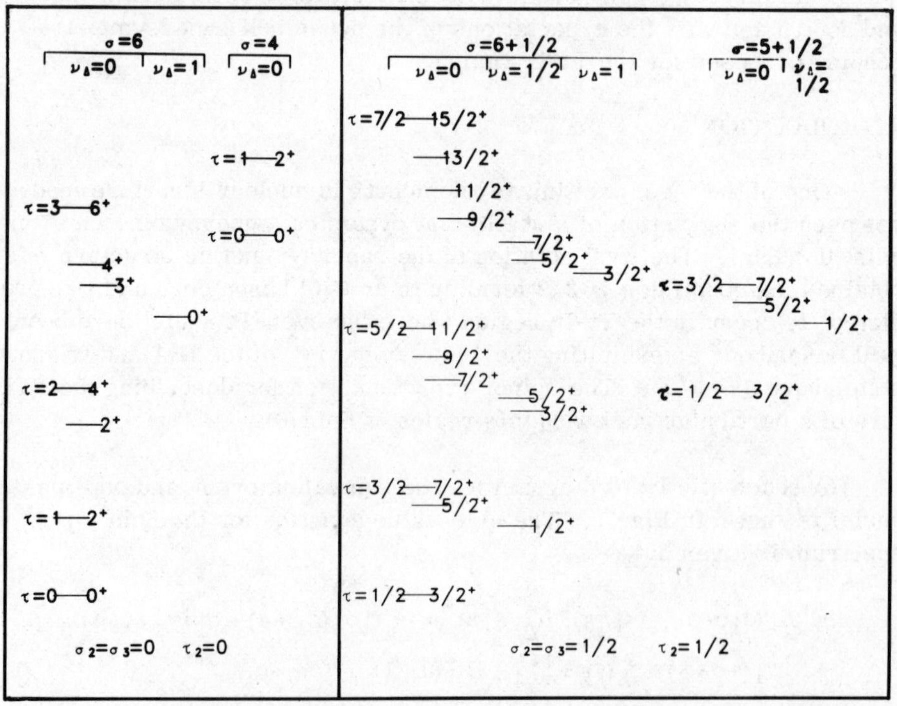

Fig. 1. Typical Spin (6) spectra for even and odd-mass nuclei.

CHARGED PARTICLE TRANSFER STRENGTHS

As in the case for the O(6) limit of the IBA (see Ref. 4, 5) there exist analytical expressions for two-neutron transfer strengths for supersymmetric systems. For Pt, Ir (t,p) reactions, where are investigating N→N-1, Iachello has obtained[1]

$$I^{even}(N\rightarrow N-1) = \alpha_\nu^2 \frac{N_\nu(N+3)}{2(N+1)}\left(\Omega_\nu - (N_\nu - 1) - \frac{(N-2)(N_\nu-1)}{2N}\right) \quad (2)$$

$$I^{odd}(N\rightarrow N-1) = \alpha_\nu^2 \frac{N_\nu(N+4)}{2(N+2)}\left(\Omega_\nu - (N_\nu - 1) - \frac{(N-2)(N_\nu-1)}{2N}\right) \quad (3)$$

where N, N_ν, and Ω_ν refer to the total number of bosons and the degeneracy of the shell as in eq. 4 of Ref. 4. In the Pt-Ir region where N ≈ 7, eq. 2,3 predict essentially equal strengths for Ir (t,p) and Pt (t,p). As in O(6) nuclei no excited L = 0 strength would be observed.[4]

The experimental enhancement factors obtained from our present Pt, Ir (t,p) measurements[5,6] are summarized in Table 1 and are compared to the supersymmetry predictions. The agreement between the empirical and predicted strengths appears exceptional. However, problems do exist. As expected, essentially no excited L = 0 strength is observed in the ^{193}Ir (t,p)^{195}Ir reaction. However, two $3/2^+$ states above 1 MeV are strongly (8-15% of g.s. strength) populated in ^{193}Ir, possibly indicating the emergence of another degree of freedom. Comparing Ir (t,p) strengths to those in Os (t,p), the observed[5] Os strengths are well below the Pt and Ir

TABLE 1. Pt, Ir (t,p) Enhancement Factors and Supersymmetry Predictions

	Reaction	σ_{exp} a) (μb/sr)	ϵ_{exp} b)	$\epsilon_{s.s.}$
N = 7	^{194}Pt (t,p) ^{196}Pt	366(7)	12.7(.4)	≡12.7
	^{193}Ir (t,p) ^{195}Ir	347(2)	12.4(1.2)	12.3
N = 8	^{192}Pt (t,p) ^{194}Pt	398(105)	12.7(1.6)	13.6
	^{191}Ir (t,p) ^{193}Ir	327(3)	13.15(1.3)	13.2

a) Experimental cross sections from Pt (Ref. 5) and Ir (Ref. 6) measurements.
b) Experimental enhancement factors given by eq. 1 of Ref. 4.
c) Predictions from eq. 2,3 normalized to ^{194}Pt (t,p) experimental value.

values, possibly indicating that the expected Os-Ir-Pt supermultiplet is not being fully realized. The entire examination of Pt-Ir-Os strengths is further complicated if one tries to incorporate (p,t) results as well. Although the experimental methods in obtaining (p,t) strengths have not been as consistent as our present (t,p) measurements, there is a clear indication that the Ir (p,t) strengths[7] are far below the Pt (p,t) strengths[8] and possibly even below the Os (p,t) strengths.[9] A more consistent measurement of Pt, Ir, Os (p,t) reactions is necessary to fully understand two-neutron transfer strengths in this region.

Possibly the most unique aspect of a supersymmetry framework is that transitions between odd- and even-mass nuclei occur with the same

TABLE 2.
Low-Lying $d_{3/2}$ and $s_{1/2}$ Strengths in $^{194,196,198}Pt(\vec{t},\alpha)^{193,195,197}Ir$ [a)]

Final Nucleus	E_x (keV)	J^π	$S^{b)}$	S_{rel}	$S_{s.s.}$ c)
^{193}Ir	0	$3/2^+$	1.6	≡1.00	≡1.00
N = 7	180	$3/2^+$	0.11	0.07	0
	460	$3/2^+$	1.1	0.69	0.64
	73	$1/2^+$	0.5(3)[d)]		
^{195}Ir	0	$3/2^+$	2.1	≡1.00	≡1.00
N = 6	234	$(3/2^+)$	0.33	0.16	0
	287	$3/2^+$	0.49	0.23	0.60
	70	$1/2^+$	0.75		
^{197}Ir	0	$3/2^+$	3.5		
N = 5	52	$1/2^+$	1.2		

a) Pt $(\vec{t},\alpha)$ measurements of Ref. 10.
b) Spectroscopic strengths obtained in Ref. 10 using optical model parameters of Ref. 11 for Pb $(\vec{t},\alpha)$.
c) Spectroscopic strengths predicted by eq. 4. Only the predictions for $^{193,195}Ir$ are directly compared to experiment. Similar strengths would be expected for ^{197}Ir, but no low-lying $3/2^+$ state other than the ground state was populated
d) Our best attempt to obtain the $s_{1/2}$ strength from the unresolved $1/2^+$ - $11/2^-$ doublet at ~79 keV in ^{193}Ir.

importance as transitions within one nucleus. To probe the single-nucleon transfer strengths, we have studied the Pt $(\vec{t},\alpha)$ Ir reactions[10] for enriched ^{194,196,198}Pt targets using a 17 MeV polarized triton beam.

For the supersymmetry based on O(6) bosons and $j = 3/2$ fermions, two $J^{\pi} = 3/2^{+}$ states should be populated[1] in Pt $(\vec{t},\alpha)$ Ir reactions, the ground state with $\sigma = N + \frac{1}{2}$ and the $\tau = 0$ excited $3/2^{+}$ state with $\sigma = N - \frac{1}{2}$, with the ratio of the spectroscopic strengths, S,[1]

$$\frac{S(\sigma = N - 1/2)}{S(\sigma = N + 1/2)} \cong \frac{N}{N + 4} \qquad (4)$$

No other state should be populated if a single $j = 3/2$ orbital is responsible for the observed spectrum. The comparison between our spectroscopic strengths and the predictions based on eq. 4 is given in Table 2. Immediately one sees that ^{195}Ir and ^{197}Ir do not follow the supersymmetry predictions in that the observed distributions of $d_{3/2}$ strengths are in clear disagreement. In addition, considerable low-lying $s_{1/2}$ strength is observed. On the other hand, the distribution of spectroscopic strengths in ^{193}Ir are in good agreement with the supersymmetry predictions, both in the distribution of $d_{3/2}$ strength and the probable $s_{1/2}$ strength.

CONCLUSIONS

Based essentially on single-particle transfer measurements, we have shown that a supersymmetry structure does not apply to all Ir nuclei. However, both a good O(6) boson description for the even-A nucleus and a single $j = 3/2$ orbital for the odd-A nucleus are needed to realize a supersymmetry scheme. Therefore, a discussion of the validity of this new approach in this region should be restricted to ^{191,193}Ir. The realization that a dynamical supersymmetry applies to nuclei would be the first manifestation of a supersymmetry in nature. It is, therefore, important to fully establish the degree to which a breakdown of the supersymmetry in nuclei may be occurring, rather than discarding the model because it does not reproduce all possible nuclear properties.

ACKNOWLEDGMENTS

It is with extreme gratitude that I thank Professor F. Iachello for invaluable and essential conversations concerning the predictions of the dynamical supersymmetry. I am also indebted to my colleagues at Los Alamos and D. G. Burke of McMaster for their assistance in the data acquisition and analysis reported here. I would also like to acknowledge with gratitude stimulating conversations with M. Vergnes, J. Wood, I. Bars, and B. Balantekin.

REFERENCES

*Work supported by U. S. Department of Energy.

† Present address: Wright Nuclear Structure Laboratory, Yale University, New Haven, Connecticut 06511.

1. F. Iachello, Phys. Rev. Lett. 44 (1980) 772; and private communication.
2. R. F. Casten and J. A. Cizewski, Nucl. Phys. A309 (1978) 477, and references therein.
3. A. Arima and F. Iachello, Phys. Rev. Lett. 40 (1978) 385.
4. J. A. Cizewski, in these proceedings.
5. J. A. Cizewski, et al., Phys. Lett. 88B (1979) 207, and to be published.
6. J. A. Cizewski, E. R. Flynn, R. E. Brown, and J. W. Sunier, Bull. Am. Phys. Soc. 25 (1980) 740, and to be published.
7. G. Lovhoiden, et al., Nucl. Phys. A302 (1978) 51; G. L. Struble, et al., Phys. Rev. C20 (1978) 927.
8. P. T. Deason, et al., Phys. Rev. C20 (1979) 927.
9. H. L. Sharma and N. M. Hintz, Phys. Rev. C13 (1976) 2288.
10. J. A. Cizewski, E. R. Flynn, R. E. Brown, J. W. Sunier and D. G. Burke, to be published.
11. E. R. Flynn, et al., Phys. Rev. Lett. 36 (1976) 79.

CONCLUDING REMARKS

Herman Feshbach

Physics Department
Massachusetts Institute of Technology
Cambridge, Mass. 02139

The promise of the Interacting Boson Model sensed and to some extent articulated explicitly in the first workshop has been amply confirmed in this workshop. As a model, it has progressed far enough in explaining and correlating data that one must now consider its limitations more carefully as well as formulate implied generalizations and extensions. It has yielded new insights into the behavior of low lying levels, and indeed has generated a renaissance of the field of Nuclear Spectroscopy. It is being extended to odd nuclei. At another level, it has been possible to more sharply formulate the approximation leading from, say, the shell model to the interacting boson picture. Connections with the current geometrical approach have been established through the derivation of a partial differential equation in five variables similar to the Bohr Hamiltonian. In the first workshop, the possibility that new dynamic symmetries could be uncovered using the algebraic and mapping methods of the IBM was noted. A striking example of this was presented to this workshop by Iachello when he showed that supersymmetry is a possible symmetry which a nuclear spectrum can exhibit. He and John Wood provided us with examples.

The question we now ask has correspondingly changed. We no longer ask if the s and d boson description is adequate. The question has become: What is the range of phenomena for which only s and d bosons are needed? R. Casten in his broad review showed us how the IBM provided a successful interpretation of the low lying spectra of a wide variety of nuclei. But there are deviations.

The sources of most of these seem to be fairly clear. The simplest one, for even-even nuclei, indicates that including the $\ell = 4^+$ boson can on occasion be advantageous, as noted by Barrett

Gelberg and van Isacker. Another possibility indicates the need for another s and d boson, s' and d', for some nuclei. In the geometric model this option might correspond to nuclei with two minima in their potential energy as a function of deformation (see Dieperink's contribution to this meeting). To some extent the effect of these additional bosons can be taken into account by renormalization, as discussed by Otsuka and Barrett, a process very similar to that which occurs in shell model theory. But as in shell model theory, renormalization as expressed in terms of effective operators may not be sufficient and one may have to expand the Hilbert space in terms of which the system is to be described. The additional states are sometimes referred to as the "intruder" states. Beside those which can be generated by adding an $\ell = 4^+$ boson, Zylicz also points to the possible effect of the collective 3^- states.

The present IBM restricts its description to that of the valence nucleons and holes. Moreover, since the IBM assumes boson number conservation, there are a maximum number of bosons in a valence shell. This leads to a saturation phenomenon in, for example, the quadrupole moments and spins as the number of nucleons of a particular type increase, a saturation which does not seem to be observed (see Gelberg and Lieb). This suggests again the need in this regime to enlarge the space to include more than just s and d bosons. Examples certainly of importance when one considers nuclei near the closed shell are the core-excitation states as mentioned by van Isacker.

High spin states, as noted by Cline and Stackel on the basis of Coulomb excitation studies, form a related area where the IBM may require extension. High spin states in the pure IBM model would have to be formed from several bosons. As long as the required boson number does not approach the saturation limit, multi boson states are possible, descriptions of states with large values of the spin. However, one does know that, in some cases, the high spin states are generated from high spin single particle orbits. In these cases it is necessary to enlarge the Hilbert space employed to include these critical orbitals. This has been done in a paper reported by Arima in which the problem of back bending is treated, putting the extra nucleons on "top of the IBM".

One surprise was provided by the results of Clement, who showed that the IBM successfully describes the spectrum and transition probabilities of relatively light nuclei. This was not anticipated as the model was thought to be restricted to nuclei with large valence shells. One is therefore left with the problem of understanding the underlying reasons for this success.

One of the recent extensions of the model and theory has been to odd nuclei. The model is referred to as the IBFM (Interacting Boson-Fermion Model). The elementary excitations include in addition the s and d boson, that of the odd nucleon. An important ingredient is the exchange term discussed at a microscopic level by Talmi, which is clearly needed to fit the experimental data. Results of the calculations and comparison of this model with data (see Scholten, Lo Bianco, Wood, Cizewski, and Casten) have been highly encouraging. But a good deal remains to be done, particularly with respect to electromagnetic transitions. Comparisons need to be made with the Nilsson model and the model of Meyer Ten Vehn.

As a final remark with respect to phenomenology, one should call attention to the expansion of the tests of IBM to the use of particle probes. Much of the recent data has come from neutron capture and Coulomb excitation studies. But in addition one now has data from two neutron transfers (J. Cizewski) including data from polarized projectiles (tritons). In the first workshop attention was called by Dieperink to the importance of inelastic electron scattering, while in this conference Lee pointed to the possible advantage of using inelastic pion scattering.

Turning now to microscopic theory, that is theory which attempts to relate the IBM to a more fundamental underlying theory, we refer to these developments as involving the IBA (the Interacting Boson Approximation). The "more fundamental" theory is for the most part the nuclear shell model. Otsuka, for example, discussed in some detail the multi-j degenerate orbits, explicitly carrying out the mapping into the s-d boson space. Broglia pointed out that the occupancy of the Nilsson orbitals for the IBA differs considerably from that of the Bohr-Mottelson unified model, which has the consequence that the two models have differing charge and mass distributions. Measurement of the charge distribution, for example, is called for. Actually one can expect that neither model will provide precise fits. It will of course be of interest to see which comes closer, and what changes are required. In the IBA, will it be necessary to include the $\ell = 4^+$ boson? A discussion of "nuclear field theory" and its relation to the IBA was given by Bortignon and Vitturi with applications presented by Maglione. An important advance in this area is provided by a paper by C.L. Wu and D.H. Feng, presented by the latter, in which the empirical diagrammatic rules of nuclear field theory are derived. Their method, using the "boson-fermion hybrid" representation, is unusual and for me will require further study.

There are other procedures beside that of the IBA for introducing boson amplitudes for the description of interacting fermion systems. Tamura, Weeks and Kishimoto (BET) have developed one such method, Belyaev and Zelevinski still another (BZM). Arima has made a numerical comparison of all three using Ginocchio's solvable model. The results indicated an equally good match by IBM and BZM but a somewhat inferior one for BET.

Finally I would like to draw your attention to the presentations made by Dieperink, Ginocchio and Gilmore. These authors discussed the reduction of the IBM in the classical limit to a classical partial differential equation of the Bohr type. As Gilmore emphasized, this is always possible if H is expressible in terms of the generators of a compact Lie group. Specific results are obtained not only for the potential energy but also for the inertial parameters. Examples using the IBM, developed by Ginocchio and Kirson, were reported by the former. At this point it again becomes possible to make a comparison with the Bohr Hamiltonian, in the cases where the parameters of the latter have been evaluated in a microscopic theory such as that of Kumar and Baranger.

In comparing this second workshop with the first one notices the rapid pace of the development of the interacting boson picture in both breadth and sophistication. Recalling that the shell model predicts an astronomical number of states, it is remarkable in the sense that the selection of an infinitesimal fraction of these on the basis of their dynamical symmetry properties suffices to provide, through the IBM, an understanding of the properties of low lying levels of both even and odd nuclei. One possible direction of development, that has not yet occurred to any great extent, is suggested by the possibility that the symmetries being used might also generate doorway state resonances to be seen perhaps as giant resonances. If these exist one would have a new way of examining interesting features in the highly excited part of the nuclear spectrum. Another suggestion, indeed, suggested by the recent discovery of the giant Gamow-Teller resonances as well as the isobar doorway model for pion scattering, is the possibility of unnatural parity bosons 0^-, 1^+, 2^- etc. corresponding to axial vector currents in nuclei. Finally, an overall question which needs to be answered asks for those essential features of the nuclear forces which are required for the validity of the interacting boson picture.

PARTICIPANTS

K. ALLAART	Natuurkundig Laboratorium Vrije Universiteit 1081 HV AMSTERDAM, The Netherlands
A. ARIMA	Department of Physics University of Tokyo Bunkyo-ku,TOKYO 113, Japan
A. BÄCKLIN	Tandem Accelerator Laboratory Box 533 S - 75121 UPPSALA, Sweden
M. BALDO	Istituto di Fisica Università di Catania I - 95129 CATANIA, Italy
B. BARRETT	Department of Physics University of Arizona TUCSON, Arizona 85721, USA
P. BORTIGNON	Istituto di Fisica Università di Padova I - 35100 PADOVA, Italy
R.A. BROGLIA	Niels Bohr Institute University of Copenhagen DK - 2100 COPENHAGEN Ø, Denmark and Nuclear Division Oak Ridge National Laboratory OAK RIDGE, Tennessee 37380, USA
R.F. CASTEN	Physics Division Brookhaven National Laboratory UPTON, New York 11973, USA
J. CIZEWSKI	Los Alamos Scientific Laboratory P.O. Box 1663 LOS ALAMOS, New Mexico 87545, USA

H. CLEMENT

Sektion Physik-Universität München
D - 8046 GARCHING, West-Germany

D. CLINE

Nuclear Structure Laboratory
University of Rochester
ROCHESTER, New York 14627, USA

P.J. DALY

Department of Chemistry
Purdue University
WEST-LAFAYETTE, Indiana 47907, USA

A.E.L. DIEPERINK

Kernfysisch Versneller Instituut
University of Groningen
9747 AA GRONINGEN, The Netherlands

P. FEDERMAN

Instituto de Fisica
UNAM
MEXICO 20, D.F., Mexico

D. FENG

Niels Bohr Institute
University of Copenhagen
DK - 2100 COPENHAGEN Ø, Denmark
and
Physics Department
Drexel University
PHILADELPHIA, Pennsylvania 19104, USA

H. FESHBACH

Department of Physics
Massachusetts Institute of Technology
CAMBRIDGE, Massachusetts, 02139, USA

A. FRANK

Centro de Estudios Nucleares
Circuito Exterior C.U.
MEXICO 20 D.F., Mexico

A. GELBERG

Institüt für Kernphysik der
Universität zu Köln
D - 500 KÖLN 41, West-Germany

R. GILMORE

Institute for Defense Analysis
Systems Evaluation Division
ARLINGTON, Virginia 22202, USA

J. GINOCCHIO

Los Alamos Scientific Laboratory
P.O. Box 1663
LOS ALAMOS, New Mexico 87545, USA

G. SCHARFF-GOLDHABER
Physics Department
Brookhaven National Laboratory
UPTON, New York 11973, USA

K. HEYDE
Laboratorium voor Kernfysica
Rijksuniversiteit Gent
B - 9000 GENT, Belgium

F. IACHELLO
Kernfysisch Versneller Instituut
University of Groningen
9747 AA GRONINGEN, The Netherlands
and
Physics Department
Yale University
NEW HAVEN, Connecticut 06520, USA

H. LEE
Physics Division
Argonne National Laboratory
ARGONNE, Illinois 60439, USA

K. LIEB
Institut für Kernphysik
Universität Göttingen
D - 3400 GÖTTINGEN, West-Germany

G. LOBIANCO
Istituto di Scienze Fisiche
Università di Milano
I - 20133 MILANO, Italy

E. MAGLIONE
Niels Bohr Institute
University of Copenhagen
DK - 2100 COPENHAGEN Ø, Denmark
and
Istituto Nazionale di Fisica Nucleare
Laboratori di Legnaro
LEGNARO, Italy

T. OTSUKA
Japan Atomic Energy Research Institute
Tokai Research Establishment
TOKAI-MURA, IBARAKI-KEN 391-11, Japan

M. SAKAI
Institute for Nuclear Study
University of Tokyo
TANASHI, TOKYO 188, Japan

J. SAU
Institut de Physique Nucléaire
Lyon-1
VILLEURBANNE, Cedex, France

O. SCHOLTEN
Kernfysisch Versneller Instituut
University of Groningen
9747 AA GRONINGEN, The Netherlands

H. SEYFARTH
Institut für Kernphysik der
Kernforschungsanlage Jülich
D - 5170 JÜLICH 1, West-Germany

P. SONA
Istituto Nazionale di Fisica
Nucleare
Largo E. Fermi 2
I - 51025 FIRENZE, Italy

J. STACHEL
Institüt für Kernphysik
Johannes Gutenberg Universität
D - 6500 MAINZ, West-Germany

I. TALMI
Physics Department
The Weizmann Institute of Science
REHOVOT, Israel

P. VAN ISACKER
Instituut voor Nukleaire Wetenschappen
Rijksuniversiteit Gent
B - 9000 GENT, Belgium

M. VERGNES
Institut de Physique Nucléaire
B.P. 1
F - 91406 ORSAY, France

A. VITTURI
Istituto di Fisica
Università di Padova
I - 35100 PADOVA, Italy

P. VON BRENTANO
Institut für Kernphysik der
Universität zu Köln
D - 500 KÖLN 41, West-Germany

J. WOOD
Physics Department
Georgia Institute of Technology
ATLANTA, Georgia 30332, USA

T.S. YANG
Sektion Physik
Universität München
D - 8046 GARCHING, West-Germany

J. ZYLICZ
Institute of Experimental Physics
University of Warsaw
PL - 04-643 WARSZAWA, Poland

INDEX